CW01509962

Human–Wildlife Interactions

Turning Conflict into Coexistence

Human–wildlife conflict (HWC) is one of the most complex and urgent issues facing wildlife management and conservation today. Originally focused on the ecology and economics of wildlife damage, the study and mitigation of HWC has gradually expanded its scope to incorporate the human dimensions of the whole spectrum of human–wildlife relationships, from conflict to coexistence. Having the conflict-to-coexistence continuum as its leitmotiv, this book explores a variety of theories and methods currently used to address human–wildlife interactions, illustrated by case studies from around the world. It presents some key concepts in the field, such as values, emotions, social identity and tolerance, and a variety of insights and solutions to turn conflict into coexistence, from individual level to national scales, including conservation marketing, incremental and radical innovation, strategic planning and socio-ecological systems. This volume will be of interest to a wide range of readers, including academics, researchers, students, practitioners and policy-makers.

BEATRICE FRANK is the Social Science Specialist for Capital Regional District Regional Parks, Canada and an adjunct professor at the University of Victoria. In the last ten years, she has focused on better defining tolerance and coexistence and developed the conflict-to-coexistence concept proposed in this book, which she is furthering in her most recent research and publications on wildlife and protected areas.

JENNY A. GLIKMAN is Associate Director of Community Engagement at the San Diego Zoo's Institute for Conservation Research, and board member of the International Union for Conservation of Nature (IUCN) Task Force on Human–Wildlife Conflict. As a social scientist, she focuses on understanding the relationships between humans and wildlife. Her work ranges from studying and addressing human–wildlife interactions, to exploring various aspects of local consumers of wildlife products in several countries.

SILVIO MARCHINI is a research associate at the Wildlife Conservation Research Unit (WildCRU) of the University of Oxford, UK, and at the University of São Paulo, Brazil, and a conservation fellow at Chester Zoo, UK. He is a board member of the IUCN Task Force on Human–Wildlife Conflict and of the IUCN Conservation Planning Specialist Group. His current work focuses on ways to upscale the analysis and management of human–wildlife conflicts.

Conservation Biology

Executive Editor

Alan Crowden – freelance book editor, UK

This series aims to present internationally significant contributions from leading researchers in particularly active areas of conservation biology. It focuses on topics where basic theory is strong and where there are pressing problems for practical conservation. The series includes both authored and edited volumes and adopts a direct and accessible style targeted at interested undergraduates, postgraduates, researchers and university teachers.

1. *Conservation in a Changing World*, edited by Georgina Mace, Andrew Balmford and Joshua Ginsberg 0 521 63270 6 (hardcover), 0 521 63445 8 (paperback)
2. *Behaviour and Conservation*, edited by Morris Gosling and William Sutherland 0 521 66230 3 (hardcover), 0 521 66539 6 (paperback)
3. *Priorities for the Conservation of Mammalian Diversity*, edited by Abigail Entwistle and Nigel Dunstone 0 521 77279 6 (hardcover), 0 521 77536 1 (paperback)
4. *Genetics, Demography and Viability of Fragmented Populations*, edited by Andrew G. Young and Geoffrey M. Clarke 0 521 782074 (hardcover), 0 521 794218 (paperback)
5. *Carnivore Conservation*, edited by John L. Gittleman, Stephan M. Funk, David Macdonald and Robert K. Wayne 0 521 66232 X (hardcover), 0 521 66537 X (paperback)
6. *Conservation of Exploited Species*, edited by John D. Reynolds, Georgina M. Mace, Kent H. Redford and John G. Robinson 0 521 78216 3 (hardcover), 0 521 78733 5 (paperback)
7. *Conserving Bird Biodiversity*, edited by Ken Norris and Deborah J. Pain 0 521 78340 2 (hardcover), 0 521 78949 4 (paperback)
8. *Reproductive Science and Integrated Conservation*, edited by William V. Holt, Amanda R. Pickard, John C. Rodger and David E. Wildt 0 521 81215 1 (hardcover), 0 521 01110 8 (paperback)
9. *People and Wildlife, Conflict or Coexistence?*, edited by Rosie Woodroffe, Simon Thirgood and Alan Rabinowitz 0 521 82505 9 (hardcover), 0 521 53203 5 (paperback)
10. *Phylogeny and Conservation*, edited by Andrew Purvis, John L. Gittleman and Thomas Brooks 0 521 82502 4 (hardcover), 0 521 53200 0 (paperback)
11. *Large Herbivore Ecology, Ecosystem Dynamics and Conservation*, edited by Kjell Danell, Roger Bergstrom, Patrick Duncan and John Pastor 0 521 83005 2 (hardcover), 0 521 53687 1 (paperback)

Human–Wildlife Interactions

Turning Conflict into Coexistence

Edited by

BEATRICE FRANK
Capital Regional District Regional Parks

JENNY A. GLIKMAN
Institute for Conservation Research, San Diego Zoo Global

SILVIO MARCHINI
*University of São Paulo and Wildlife Conservation Research Unit (WildCRU),
Department of Zoology, University of Oxford*

CAMBRIDGE
UNIVERSITY PRESS

University Printing House, Cambridge CB2 8BS, United Kingdom

One Liberty Plaza, 20th Floor, New York, NY 10006, USA

477 Williamstown Road, Port Melbourne, VIC 3207, Australia

314–321, 3rd Floor, Plot 3, Splendor Forum, Jasola District Centre,
New Delhi – 110025, India

79 Anson Road, #06–04/06, Singapore 079906

Cambridge University Press is part of the University of Cambridge.

It furthers the University's mission by disseminating knowledge in the pursuit of
education, learning, and research at the highest international levels of excellence.

www.cambridge.org
Information on this title: www.cambridge.org/9781108416061
DOI: 10.1017/9781108235730

First published 2019

Printed in the United Kingdom by TJ International Ltd, Padstow Cornwall

A catalogue record for this publication is available from the British Library.

Library of Congress Cataloging-in-Publication Data
NAMES: Frank, Beatrice, 1977- editor. | Glikman, Jenny A., 1979- editor. | Marchini,
    Silvio, 1968- editor.
TITLE: Human-wildlife interactions : turning conflict into coexistence / edited by
    Beatrice Frank, Jenny A. Glikman, Silvio Marchini.
OTHER TITLES: Human-wildlife interactions (Cambridge University Press)
DESCRIPTION: Cambridge, United Kingdom ; New York, NY : Cambridge University
    Press, 2019. | Series: Conservation biology ; 23 | Includes bibliographical
    references.
IDENTIFIERS: LCCN 2018042036 | ISBN 9781108416061 (hardback) |
    ISBN 9781108402583 (paperback)
SUBJECTS: LCSH: Human ecology. | Human-animal relationships. | Wildlife
    conservation. | Wildlife management. | BISAC: NATURE / Ecology.
CLASSIFICATION: LCC GF50 .H88 2019 | DDC 333.95/416–dc23
LC record available at https://lccn.loc.gov/2018042036

ISBN 978-1-108-41606-1 Hardback
ISBN 978-1-108-40258-3 Paperback

To Alan Rabinowitz, inspirational leader in conservation, who provided us with a visionary and inspired path for coexisting with wildlife

Contents

Contributors

SHELLEY M. ALEXANDER Department of Geography, University of Calgary, Calgary, AB, Canada

JEREMY T. BRUSKOTTER School of Environment and Natural Resources, The Ohio State University, Columbus, OH, USA

NEIL H. CARTER College of Innovation and Design, Boise State University, Boise, ID, USA

SALISHA CHANDRA Lion Guardians, Nairobi, Kenya

SUSAN G. CLARK School of Forestry & Environmental Studies and Institution for Social and Policy Studies, Yale University, New Haven, CT, USA

PEDRO L. P. CORREA Department of Computer Engineering and Digital Systems, University of São Paulo, São Paulo, Brazil

ROBERT CRABTREE Yellowstone Ecological Research Center, Bozeman, MT, USA

ALIA M. DIETSCH School of Environment and Natural Resources, The Ohio State University, Columbus, OH, USA

STEPHANIE DOLRENRY Lion Guardians, Nairobi, Kenya; and Department of Biological Sciences, University of Cape Town, Rondebosch, Cape Town, South Africa

LEO R. DOUGLAS Department of Ecology, Evolution and Environmental Biology, Columbia University, New York, NY, USA

DIANNE L. DRAPER Department of Geography, University of Calgary, Calgary, AB, Canada

KATIA M. P. M. B. FERRAZ Luiz de Queiroz College of Agriculture, University of São Paulo, São Paulo, Brazil

BEATRICE FRANK Capital Regional District Regional Parks, Victoria, BC, Canada

SUNETRO GHOSAL STAWA and Norwegian University of Life Sciences, Ås, Norway

JENNY A. GLIKMAN Institute for Conservation Research, San Diego Zoo Global, Escondido, CA, USA

MEREDITH L. GORE Department of Fisheries and Wildlife, Michigan State University, East Lansing, MI, USA

THAÍS GUIMARÃES-LUIZ Department of Wildlife, The Secretariat for the Environment of the State of São Paulo, São Paulo, Brazil

REBECCA G. HARVEY Department of Wildlife Ecology and Conservation, University of Florida, Davie, FL, USA

LEELA HAZZAH Lion Guardians, Nairobi, Kenya; Department of Human Dimensions in Natural Resources, Colorado State University, Fort Collins, CO, USA; and Department of Biological Sciences, University of Cape Town, Rondebosch, Cape Town, South Africa

MAARTEN JACOBS Department of Environmental Sciences, Wageningen University, Wageningen, The Netherlands

HANNAH JAICKS Environmental Psychology Program, the City University of New York, New York, NY, USA

BJØRN P. KALTENBORN Norwegian Institute for Nature Research, Lillehammer, Norway

GABRIEL KARNS College of Food, Agriculture, and Environmental Science, The Ohio State University, Columbus, OH, USA

LUCY E. KING Department of Zoology, University of Oxford, Oxford, UK; Save the Elephants, Nairobi, Kenya

SIDDHARTHA KRISHNAN Ashoka Trust for Research in Ecology and the Environment (ATREE), Bengaluru, Karnataka, India

JOHN D. C. LINNELL Norwegian Institute for Nature Research, Trondheim, Norway

MICHELLE L. LUTE Department of Biology, Texas State University, San Marcos, Texas, USA

DAVID W. MACDONALD Wildlife Conservation Research Unit (WildCRU), Department of Zoology, University of Oxford, Oxford, UK

MICHAEL J. MANFREDO Department of Human Dimensions of Natural Resources, Colorado State University, Fort Collins, CO, USA

SILVIO MARCHINI Luiz de Queiroz College of Agriculture, University of São Paulo, São Paulo, Brazil; Wildlife Conservation Research Unit (WildCRU), Department of Zoology, University of Oxford, Oxford, UK; and Chester Zoo, Chester, UK

FRANK J. MAZZOTTI Department of Wildlife Ecology and Conservation, University of Florida, Davie, FL, USA

RONALDO MORATO National Center for Research and Conservation of Mammalian Carnivores (CENAP), Chico Mendes Institute for Biodiversity Conservation, Atibaia, Brazil

JOSHUA MORSE Rubenstein School of Environment and Natural Resources, University of Vermont, Burlington, VT, USA

MICHAEL PAUL NELSON Forest Ecosystems & Society, Oregon State University, Corvallis, OR, USA

OMAR OHRENS Nelson Institute for Environmental Studies, University of Wisconsin-Madison, Madison, WI, USA

ALAN RABINOWITZ Panthera, New York, NY, USA

ERIN P. RILEY Department of Anthropology, San Diego State University, San Diego, CA, USA

RYO SAKURAI College of Policy Science, Ritsumeikan University, Ibaraki, Osaka, Japan

FRANCISCO SANTIAGO-ÁVILA Nelson Institute for Environmental Studies, University of Wisconsin-Madison, Madison, WI, USA

KETIL SKOGEN Norwegian Institute for Nature Research (NINA), Oslo, Norway

SILJE SKULAND Consumption Research Norway, Oslo, Norway

KRISTINA SLAGLE School of Environment and Natural Resources, The Ohio State University, Columbus, OH, USA

DOUG SMITH Yellowstone Wolf Project, Yellowstone National Park, WY, USA

CARL D. SOULSBURY School of Life Sciences, University of Lincoln, Lincoln, UK

LEEANN SULLIVAN Department of Human Dimensions of Natural Resources, Colorado State University, Fort Collins, CO, USA

TARA L. TEEL Department of Human Dimensions of Natural Resources, Colorado State University, Fort Collins, CO, USA

ADRIAN TREVES Nelson Institute for Environmental Studies, University of Wisconsin-Madison, Madison, WI, USA

BROOKE TULLY Independent Trainer and Consultant, Poughkeepsie, NY USA

JERRY J. VASKE Human Dimensions of Natural Resources, Colorado State University, Fort Collins, CO, USA

DIOGO VERÍSSIMO Department of Zoology, University of Oxford, Oxford, UK

JOHN VUCETICH School of Forest Resources and Environmental Sciences, Michigan Technological University, MI, USA

PIRAN C. L. WHITE Environment Department, University of York, York, UK

ALEXANDRA ZIMMERMANN Wildlife Conservation Research Unit (WildCRU), Department of Zoology, University of Oxford, Oxford, UK

Foreword

ROSIE WOODROFFE AND ALAN RABINOWITZ

Throughout the world, people have wildlife for neighbours. Like their human neighbours, some of these wildlife species are beloved community members, while others are downright disruptive. Wildlife can be especially hard to live with if they kill people's domestic animals or destroy their crops. Some species – like lions, elephants and bears – can even threaten human lives on occasion.

For hundreds of years, people viewed such wildlife as adversaries and sought to reduce their impact by killing them. Species like rats and coyotes thrived nonetheless; others did not. Tasmania's thylacine, extirpated due to its predation on sheep, is among the most famous victims; its analogue in the South Atlantic, the Falkland Island wolf, met the same fate half a century earlier. Other species, like prairie dogs, persisted in tiny fractions of their historic ranges. For species as dissimilar as hen harriers and African wild dogs, it is deliberate killing by people which prevents population recovery and sets populations on a path to local extinction.

In 2005 we edited, together with Simon Thirgood, a book entitled *People and Wildlife, Conflict or Coexistence?* which explored how to conserve wildlife species that make difficult neighbours. We recognized that living with wildlife – even endangered wildlife – can have real and serious impacts on people's lives and livelihoods, and we tried to identify the most promising ways to mitigate these harmful impacts while still conserving the wildlife species. We had expected to draw on experience from managing more abundant species, but such experience was surprisingly scant: in most settings where wildlife threatened human lives or livelihoods, the response was to kill the wildlife, and non-lethal approaches had received little attention. For example, while there have now been many trials of non-lethal ways to prevent elephants from damaging crops, it was 2018 before the first such study of garden slugs was initiated. Curiously, lessons learned from managing endangered species may prove useful for the management of more abundant species.

The contributors to our book were mostly trained as biologists, and we focused mainly on technical solutions. We reasoned that an important first step to resolving human–wildlife conflict was to reduce the impact of wildlife on people, and so our book evaluated tools to reduce damage, like herding and crop-guarding, and approaches to offset the costs of such damage, like ecotourism and compensation. But human behaviour is complex, and fostering coexistence with wildlife demands more than adjusting a balance sheet. That is where this new book comes in. Our contributors were mostly experts on the 'wildlife' side of human–wildlife conflict, but this book bursts with expertise on the 'human' side. It covers topics like psychology, essential to understanding people's perceptions of their interactions with wildlife, and approaches like marketing and innovation, which can help alter those perceptions in ways likely to benefit both people and wildlife.

Perceptions of coexistence with wildlife are changing rapidly. When we first proposed our book, one reviewer marvelled at the paradox of a book about 'conserving pests' and another questioned the need for yet another book about 'pest management'. Our book marked a step away from such perceptions, away from conflict and towards coexistence. This book takes the next step, emphasizing the need to learn from positive interactions between people and wildlife to resolve more negative interactions, and tackling coexistence not only with the threatened species that we considered, but also with more abundant species that might indeed be termed pests in some settings.

It is exciting to see the concept of coexistence between people and wildlife taking hold, with new and diverse perspectives on how it may be achieved. We wish that our friend and co-editor Simon Thirgood[1] was still here to see it. Every step towards coexistence adds to his legacy, and builds our hope for the future of people and wildlife.

---

[1] While finalizing this foreword, Alan Rabinowitz passed away, leaving behind a path to further develop a visionary and inspired advocacy for enhancing coexisting with wildlife.

Preface

When we talk about human–wildlife interactions, negative experiences are often most forefront, including situations that involve the injury or death of humans and non-human animals by wild animals, and those that cause damage to crops or to other human belongings. Discourses around human–wildlife conflicts have, indeed, grabbed the attention of practitioners, managers and researchers in the last few decades since this type of conflict is considered to be one of the most critical challenges facing wildlife conservation. However, human–wildlife interactions go beyond the negative impacts that species have on the environment or humans and their belongings. In fact, the research emphasis has recently expanded from a focus on conflicts to include the broad spectrum of interactions between people and wildlife that range from negative to neutral to positive.

In response to this timely and emerging trend, we aim to build upon Woodroffe, Thirgood and Rabinowitz's 2005 book, *People and Wildlife, Conflict or Coexistence?*, by placing a greater emphasis on the coexistence aspect of human–wildlife interactions. Our goal is to contribute to the development of more comprehensive frameworks and approaches, and present a variety of perspectives and solutions that emphasize positive rather than negative aspects of human–wildlife interactions. To foster the inclusion of coexistence in human–wildlife interactions, this book uses the *conflict-to-coexistence continuum* (Frank 2016) as a leitmotiv. This continuum spans from negative to positive attitudes and/or behaviours, which defines the different degrees of conflict and coexistence that characterize human–wildlife interactions. The book also encourages complementary as well as competing perspectives on human–wildlife interactions, as the intent of this contribution is to propose new directions for how best to include positive interactions in human–wildlife research and conservation.

We are thankful and feel very fortunate to have worked with a remarkable team of renowned experts on this book. Together, we have explored and discussed the conceptual and practical implications, as well as the innovative aspects of shifting the paradigm to encompass human–wildlife coexistence. The first five chapters of this book focus on conceptualizing coexistence and analysing human–wildlife interactions from an anthropocentric standpoint (Chapter 1). We also look at values (Chapter 2), social identity (Chapter 3), emotions (Chapter 4) and tolerance (Chapter 5) from a positive interaction perspective. Novel theories and frameworks are described to better encompass and assess the role played by these factors in shaping human–wildlife relationships. The following nine chapters (Chapters 6–14) portray a series of case studies that include different degrees of neutral to positive interactions between humans and wildlife. The case studies apply frameworks for coexisting with urban wildlife (Chapter 6), governance of long-distance migration (Chapter 8) and effectiveness and acceptability of interventions for coexistence with predators (Chapter 12). Understanding the place wildlife holds in different landscapes is another key theme explored through the lens of carnivores (Chapter 7). The conflict-to-coexistence concept is discussed further by looking at problematic species in Japan (Chapter 9), Indonesia and the USA (Chapter 10) and invasive species in the USA (Chapter 13). The use of beehives as barriers to reduce crop raiding by African elephants (*Loxodonta africana*) (Chapter 11) highlights an innovative approach towards increasing positive interactions with wildlife. To conclude, we offer an overview focused on European institutions working at different scales to foster human–wildlife coexistence for large herbivores and carnivores (Chapter 14). Many of the examples given in this book are biased towards terrestrial wildlife (e.g. elephants, large carnivores and monkeys) due to a historical focus on these taxa. However, the concepts and outcomes of these case studies are broadly applicable to other taxa, and may provide inspiration and solutions to improve the understanding and assessment of neutral to positive interactions with wildlife. In addition, they provide insights into stakeholder engagement and power distribution that can be extrapolated to any wildlife species and are relevant to anyone working on human–wildlife interactions. The last four chapters (Chapters 15–19) depict pathways to make coexistence happen. A series of topics developed in other disciplines, including world-view perspectives (Chapter 15), conservation marketing (Chapter 16), incremental and radical innovation (Chapter 17), socio-ecological and landscape approaches (Chapter 18) and conservation planning (Chapter 19) are discussed from a social science standpoint and applied to foster coexistence with wildlife. A world map (Figure 0.1) that includes all of the topics, species and locations addressed in this book is presented in this preface to offer the reader a better understanding of the broad, diverse and innovative ideas our team has explored, applied and implemented to shift the focus of human dimensions discourse from conflict to coexistence. These new perspectives further push the conflict-to-coexistence continuum and challenge its scale and singular dimension to better encompass the multifaceted nature of human–wildlife interactions. Following the example of Woodroffe et al. (2005), we close the book by offering our conclusions and hopes for a future that entails coexistence with wildlife.

Human–wildlife interactions – and clashes of opinions among stakeholders regarding how to deal with the encounters – will grow as the boundaries between the space in which humans and wildlife exist become more blurred. Through the innovative conceptual and applied contributions in this book, we aim to catalyse a paradigm shift from discourse on human–wildlife conflict to dialogue promoting human–wildlife interactions and coexistence. Exploring when and why people start to accept wildlife in their proximity is key to this endeavour and represents a pathway by which we may begin to shift toward mechanisms that enhance the willingness to coexist with wild species.

References

Frank, B. (2016). Human–wildlife conflicts, the need to include tolerance and coexistence: An introductory comment. *Society & Natural Resources,* 29(6), 738–43.

Woodroffe, R., Thirgood, S. & Rabinowitz, A. (2005). *People and Wildlife, Conflict or Coexistence?* Cambridge: Cambridge University Press.

Beatrice Frank
Jenny A. Glikman
Silvio Marchini

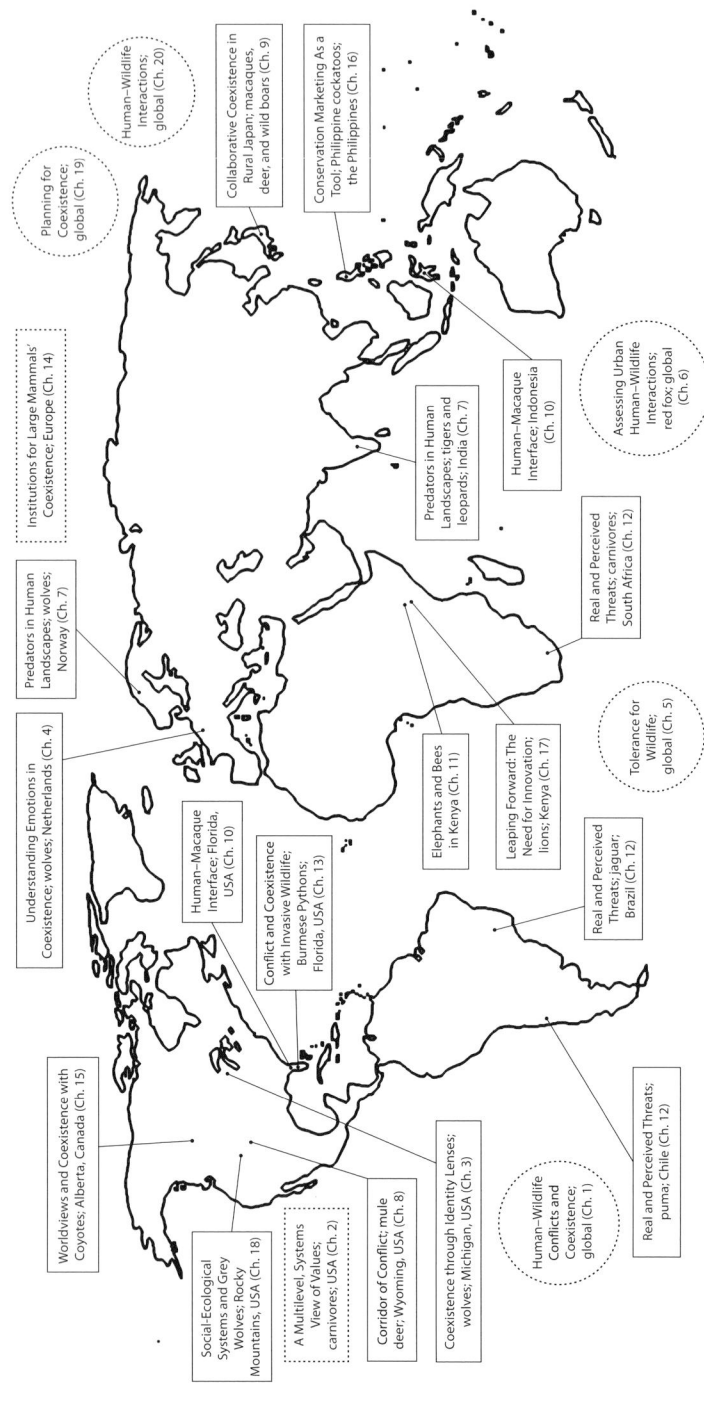

Figure 0.1 World map showing all the topics, species and locations addressed throughout the book. Map reproduced from Wikimedia and created by Jillian Knox using Illustrator.

Human–Wildlife Interactions; global (Ch. 20)

Planning for Coexistence; global (Ch. 19)

Collaborative Coexistence in Rural Japan; macaques, deer, and wild boars (Ch. 9)

Conservation Marketing As a Tool; Philippine cockatoos; the Philippines (Ch. 16)

Institutions for Large Mammals' Coexistence; Europe (Ch. 14)

Predators in Human Landscapes; tigers and leopards; India (Ch. 7)

Human–Macaque Interface; Indonesia (Ch. 10)

Assessing Urban Human–Wildlife Interactions; red fox; global (Ch. 6)

Predators in Human Landscapes; wolves; Norway (Ch. 7)

Real and Perceived Threats; carnivores; South Africa (Ch. 12)

Understanding Emotions in Coexistence; wolves; Netherlands (Ch. 4)

Elephants and Bees in Kenya (Ch. 11)

Leaping Forward: The Need for Innovation; lions; Kenya (Ch. 17)

Tolerance for Wildlife; global (Ch. 5)

Human–Macaque Interface; Florida, USA (Ch. 10)

Conflict and Coexistence with Invasive Wildlife; Burmese Pythons; Florida, USA (Ch. 13)

Real and Perceived Threats; jaguar; Brazil (Ch. 12)

Worldviews and Coexistence with Coyotes; Alberta, Canada (Ch. 15)

Social-Ecological Systems and Grey Wolves; Rocky Mountains, USA (Ch. 18)

A Multilevel, Systems View of Values; carnivores; USA (Ch. 2)

Corridor of Conflict; mule deer; Wyoming, USA (Ch. 8)

Coexistence through Identity Lenses; wolves; Michigan, USA (Ch. 3)

Human–Wildlife Conflicts and Coexistence; global (Ch. 1)

Real and Perceived Threats; puma; Chile (Ch. 12)

Human–Wildlife Conflicts and the Need to Include Coexistence

BEATRICE FRANK AND JENNY A. GLIKMAN

How societies view wildlife determines the outcome of human–wildlife interaction and, depending on the context, translates into a coexistence, neutral or conflict situation. Throughout history, the social meaning of wildlife has changed, shaping the role and the place wildlife hold in different societies, from beloved pets cherished at home (e.g. dogs) to despicable vermin to be eradicated from the wild (e.g. wolves). For example, white-tailed deer (*Odocoileus virginianus*) coexist and are often tolerated within urban human settlements in North America. Yet those species perceived as a threat (e.g. coyotes *Canis latrans*), or pest (e.g. raccoon *Procyon lotor*) or with deep-rooted social meaning, as in the case of the *big bad wolf* of Little Red Riding Hood fame in Western cultures, can be rejected by society, potentially turning an encounter with such species into a conflict situation (Varga 2009). Humans may accept, or not, a wildlife species depending on how that wildlife species is defined, and where a particular society draws the line between humans and wildlife spaces (Knight 2000; Philo & Wilbert 2000; Creager & Jordan 2002). In this chapter, we first describe the rise of physical and figurative boundaries between humans and wildlife, and how these have influenced the rise of human–wildlife conflict. Text Box 1.1 is included to provide a historical perspective of the changing relationship between humans and nature from an animal geography standpoint. This chapter also explores human–wildlife interactions and coexistence and introduces the conflict-to-coexistence continuum concept. The final section of this chapter focuses both on turning conflict into coexistence and on how the conflict-to-coexistence continuum concept can help researchers and practitioners better understand and address human–wildlife interactions.

Text Box 1.1 A Changing Relationship between Humans and Nature

Animal geography, a subfield of human geography, studies human–animal relations in terms of space and place, and critically interrogates the relationship between humans and other species (Johnson 2008). It describes the ways individuals, social groups and societies organize, perceive, provide meanings for and communicate about nature and wildlife. Throughout history, as critically pinpointed in the animal geography literature, humans have used an anthropocentric perspective to separate themselves from nature and wildlife (Oliver & Johnston 2000). Early nomadic and hunting societies were an integral component of their environment until they gradually settled, embraced an agricultural lifestyle and domesticated wildlife (Oelschlaeger 1991; Ingold 1994; Manning & Serpell 1994; Emel et al. 2002; Kruuk 2002). During this agricultural transition humans started to selectively include certain wildlife species within their realm, converting them into pets (e.g. dogs, cats) or livestock (e.g. horses, cattle) (Leighly 1963), while excluding others either because they were unable to be domesticated (e.g. antelopes) or were competing predators (e.g. carnivores) (Lee-Thorp et al. 2000; Berger & McGraw 2007). The growing separation of humans from nature enabled humans, in several societies, to believe that they were the centre of the world, and all other species were re-organized by humans as domesticated animals and/or wildlife species (Evernden 1992; Tovey 2003).

Over the following centuries, the idea that modifying nature and taming wildlife was essential to satisfy human needs and to subdue the perceived dark, undesired side of wilderness was strengthened by Western religious dogmas (Cronon 1995; Wolch & Emel 1998; Nash 2001). The rise of scientific reasoning further shaped this anthropocentric perspective by separating humans from the subjects to be studied (the true shape of nature could only be understood through scientific approaches), with humans no longer being part of nature and becoming sovereign over nature and all other species (Oelschlaeger 1991; Evernden 1992; Manning & Serpell 1994). With the development of the Romantic movements, and the start of industrialization and urbanization, the human anthropocentric perspective of nature became twofold. On one side, nature turned into a source of salvation from human society and a respite from the pressure of modern life (Manning & Serpell 1994; Cronon 1995; Nash 2001). As advocated by Henry D. Thoreau and John Muir, nature needed to be preserved for its beauty, spiritual truth, innocence and purity (Oelschlaeger 1991; Manning & Serpell 1994; Nash 2001). On the other side, the rapid human population growth, industrialization and urbanization created pressures for the allocation and use of resources to increase human wealth and livelihoods. By advocating for the conservation of nature through planned use and renewal, Gifford Pinchot introduced the idea of the wise and economically efficient employment of resources, including wildlife (Rothman 2000; Nash 2001). As a result, a division between preserving and exploiting nature and wildlife took root – shaping the still ongoing debate of nature and wildlife *as only for use* versus *to be cared for* (Evernden 1992; Rothman 2000; Nash 2001).

Over the last centuries, nature and wildlife have become commodities scarce consumption goods and objects with a market value advertised by media commercials, sold as gadgets in malls, consumed in park resorts and commercialized through many other venues (Price 2000; Polanyi 2001). The definition of nature as a commodity, however, conflicts with the idea of nature as pristine and untouched, worth protecting and an emblem of environmental movements. The contradictions inherent in rationalizing nature and wildlife from an anthropomorphized perspective as goods to be consumed and as valuable for their own sake have played, and still play, a fundamental role in shaping societal values, beliefs and attitudes towards nature, and determine whether a human–wildlife interaction is perceived as a coexistence, neutral or conflict experience.

1.1 BOUNDARIES BETWEEN HUMANS AND WILDLIFE

Human–wildlife interactions, either conflict or coexistence, are driven by the separation between *us* – the humans – and *them* – the non-human species. This separation often represents the root cause of human–wildlife conflicts through the creation of exclusive places – either for humans (e.g. cities) or for wildlife (e.g. protected areas). Indeed, humans have increasingly drawn boundaries to keep wildlife outside the human space. Those boundaries are sometimes physical, such as fences and walls around communities, but also figurative as expressed by the concept of wilderness and protected areas, which represent places where wildlife should live (Knight 2000; Creager & Jordan 2002). Boundaries are territories of separation, as much as contact; they influence the physical and ecological features present in a place while shaping participants' values, attitudes and behaviours towards wildlife (Frank & Bath 2012). The meaning we confer to nature and the way we relate to other species will ultimately affect whether human–wildlife interactions turn into a conflict or coexistence situation.

Through the creation of boundaries, human society has further distanced itself from nature, adding complexity to the society–nature relationship and laying the groundwork for conflict with and over wildlife. Boundaries influence the relationship between humans and wildlife and add a further layer of complexity to human–wildlife interactions (Ripple et al. 2014; Liordos et al. 2017). Yet the separation through boundaries between humans and wildlife is not as clear as one might think. Indeed, boundaries become blurry in countries where people and wildlife have shared the same landscape for millennia, and where protected areas were established within human territories (e.g. Southern Europe, India) in contrast with places where natural parks have been established in remote areas or by relocating people (e.g. North America, Africa) (Woodroffe 2000; Jenkins & Keal 2004; Woodroffe et al. 2005). The expansion of human settlement near wildlife, and the arrival of species such as deer and foxes in urban places, have made it even more challenging to distinguish between human and wildlife spaces. It has become an everyday occurrence for humans and wildlife to cross boundaries and enter into each other's space, leading to increased human–wildlife interactions, and resulting in controversies among various social groups, institutions and ideologies over wildlife, their meaning and belonging.

1.1.1 The Rise of Human–Wildlife Conflicts

In the traditional anthropocentric definition of human–wildlife conflict (HWC), the human role is absent from the conflict analysis and solutions are developed by only focusing on wildlife. Indeed, HWC has been addressed by definitions centred on the competition of humans and wildlife over space, resources and livelihood (Knight 2000). Often the focus is on wildlife threatening human interests, safety and well-being (White & Ward 2010). More recently, authors have included the human dimensions in the HWC equation by adding the negative effects/actions of humans and/or wildlife on the needs of each other (Conover 2002; Madden 2004a, 2004b; Woodroffe et al. 2005; Nyhus 2016).

As initially pointed out by Peterson et al. (2010), conceptualizing HWC as direct oppositions between humans and wild animals implies that wildlife are not acting opportunistically but consciously against humans. However, in recorded history, only humans have been known to consciously target and kill non-human animals not for subsistence or survival reasons. The perception of wildlife impairing human interests, by damaging crops, livestock predation and threatening human safety, frequently motivates responses such as retaliatory killing of individual animals or persecution of entire wildlife populations (Woodroffe et al. 2005; Madden 2008; Hazzah et al. 2014; Nyhus 2016; Ravenelle & Nyhus 2017; Gebresenbet et al. 2018). With a steady growth of human populations and the connected needs for space and resources, the picture of HWC has become even more complex as humans and wildlife are forced to live in closer proximity and cross each other's boundaries constantly (Blackwell et al. 2017; Senthilkumar et al. 2017; Songhurst 2017), pushing human–wildlife interactions towards conflict.

When the balance is tipped towards conflict, species can be banned from the human realm, seen as negatively interfering with *our* material properties (e.g. causing damage to pets, livestock and crops), and/or *our* psycho-sociological well-being (e.g. fear of being attacked and reduced health) no matter what they do (Barua et al. 2013). For example, negative experiences with wildlife (Kansky et al. 2016), concerns about safety (Sponarski et al. 2015), economic costs of living with wildlife (Treves & Bruskotter 2014) and labour and opportunity costs (Dickman 2010) are some of the factors, among many others, that can lead to HWC (Woodroffe et al. 2005; Dickman 2010; Frank 2016). These conflicts may go beyond evident physical impacts, as they can trigger social and psychological backlash to those who experience them (e.g. poor attendance or performance in schools, decrease in well-being, loss of sleep for guarding crops and food insecurity) (Ogra 2008; Barua et al. 2013; Kansky et al. 2016; Nyhus 2016; Yurco et al. 2017). Such hidden impacts are difficult to measure and can be magnified by human–human conflict (HHC) over what reactions towards wildlife are acceptable under different circumstances (i.e. poaching, killing, negatively affecting species and wildlife habitats). A sense of impotence about governing systems decisions over wildlife conservation and power imbalance with other stakeholders, may result in those feeling that they bear all the cost of conservation redirecting their anger to wild species rather than addressing the power issues with other players in the decision-making process (Barua et al. 2013; Nyhus 2016). Power dynamics between humans can thus exacerbate conflict over wildlife, and turn human–wildlife conflicts into human rights and environmental justice issues. So, is it really a human–wildlife conflict, and for whom? In the next section we address this key question to portray the multiple dimensions and the challenges behind the concept of human–wildlife conflicts, thus allowing us to move from focusing on conflict to looking at interactions between humans and wildlife.

1.1.2 Is It Really a Conflict, and for Whom?

The term HWC, with its multiple implications, has commonly been used in the conservation literature (Hill 2017), but to whom does it really refer? People engaged in wildlife conservation and/or management use this concept as an operative tool that helps define when a human–wildlife interaction turns into a negative situation/experience. Yet the definition of HWC often does not fully consider the psychological dispositions of those living and sharing the landscape with wildlife. Indeed, HWC definitions used in conservation do not really take into account the following questions: How do local people perceive wildlife? Is a certain species problematic for them? Are these perceptions mutually exclusive? Is the killing the main problem, and is it caused by deep-rooted conflicts? As defined by Madden and McQuinn (2014), deep-rooted conflicts are pre-existing and often non-negotiable disagreements between different stakeholders, which can arise from contentious history between and among groups, power imbalance, opposing values and different groups' identities. Because of deep-rooted conflicts, any interaction among such groups can add meaning and emotion to each new dispute, making any type of conflict an unsolvable problem. Many other questions that include local perspective can be added to the above list. With this list of questions, we aim at making the point – is it a conflict? If so, for whom – the conservationist, the local people or for others?

It is also important to bear in mind that a *conflict* in one specific situation may not be perceived as such in another similar one due to culture, location, severity and time, among other factors. Indeed, some perceived conflicts are more about social and cultural values than about actual impacts (McIntyre et al. 2008; Soulsbury & White 2015). For example, in some societies the killing of wildlife is cultural or commercial-related, and has nothing to do with the visible material impacts and conflict we hear and/or read about in the media or in peer review articles (Hazzah et al. 2014; Dayer et al. 2017). This is the case of killing lions in Maasai culture (see Chapter 17), or of the notorious Cecil – the lion that nowadays symbolizes trophy hunting. Neither of these examples has anything to do with the traditional definition of HWC (Nelson et al. 2016).

To overcome the *for whom* challenge, recent literature has shifted towards recognizing how, in many cases, the HWC framework hides a human–human component; a friction between different stakeholders over different interests, including how to protect and conserve wildlife (Manfredo & Dayer 2004; Peterson et al. 2010; Redpath et al. 2015; Bhatia et al. 2017; Madden & McQuinn 2017). To distinguish explicitly the different factors of HWC, Young et al. (2010) suggested extracting and separating the two components of HWC: (1) the impacts caused by wildlife on humans, and (2) the HHC between those defending pro-wildlife positions and those defending other positions. Because of the conservation-oriented perspective of HWC, the traditional approach to solve conflicts has focused on the first component by reducing the tangible side of impacts caused by wildlife through bio-physical and economic solutions (Madden 2008; Pooley et al. 2017). Examples of interventions to mitigate conflict include lethal and non-lethal wildlife management measures, technical fixes for preventing damage such as building fences, as well as financial instruments to offset the direct impact of wildlife on human belongings (for reviews see Eklund et al. 2017; van Eeden et al. 2018). However, these wildlife management efforts have, in some instances, failed to reduce HWC as they have fallen short of considering the second component of HWC, the social drivers and root causes of HHC (Dickman 2010; Young et al. 2010; Draheim et al. 2015).

Often HHC has been completely overlooked as conservationists, managers and decision-makers position themselves as neutral subjects in the HWC. Such a perspective fails to recognize that: (1) researchers, conservationists and other decision-makers hold value-laden and socially constructed perspectives about nature, and (2) local people are the ones bearing the consequences of living with wildlife; they are the ones sharing the landscapes and often coexisting with the species the conservationists want to protect. This separation between conservationists' perspective/strategies and the *other* stakeholders has been documented in the literature through various case studies (e.g. Logsdon et al. 2015; van Heel et al. 2017). A typical example of management efforts that have failed to reduce HWC is around large carnivores, in Scandinavian countries (Skogen & Krange 2003; Bisi et al. 2010) and in North America (Lute & Gore 2014; Browne-Nuñez et al. 2015) where people perceived the presence of wolves as imposed by the authorities, or in Tanzania, where tribes believe lions are sent by their rivals to jeopardize their communities' safety and well-being (Dickman & Hazzah 2016). The separation and divergence in views on how to manage natural resources can hinder support for conservation (Hill 2017; Madden & McQuinn 2017). Opposition towards conservation can indeed become even more severe when local communities perceive that their own needs are being subordinated to those of wildlife (Madden 2008; Songhurst 2017). As a result, conservation interventions find local resistance or fail, as they do not build trust and transparency between groups interested in the human–wildlife interaction, hence addressing the deep-rooted reasons behind HWC (Madden & McQuinn 2014, 2017; Dayer et al. 2017; Hill 2017).

1.2 RECONCILIATION AND COEXISTENCE WITH WILDLIFE

Through their pioneering research, Madden (2004a, 2004b) and Woo-droffe et al. (2005) have initiated a shift in wildlife conservation and management perspectives, by including tolerance and coexistence in HWC and by recognizing that humans are not only part of the problem, but also part of the solution. The integration of tolerance and coexistence into the HWC discourse has contributed to the alleviation of the imme-diate feeling and perceived perspective of antagonism and separation between humans and wildlife (Peterson et al. 2010; Frank 2016; Hill 2017). Accordingly, an increased importance has been given to the concepts of tolerance and coexistence, and their use and role in refram-ing conservation challenges and opportunities (Frank 2016; Pooley et al. 2017). The inclusion of tolerance and coexistence in HWC is helping conservationists to recognize that wildlife can thrive in human land-scapes, and that most of the time people do live with wildlife and experience impacts or compete for space without calling such inter-actions conflicts. For example, some people will tolerate losing part of their crop or some livestock to wildlife as part of the risks of farming and cultural benefits perceived from wildlife (Goodale et al. 2015).Thus, tolerance and coexistence are more than the ability of human and wildlife to co-occur in the same place, often at the same time. Tolerance and coexistence are about the ability of humans and wildlife to interact, and through those interactions build a community that is integrated, and can cope with moderate and manageable competition (López-Bao et al. 2015; Soulsbury & White 2015; Carter & Linnell 2016; Chapron & López-Bao 2016). For example, in Romania, as well as in other European countries (e.g. Croatia, Italy), where brown bears have successfully co-inhabited the landscape with humans, learning through positive experi-ence plays an important role for coexistence (Majić et al. 2011; Glikman et al. 2012; Dorresteijn et al. 2016).

The terms coexistence and tolerance are becoming increasingly popular in human–wildlife interaction literature (Nyhus 2016; Hill 2017). Yet the meanings of tolerance and especially of coexistence are generally used implicitly, and therefore are not defined in conservation literature (e.g. Karanth & Chellam 2009; Ripple et al. 2014). Indeed, the meanings of tolerance and coexistence remain unclear, especially as they are used to describe attitudes and behaviours across social and natural science perspectives (Treves & Bruskotter 2014; Carter & Linnell 2016; Inskip et al. 2016). In recent articles, coexistence has been conceptualized and operationalized for research and conservation use. According to Frank (2016) 'coexistence takes place when the interests of humans and wildlife are both satisfied, or when a compromise is negotiated to allow the existence of both humans and wildlife together' (Frank 2016, p. 739). Chapron and López-Bao (2016) describe the term of coexistence from an ecological community perspective, using as an example large carnivore populations roaming free in the human-made European landscape. They argue that coexistence happens when species have different ecological niches and do moderately compete for the same resources. These authors question the feasibility of coexistence by wondering if *super predators* like humans who alter ecological and evolutionary processes globally (Darimont et al. 2015) have the ability to become less competitive and differentiate their niche to avoid conflict with other species, especially other carnivores, which share our same need for space and resources. Carter and Linnell (2016) further advance the definition to a broader landscape level by stating that coexistence arises in dynamic and sustainable socio-ecological systems where humans and wildlife are integrated and co-adapt to living together in space and over time. From this point of view humans and wildlife are mutually adaptable – they co-adapt – when they 'are able to change their behaviour, learn from experience, and pursue their own interests with respect to each other' (Carter & Linnell 2016, p. 577). Morehouse and Boyce (2017) offer another interpretation of coexistence, which occurs when wildlife share the same landscape with humans without impacting human safety, property or rights. Within shared landscapes, effective institutions ensure the presence of wildlife in the long term while fostering social legitimacy through dialogue with and between groups, and by pursuing a tolerable level of wildlife-related risks. These definitions look at coexistence at different scales, yet unify multidisciplinary perspectives and consider human–wildlife and human–human interactions.

Interestingly, there is more conceptual variation in the conservation literature around tolerance than coexistence. Indeed, depending on the context, tolerance is being defined either as a behaviour or behavioural intention (e.g. Hazzah et al. 2009; Marchini & Macdonald 2012; Bruskotter et al. 2015; Gebresenbet et al. 2018) or as an attitude (e.g. Manfredo & Dayer 2004; Zimmermann et al. 2005; Treves 2012; Lindsey et al. 2013; Harvey et al. 2017). Adopting the definition of tolerance as an attitude or mindset that signals an intention, coexistence is then referred to as a state or an array of behaviours (Treves 2012; Frank 2016; Harvey et al. 2017). As stressed by Treves (2012) tolerance may not translate in behaviours. Indeed, even when wildlife causes impacts, humans can still tolerate them if they perceive some sort of benefit (e.g. spiritual, economic) (Madden 2004a, 2008; Goodale et al. 2015). Hence, tolerance is described as the passive acceptance of a wildlife population (Treves & Bruskotter 2014), which depends upon the risk–benefit beliefs people have towards a species and the related perceptions of control over hazard, social trust, conflicts among groups and the effect such species generates (Bruskotter & Wilson 2014; Inskip et al. 2016; see Chapter 5). Based on this, some researchers use the terms *acceptance* and *tolerance* synonymously (Bruskotter & Fulton 2012; Inskip et al. 2016). Recent research further indicates that outer variables (i.e. experience) and inner variables (i.e. value orientations, empathy, taxonomic bias, personal norms, emotions) drive perceptions of the costs-benefits of living with wildlife (see Chapters 2 and 4). Such description of tolerance acknowledges that people sharing the landscape with wildlife will bear added costs – physical as much as psychological – of living with wild species and yet still be willing to have wildlife in their proximity (Kansky et al. 2016). Tolerance is about people not interfering with or harming species, and bearing the costs/risks of sharing the landscape with wild animals, as much as it is about accepting feelings, habits, beliefs or behaviours differing from, or conflicting with, one's own. Even when wildlife cause conflicts, people can be tolerant towards them if the species are perceived as beneficial to the personal, spiritual, cultural, economic, social or political well-being of society. 'Tolerance can also be the result of adjustment, for instance, when local residents would be willing to accept damage caused by wildlife up to a certain threshold' (Frank 2016, p. 740).

1.3 THE CONFLICT-TO-COEXISTENCE CONTINUUM

For some, the discussion around conflict or coexistence may be a matter of semantics. However, while working towards conservation solutions, focusing on mechanisms of coexistence represents a more positive approach than simply mitigating conflicts. As stressed by Peterson et al. (2010) the words we use while describing human–wildlife interactions matter, as defining an interaction through a positive label may help focus on affinity and empathy between human and wildlife while still acknowledging that such species compete over space and limited resources. However, shifting from labelling human–wildlife conflict to human–wildlife coexistence may not be enough. There is a need to consider conflict and coexistence as they relate to each other.

To foster the inclusion of tolerance and coexistence in human–wildlife research, Frank (2016) introduced the conflict-to-coexistence continuum. This continuum spans from negative to positive attitudes and/or behaviours, which defines the different degrees of conflict and coexistence that characterize human–wildlife interactions. On the extreme end of the conflict side of the continuum, negative attitudes/ behaviours can result in retaliatory killing of wildlife, support for eradication policies, and/or the sabotage of species conservation. Moving away from this end position, attitudes/behaviours become less negative and/or extreme, with people disagreeing and opposing species management and conservation, but likely not taking extreme actions against wildlife. This section includes support for wildlife management that welcomes lethal control or species population management through relocation and/or selective killing of problematic individuals. In contrast, killing is not a retaliatory action done by affected stakeholders or public, but a management intervention undertaken by wildlife agencies or elected institutions based on public requests/needs. The continuum moves then towards neutral or mixed attitudes/behaviours, where individuals do not strictly fall into the conflict or coexistence section of the continuum. People within this part of the continuum may not be interested in wildlife and thus remain indifferent towards wildlife issues. Passive tolerance and coexistence characterize this section, which is followed by the positive end of the continuum. In this last section of the continuum, attitudes/behaviours span 'from full integration and respect for wildlife within the human landscape to deep affiliation with nature and willingness to forgo one's own interests to further those of wildlife' (Frank 2016, p. 740). Humans favouring the needs of wildlife (e.g. the development and maintenance of strict nature reserves and wilderness areas, donating for wildlife conservation and transforming their private land in covenants) represent some examples of the extreme positive side of the continuum.

'Conflict, coexistence, and tolerance are context-laden and vary across human–wildlife interactions. Their meaning can take different nuances depending on the socio-cultural background, types of conservation law enforcement, economic benefits, and other aspects of societies living with wildlife' (Frank 2016, p. 741). For example, coexistence with carnivores has been shown to be more widespread in emerging countries where people are accustomed to living with wild species and accepting the risks of sharing the landscape with predators as part of their lifestyle (Inskip et al. 2016). Indeed conflict and coexistence are not locked to a fixed point along the continuum. Individual attitudes/behaviours towards a species may change over time, across space and in degree, with individuals shifting along the continuum as their interactions with wildlife evolve. The continuum represents a framework to help us in reasoning about the relationship human and species have, and how human–wildlife and human–human interactions may favour a shift in position along the continuum. Changes can be fostered through culture, location of residence, emotions, world-views and social identity – all themes explored in this book to further develop this concept and test its validity as a working tool.

The conflict-to-coexistence continuum will represent the leitmotiv of this book, to help us think about these concepts not only as antonyms, but as connected themes characterized by the complexity of dealing with human–wildlife and human–human interactions. To work towards solutions and applications that maximize conservation success, it is necessary to better portray the multiple dimensions influencing human–wildlife relationships by drawing attention to other possible, complementary or competing pathways by which we may address human–wildlife interactions (e.g. minimizing damage, mitigating social conflict, promoting coexistence, planning for conservation, optimizing wildlife value). Exploring how conflicts can be reduced to the point where people start to accept wildlife in their proximity is critical to this endeavour, and represents a pathway by which we may begin to shift towards mechanisms that enhance coexistence and tolerance towards wildlife.

1.4 TURNING CONFLICT INTO COEXISTENCE

Over time, human–wildlife interactions have increased and diversified, as has the discourse around them. Humans went from being mainly prey in prehistoric times to being defined as hyper and super predators (Darimont et al. 2015; Chapron & López-Bao 2016). In this new anthropocentric conservation paradigm, humans have been recognized as manipulating and shaping the landscape to a dramatic extent, one that is causing the mass extinctions of species, the diffusion of invasive species and the change of climate among many environmental challenges. While humans are the cause of many environmental changes, they can also become the solution to such challenges – a vision that can offer prolific innovation potential for conservation (Kueffer & Kaiser-Bunbury 2014; Fisher 2016).

Humans are now compelled to accept that wildlife is part of their backyard (Soulsbury & White 2015), and are learning to live with wildlife in many ways (Carter & Linnell 2016; Gupta et al. 2017). As the relationship between humans and wildlife evolves, so does the experience of their encounters. The human–wildlife relationship is not a static condition. Quite the contrary, the relationship is based on the evolving definition of what nature means to society, where humans and wildlife belong and of the dynamic interactions humans–species have, which can at times induce conflict–to–coexistence reactions (Morzillo et al. 2014; Frank 2016; Yurco et al. 2017). As Yurco et al. (2017) pointed out, conflict is neither merely present or absent, but consistently negotiated through daily experience and occurring at certain times in certain spaces. The real challenge is how to catalyse a paradigm shift from human–wildlife conflict discourse to human–wildlife interactions and coexistence dialogue for a more positive and inclusive relation with wildlife and nature. We believe that this book and its authors have embarked in this discussion, and are pushing the discourse of human–wildlife interactions in new and exciting directions.

1.5 RECOMMENDATION AND FUTURE DIRECTIONS

- Understanding that societies have conferred a meaning to nature and wildlife and defined the physical and psychological space for human and species is key when looking at human–wildlife interactions as often conflicts are deep-rooted in the human–nature separation concept, in human–human conflicts and other reasons that go beyond just the species at hand.
- Human–wildlife interactions need to be considered as context-specific, multidimensional and complex. It is important to consider neutral-to-positive interactions in order to really evaluate the extent/severity of the conservation challenge and provide solutions that go beyond just minimizing conflicts and for whom.
- A cohesive understanding of what coexistence and tolerance concepts mean and entail represents an important step forward for wildlife conservation and management success. Such an understanding can shift practitioner approaches towards promoting positive behaviours and attitudes rather than addressing just negative interactions with

wildlife, thus enhancing humans' willingness to coexist and tolerate wildlife in the long term.

- The conflict-to-coexistence continuum helps reasoning about conflict-to-coexistence attitudes/behaviours. Such an understanding is key to develop policy and management approaches that are tailored to the specific intensity and form of human–wildlife interaction of interest.
- Future research should showcase coexistence and tolerance. It is indeed important to start conducting research that documents positive attitudes/behaviour and explore factors (i.e. values, culture and location of residence) that foster positive psychological dispositions and coexistence towards wildlife.

1.6 References

Barua, M., Bhagwat, S. A. & Jadhav, S. (2013). The hidden dimensions of human–wildlife conflict: Health impacts, opportunity and transaction costs. *Biological Conservation*, 157, 309–16.

Berger, L. R. & McGraw, W. S. (2007). Further evidence for eagle predation of, and feeding damage on, the Taung child. *South African Journal of Science*, 103, 496–8.

Bhatia, S., Redpath, S. M., Suryawanshi, K. & Mishra, C. (2017). The relationship between religion and attitudes toward large carnivores in Northern India? *Human Dimensions of Wildlife*, 22(1), 30–42.

Bisi, J., Liukkonen, T., Mykra, S., Pohja-Mykra, M. & Kurki, S. (2010). The good bad wolf–wolf evaluation reveals the roots of the Finnish wolf conflict. *European Journal of Wildlife Research*, 56, 771–9.

Blackwell, B. F., DeVault, T. L., Fernandez-Juricic, E., Gese, E. M., Gilbert-Norton, L. & Breck, S. W. (2017). No single solution: Application of behavioural principles in mitigating human–wildlife conflict. *Animal Behaviour*, 120, 245–54.

Browne-Nuñez, C., Treves, A., MacFarland, D., Voyles, Z. & Turng, C. (2015). Tolerance of wolves in Wisconsin: A mixed-methods examination of policy effects on attitudes and behavioural inclinations. *Biological Conservation*, 189, 59–71.

Bruskotter, J. T. & Fulton, D. C. (2012). Will hunters steward wolves? A comment on Treves and Martin. *Society & Natural Resources*, 25, 97–102.

Bruskotter, J. T., Singh, A., Fulton, D. C. & Slagle, K. (2015). Assessing tolerance for wildlife: Clarifying relations between concepts and measures. *Human Dimensions of Wildlife*, 20, 255–70.

Bruskotter, J. T. & Wilson, R. S. (2014). Determining where the wild things will be: Using psychological theory to find tolerance for large carnivores. *Conservation Letters*, 7, 158–65.

Carter, N. H. & Linnell, J. D. C. (2016). Co-adaptation is key to coexisting with large carnivores. *Trends in Ecology & Evolution*, 31(8), 575–8.

Chapron, G. & López-Bao, J. V. (2016). Coexistence with large carnivores informed by community ecology. *Trends in Ecology & Evolution*, 31(8), 578–80.

Conover, M. R. (2002). *Resolving Human–Wildlife Conflicts.* Boca Raton, FL: CRC Press.

Creager, A. N. H. & Jordan, W. C. (2002). *The Animal/Human Boundary: Historical Perspectives (Studies in Comparative History).* Rochester, NY: University of Rochester Press.

Cronon, W. (1995). The trouble with wilderness; or, getting back to the wrong nature. In W. Cronon, ed., *Uncommon Ground: Rethinking the Human Place in Nature.* New York, NY: W.W. Norton, pp. 69–90.

Darimont, C. T., Fox, C. H., Bryan, H. M. & Reimchen, T. E. (2015). Human impacts: The unique ecology of human predators. *Science*, 349(6250), 858–60.

Dayer, A. A., Williams, A., Cosbar, E. & Racey, M. (2017). Blaming threatened species: Media portrayal of human–wildlife conflict. *Oryx*, 1–8.

Dickman, A. J. (2010). Complexities of conflict: The importance of considering social factors for effectively resolving human–wildlife conflict. *Animal Conservation*, 13, 458–66.

Dickman, A. J. & Hazzah, L. (2016). Money, myths and man-eaters: Complexities of human–wildlife conflict. In F. M. Angelici, ed., *Problematic Wildlife: A Cross-Disciplinary Approach.* Cham, Switzerland: Springer International Publishing, pp. 339–56.

Dorresteijn, I., Milcu, A. I., Leventon, J., Hanspach J. & Fischer, J. (2016). Social factors mediating human–carnivore coexistence: Understanding thematic strands influencing coexistence in Central Romania. *Ambio*, 45(4), 490–500.

Draheim, M. M., Madden, F., McCarthy, J.-B. & Parsons E. C. M. (2015). *Human–Wildlife Conflict: Complexity in the Marine Environment.* Oxford: Oxford University Press.

Eklund, A. López-Bao, J. V. Tourani M., Chapron, G. & Frank, F. (2017). Limited evidence on the effectiveness of interventions to reduce livestock predation by large carnivores. *Scientific Reports*, 7, 2097.

Emel, J., Wilbert, C. & Wolch, J. (2002). Animal geographies. *Society & Animals*, 10(4), 407–12.

Evernden, N. (1992). *The Social Creation of Nature.* Baltimore, MD: The Johns Hopkins University Press.

Fisher, M. (2016). Whose conflict is it anyway? Mobilizing research to save lives. *Oryx*, 50(3), 377–8.

Frank, B. (2016). Human–wildlife conflicts, the need to include tolerance and coexistence: An introductory comment. *Society & Natural Resources*, 29(6), 738–43.

Frank, B. & Bath, A. J. (2012). Does it matter where people live? Wildlife management across protected area boundaries. *Journal of Science & Management of Protected Areas (SAMPAA)*, 1, 12–21.

Gebresenbet, F., Baraki, B., Yirga, G., Sillero-Zubiri, C. & Bauer, H. (2017). A culture of tolerance: Coexisting with large carnivores in the Kafa Highlands, Ethiopia. *Oryx*, 52(4), 751–60.

Glikman, J. A., Vaske, J. J., Bath, A. J., Ciucci, P. and Boitani, L. (2012). Residents' support for wolf and bear conservation: The moderating influence of knowledge. *European Journal of Wildlife Research*, 58, 295–302.

Goodale, K., Parsons, G. J. & Sherren, K. (2015). The nature of the nuisance – damage or threat – determines how perceived monetary costs and cultural benefits influence farmer tolerance of wildlife. *Diversity*, 7, 318–41.

Gupta, N., Rajvanshi, A. & Badola, R. (2017). Climate change and human–wildlife conflicts in the Indian Himalayan biodiversity hotspot. *Current Science*, 113(5), 846–7.

Harvey, R. G., Briggs-Gonzalez, V. S. & Mazzotti, F. J. (2017). Conservation payments in a social context: Determinants of tolerance and behavioural intentions towards wild cats in northern Belize. *Oryx*, 51(4), 730–41.

Hazzah, L., Borgerhoff, M. M. & Frank, L. (2009). Lions and warriors: Social factors underlying declining African lion populations and the effect of incentive-based management in Kenya. *Biological Conservation*, 142(11), 2428–37.

Hazzah, L., Dolrenry, S., Naughton, L., Edwards, C. T. T., Mwebi, O., Kearney, F. & Frank, L. (2014). Efficacy of two lion conservation programs in Maasailand, Kenya. *Conservation Biology*, 28, 851–60.

Hill, C. (2017). Introduction. Complex problems: Using a biosocial approach to understanding human–wildlife interactions. In C. M. Hill, A. D. Webber & N. E. C. Priston, eds., *Understanding Conflicts about Wildlife: A Biosocial Approach*. Oxford: Berghahn, pp. 1–14.

Ingold, T. (1994). From trust to domination: An alternative history of human–animal relations. In A. Manning & J. Serpell, eds., *Animals and Human Society: Changing Perspectives*. London: Routledge, pp. 1–22.

Inskip, C., Carter, N., Riley, S., Roberts, T. & MacMillan, D. (2016). Toward human–carnivore coexistence: Understanding tolerance for tigers in Bangladesh. *PLoS ONE*, 11(1), e0145913.

Jenkins, J. & Keal, A. (2004). *The Adirondack Atlas*. Syracuse, NY: Syracuse University Press.

Johnson, C. (2008). Beyond the clearing: Towards a dwelt animal geography. *Progress in Human Geography*, 32(5), 633–49.

Kansky, R., Kidd, M. & Knight, A. T. (2016). A wildlife tolerance model and case study for understanding human–wildlife conflicts. *Biological Conservation*, 201, 137–45.

Karanth, K. U. & Chellam, R. (2009). Carnivore conservation at the crossroads. *Oryx*, 43, 1–2.

Knight, J. (2000). *Natural Enemies: People–Wildlife Conflicts in Anthropological Perspective*. London: Routledge.

Kruuk, H. (2002). History of a conflict: Carnivores and the first hominids. In H. Kruuk, *Hunter and Hunted: Relationship between Carnivores and People*. Cambridge: Cambridge University Press, pp. 103–15.

Kueffer, C. & Kaiser-Bunbury, C. N. (2014). Reconciling conflicting perspectives for biodiversity conservation in the Anthropocene. *Frontiers in Ecology & the Environment*, 12, 131–7.

Lee-Thorp, J., Thackeray, J. F. & Van der Merwe, N. (2000). The hunters and the hunted revisited. *Journal of Human Evolution*, 39, 565–76.

Leighly, J. (1963). *Land and Life: A Selection from the Writings of Carl Ortwin Sauer*. Berkley and Los Angeles: University of California Press.

Lindsey, P. A., Havemann, C. P., Lines, R., Palazy, L., Price, A. E., Retief, T. A., Rhebergen, T. & Van der Waal, C. (2013). Determinants of persistence and tolerance of carnivores on Namibian ranches: Implications for conservation on Southern African private lands. *PLoS ONE*, 8(1), e52458.

Liordos, V., Kontsiotis, V. J., Georgari, M., Baltzi, K. & Baltzi, I. (2017). Public acceptance of management methods under different human–wildlife conflict scenarios. *Science of the Total Environment*, 579, 685–93.

Logsdon, R. A., Kalcic, M. M., Trybula, E. M., Chaubey, I. & Frankenberger, J. R. (2015). Ecosystem services and Indiana agriculture: Farmers' and conservationists' perceptions. *International Journal of Biodiversity Science, Ecosystem Services & Management*, 11(3), 264–82.

López-Bao, J. V., Kaczensky, P., Linnell, J. D. C., Boitani, L. & Chapron, G. (2015). Carnivore coexistence: Wilderness not required. *Science*, 348, 871–72.

Lute, M. L. & Gore, M. L. (2014). Stewardship as a path to cooperation? Exploring the role of identity in intergroup conflict among Michigan wolf stakeholders. *Human Dimensions of Wildlife*, 19(3), 267–79.

Madden, F. (2004a). Creating coexistence between humans and wildlife: Global perspectives on local efforts to address human–wildlife conflict. *Human Dimensions of Wildlife*, 9, 247–57.

Madden, F. (2004b). Can traditions of tolerance help minimize conflict? An exploration of cultural factors supporting human–wildlife coexistence. *Policy Matters*, 13, 234–41.

Madden, F. (2008). The growing conflict between humans and wildlife: Law and policy as contributing and mitigating factors. *Journal of International Wildlife Law & Policy*, 11, 189–206.

Madden, F. & McQuinn, B. (2014). Conservation's blind spot: The case for conflict transformation in wildlife conservation. *Biological Conservation*, 178, 97–106.

Madden, F. & McQuinn, B. (2017). Conservation conflict transformation: Addressing the missing link in wildlife conservation. In C. M. Hill, A. D. Webber & N. E. C. Priston, eds., *Understanding Conflicts about Wildlife: A Biosocial Approach*. Oxford: Berghahn, pp. 148–69.

Majić, A., Marino, A., Huber, D. & Bunnefeld, N. (2011). Dynamics of public attitudes toward bears and the role of bear hunting in Croatia. *Biological Conservation*, 144(12), 3018–27.

Manfredo, M. J. & Dayer, A. A. (2004). Concepts for exploring the social aspects of human–wildlife conflict in a global context. *Human Dimensions of Wildlife*, 9, 1–20.

Manning, A. & Serpell, J. (1994). *Animals and Human Society: Changing Perspectives*. London: Routledge.

Marchini, S. & Macdonald, D. W. (2012). Predicting ranchers' intention to kill jaguars: Case studies in Amazonia and Pantanal. *Biological Conservation*, 147, 213–21.

McIntyre, N., Moore, J. & Yuan, M. (2008). A place-based, values-centered approach to managing recreation on Canadian crown lands. *Society & Natural Resources*, 21, 657–70.

Morehouse, A. T. & Boyce, M. S. (2017). Troublemaking carnivores: Conflicts with humans in a diverse assemblage of large carnivores. *Ecology and Society*, 22(3), art. 4.

Morzillo, A., de Beurs, K. & Martin-Mikle, C. (2014). A conceptual framework to evaluate human–wildlife interactions within coupled human and natural systems. *Ecology & Society*, 19, art. 44.

Nash, R. (2001). *Wilderness and the American Mind*, 3rd edn. New Haven, CT: Yale University Press.

Nelson, M. P., Bruskotter, J. T., Vucetich, J. A. & Chapron, G. (2016). Emotions and the ethics of consequence in conservation decisions: Lessons from Cecil the Lion. *Conservation Letters*, 9(4), 302–6.

Nyhus, P. J. (2016). Human–wildlife conflict and coexistence. *Annual Review of Environment & Resources*, 41, 143–71.

Oelschlaeger, M. (1991). *The Idea of Wilderness: From Prehistory to the Age of Ecology*. Binghamton, NY: Vail-Ballou Press.

Ogra, M. V. (2008). Human–wildlife conflict and gender in protected area borderlands: A case study of costs, perceptions, and vulnerabilities from Uttarakhand (Uttaranchal), India. *Geoforum*, 39, 1408–22.

Oliver, P. & Johnston, H. (2000). What a good idea! Ideology and frames in social movement research. *Mobilization: An International Journal*, 4, 37–54.

Peterson, M. N., Birckhead, J. L., Leong, K., Peterson, M. J. & Peterson. T. R. (2010). Rearticulating the myth of human–wildlife conflict. *Conservation Letters*, 3, 74–82.

Philo, C. & Wilbert, C. (2000). *Animal Spaces, Beastly Places: New Geographies of Human–Animal Relations*. London: Routledge.

Polanyi, K. (2001). *The Great Transformation: The Political and Economic Origins of Our Time*. Boston, MA: Beacon Press.

Pooley, S., Barua, M., Beinart, W., Dickman, A., Holmes, G., Lorimer, J., Loveridge, A. J., Macdonald, D. W., Marvin, G., Redpath, S., Sillero-Zubiri, C., Zimmermann, A. & Milner-Gulland, E. J. (2017). An interdisciplinary review of current and future approaches to improving human–predator relations. *Conservation Biology*, 31, 513–23.

Price, J. (2000). *Flight Maps: Adventures with Nature in Modern America*. New York: Basic Books.

Ravenelle, J. & Nyhus, P. J. (2017). Global patterns and trends in human–wildlife conflict compensation. *Conservation Biology*, 31(6), 1247–56.

Redpath, S. M., Bhatia, S. & Young, J. C. (2015). Tilting at wildlife – reconsidering human–wildlife conflict. *Oryx*, 49(2), 222–5.

Ripple, W. J., Estes, J. A., Beschta, R. L., Wilmers, C. C., Ritchie, E. G., Hebblewhite, M., Berger, J., Elmhagen, B., Letnic, M., Nelson, P. M., Schmitz, O. J., Smith, D. W., Wallach, A. D. & Wirsing, A. J. (2014). Status and ecological effects of the world's largest carnivores. *Science*, 343, 1241484.

Rothman, H. (2000). *Saving the Planet: The American Response to the Environment in the Twentieth Century*. Chicago: Ivan R. Dee.

Senthilkumar, K., Mathialagan, P., Sabarathnam, V. E. & Manivannan, C. (2017). Development of perception test for human–wildlife conflict. *International Journal of Current Microbiology & Applied Sciences*, 6(6), 817–24.

Skogen, K. & Krange, O. (2003). A wolf at the gate: The anti-carnivore alliance and the symbolic construction of community. *Sociologia Ruralis*, 43, 309–25.

Songhurst, A. (2017). Measuring human–wildlife conflicts: Comparing insights from different monitoring approaches. *Wildlife Society Bulletin*, 41(2), 351–61.

Soulsbury, C. D. & White, P. C. L. (2015). Human–wildlife interactions in urban areas: A review of conflicts, benefits and opportunities. *Wildlife Research*, 42, 541–53.

Sponarski, C. C., Vaske, J. J. & Bath, A. J. (2015). Differences in management action acceptability for coyotes in a National Park. *Wildlife Society Bulletin*, 39, 239–47.

Tovey, H. (2003). Theorizing nature and society in sociology: The invisibility of animals. *Sociologia Ruralis*, 43(3), 196–215.

Treves, A. (2012). Tolerant attitudes reflect an intent to steward: A reply to Bruskotter and Fulton. *Society & Natural Resources*, 25, 103–4.

Treves, A. & Bruskotter, J. (2014). Tolerance for predatory wildlife. *Science*, 344, 476–7.

van Eeden, L. M., Crowther, M. S., Dickman, C. R., Macdonald, D. W., Ripple, W. J., Ritchie, E. G. & Newsome, T. M. (2018). Managing conflict between large carnivores and livestock. *Conservation Biology*, 32, 26–34.

van Heel, B. F., Boerboom, A. M., Fliervoet, J. M., Lenders, H. J. R. & van den Born, R. J. G. (2017). Analysing stakeholders' perceptions of wolf, lynx and fox in a Dutch riverine area. *Biodiversity & Conservation*, 26, 1723–43.

Varga, D. (2009). Babes in the woods: Wilderness aesthetics in children's stories and toys, 1830–1915. *Society & Animals*, 17, 187–205.

White, P. C. L. & Ward, A. I. (2010). Interdisciplinary approaches for the management of existing and emerging human–wildlife conflicts. *Wildlife Research*, 37, 623–29.

Wolch, J. & Emel, J. (1998). *Animal Geographies: Place, Politics, and Identity in the Nature–Culture Borderlands*. New York: Verso.

Woodroffe, R. (2000). Predators and people: Using human densities to interpret declines of large carnivores. *Animal Conservation*, 3, 165–73.

Woodroffe, R., Thirgood, S. & Rabinowitz, A. (2005). *People and Wildlife, Conflict or Coexistence?* Cambridge: Cambridge University Press.

Young, J. C., Marzano, M., White, R. M., McCracken, D. I., Redpath, S. M., Carss, D. N., Quine, C. P. & Watt A. D. (2010). The emergence of biodiversity conflicts from biodiversity impacts: Characteristics and management strategies. *Biodiversity & Conservation*, 19, 3973–90.

Yurco, K., King, B., Young, K. R. & Crews, K. A. (2017). Human–wildlife interactions and environmental dynamics in the Okavango Delta, Botswana. *Society & Natural Resources*, 30(9), 1112–1126.

Zimmermann, A., Walpole, M. J. & Leader-Williams, N. (2005). Cattle ranchers' attitudes to conflicts with jaguar *Panthera onca* in the Pantanal of Brazil. *Oryx*, 39(4), 406–412.

2

A Multilevel, Systems View of Values Can Inform a Move towards Human–Wildlife Coexistence

ALIA M. DIETSCH, MICHAEL J. MANFREDO, LEEANN SULLIVAN, JEREMY T. BRUSKOTTER AND TARA L. TEEL

Perceptions of human–wildlife interactions, as either conflicts or opportunities for coexistence, can lead to drastically different human responses with important outcomes for wildlife. As examples, one person may revel in the sound of a coyote (*Canis latrans*) howling nearby and happily recount his or her experience to others, helping to raise awareness of these canine creatures. In contrast, another person may report the same interaction to a wildlife management authority and demand the animal be killed due to concern for the safety of local children. The possibility of future coexistence between humans and wildlife (particularly carnivores) could be quite bleak if only those who demanded retribution after having negative interactions, either perceived or realized as conflicts, report their experiences to managing authorities. Therefore, it is important to understand the basis for people's perceptions of their interactions with wildlife and, ultimately, their beliefs about how wildlife should be treated.

Research in the social sciences suggests that values have a fundamental influence on the way we perceive and interact with the world. Values (e.g. freedom, honesty, independence) are critical guiding beliefs about how the world operates and our relationships with others (Schwartz 1992). Values can be oriented towards a particular cognitive domain, such as wildlife, in a variety of ways. Thus, to provide useful information to wildlife management agencies on public thought about wildlife, we have developed a long-term programme of research in the United States (USA) that explores the role of values, and more specifically wildlife value orientations, in explaining attitudes and behaviours in relation to myriad wildlife-related topics. We also investigate how values are shaped by the world in which we live. Evidence suggests that a contemporary shift in values is greatly affecting wildlife management in the USA, with implications for trust in wildlife managing authorities, wildlife conservation funding mechanisms, wildlife-related attitudes about topics such as endangered species and carnivore conservation, as well as participation in wildlife-dependent recreation (e.g. hunting, fishing, trapping, wildlife viewing). We therefore outline a need for multilevel models that fully investigate these important systemic processes (i.e. how values work and what influences values) if coexistence with wildlife is to be truly realized.

2.1 A SOCIAL-ECOLOGICAL CONCEPT OF VALUES

Our research programme is guided by a social-ecological systems view of values (Text Box 2.1; Manfredo et al. 2017a). Values are important because they are fundamental motivational goals that influence human thought and, ultimately, human behaviour. For managers and decision-makers involved in wildlife conservation, values research has been used to compare and contrast the foundational goals of groups involved in a particular issue, clarify the basis of social conflict among various stakeholders or between stakeholders and a wildlife management authority, and understand and predict behaviours that either contribute to or detract from possible human–wildlife coexistence.

Past research in this area has been dominated by an individual-level view of values, relying on both static and linear behavioural prediction models such as the *value-belief-norm* (Stern & Dietz 1994) and *value-attitude-behaviour* (Homer & Kahle 1988; Fulton et al. 1996) frameworks. Many of these efforts applied to wildlife have traditionally focused on determining the social acceptability of lethal control (i.e. death of wildlife) as a way to curb interactions perceived as conflict (e.g. wildlife get into trash, wildlife kill pets or threaten human safety). However, such approaches are limited for several reasons. First, they

> **Text Box 2.1 Social-Ecological Systems**
>
> Folke et al. (2010) describe social-ecological systems as integrated systems consisting of reciprocal feedbacks and interdependence between a human society and the environment in which those people live. In these systems, everything is connected in ways that are both known and unknown, so a change in one part of the system will impact other parts of the system.

focus on wildlife as the *culprit* rather than considering what may have led to the interaction (e.g. were humans feeding wildlife around their homes? Has the species in question lost significant amounts of habitat?). Second, approaches that focus primarily on defining public attitudes and behaviours with only a modest nod (or none at all) to the role of values in predicting these constructs may be irrelevant outside of the particular time and context in which the study occurred. Third, these approaches rarely explore the role of broader societal trends in shaping values, which overlooks how value shift may occur over time. A societal value shift (and consequently, diversification of values) in the USA appears to have proceeded and challenged state wildlife management authorities that have not adjusted accordingly (Decker et al. 2016).

Advances in both psychology and social-ecological systems science led us to expand beyond this individual view of values to one that is multilevel, adaptive and dynamic (Manfredo et al. 2017a). We embrace the conventional definition of values as trans-situational goals and principles that guide human behaviour (Schwartz 1992). However, values are not just bounded cognitions in the mind; they are part of a meaning system that includes our thoughts and emotions, which are reflected in everything around us (see Chapter 3 and Chapter 4). These reflections include our verbal and non-verbal symbols, communication patterns, daily routines, material culture, social institutions and the ways in which we structure and relate to our natural and social surroundings (Kitayama 2002). Accordingly, values permeate a complex multilevel social structure. Such a view suggests that social groups, organizations and societies each have their own values in addition to individually held values. Groups and organizations are emergent structures with self-perpetuating properties and influences that may not fully align with each individual who comprises those groups and organizations (Conte et al. 2007).

Values also serve an adaptive function, arising through an evolutionary process by which the values that perpetuate are those that provide selective advantage for survival (Inglehart & Welzel 2005). Within individuals, values are formed through imitative learning that begins at a very early age, such as when children emulate their caregivers' behaviour, and thus values are highly resistant to change. Across individuals, values are a critical part of the cultural transmission process unique to humans. However, values are not static entities; they can shift (or change over time) when there is significant alteration in the social-ecological context. The twentieth century witnessed extraordinary changes in social life following the Second World War with greater economic stability and increased technological innovation in industrialized countries, which led to a shift in core values within many nations worldwide (Inglehart & Welzel 2005). These global events appear to have similarly affected how values in the USA have influenced thought about wildlife in recent decades (Manfredo et al. 2009).

2.2 DOMINATION AND MUTUALISM: CORE WILDLIFE VALUES IN THE UNITED STATES

Our research has identified two core ways in which values are oriented towards wildlife in the USA. These wildlife value orientations are domination and mutualism (Fulton et al. 1996; Manfredo et al. 2009; Teel & Manfredo 2010). A domination orientation cultivates beliefs of human superiority and control over their social and environmental surroundings, and that humans have a destiny and right to subjugate others, including wildlife, to benefit themselves. Domination has been recognized as a defining value found within the USA (Kluckholn & Strodtbeck 1961; Schwartz 2006). The strong influence of domination values has been traced to Judeo-Christian religions, and particularly the Reformation (Pattberg 2007). Domination has also been identified as the basis for the emergence of science and technology, competitive capitalism and sustained erosion of the environment (Buttel & Humphrey 2002; Pattberg 2007).

Domination values extended to animals assume that humans hold power over them, and can therefore relegate animals such as wildlife to conditions (e.g. habitat loss and alteration) and roles (e.g. "*game*" species) that benefit people. Domination values create psychological distance from what is being dominated and provide justification for various beliefs and practices (Pratto 1999). For example, wildlife management in the USA, including its institutions and culture, traditionally revolves around a domination ideology, resulting in policies and practices that cater to groups of people who participate in activities that result in death or harm to wildlife so long as it benefits humans in some way. In particular, the funding structure of wildlife agencies relies heavily upon the sale of licences that allow licence holders to *take* wildlife through activities such as hunting, trapping and fishing (Decker et al. 2016), and receive some *reward*, such as food, clothing or recreational opportunities that meet psychological desires (e.g. maintain family tradition, social interaction, experience nature). Coexistence for this group might be seen as living in harmony with preferred species (e.g. game animals, such as deer and elk; livestock raised by people, including cows, sheep and chicken), and protecting those species from real or perceived threats, including attacks by carnivores.

Mutualism, in contrast, is defined by a view that humans are of relatively equal standing with animals. The latter are seen as part of an extended family, capable of human-like qualities such as trust, and deserving of rights and caring. Mutualism is rooted in egalitarian ideology that places emphasis on equality and consideration for all (Wildavsky 1991). We contend that there has been a rise of mutualism in post-industrialized countries such as the USA due, in part, to the changing nature of human contact with wildlife. A modern lifestyle leads to less risk from wildlife and less reliance on wildlife as a food source. People are also less likely to learn about wildlife as a result of direct interaction, and instead, learn more through self-selected indirect sources (e.g. television, internet), resulting in depictions of animals that are limited to those preferred by the information consumer. In this context, the tendency to anthropomorphize animals fosters the view that wildlife are like humans and thus deserving of similar consideration (Mithen 1996). This anthropomorphization is enhanced in situations where humans have an increased need for social affiliation, as is the case in urbanized societies or where there is a loss of community (Epley et al. 2007; Manfredo et al. 2009).

The mutualism wildlife value orientation that has emerged from these collective circumstances is associated with attitudes and behaviours that are in contrast to the traditional practices of wildlife management in the USA. For example, individuals with a mutualism orientation tend to be less accepting than people with a domination orientation of lethal control as a wildlife management tool (e.g. Dietsch et al. 2016). Additionally, a mutualism orientation is associated with support for restrictions on humans (e.g. trail closures, habitat protection from development) to benefit wildlife, such as endangered species, because this orientation tends to prioritize an idealized version of coexistence and living in harmony with all. However, a person with a mutualism orientation may behave in a way (e.g. feed wildlife, keep wild animals as pets) that can result in human–wildlife conflict.

2.3 ASSESSMENT OF DOMINATION AND MUTUALISM WILDLIFE VALUE ORIENTATIONS

A variety of techniques, both qualitative and quantitative, have been developed to measure wildlife value orientations (Dayer et al. 2007;

Manfredo et al. 2009; Teel & Manfredo 2010; Chase et al. 2015; McCoy et al. 2016). The most commonly employed method consists of quantitative survey assessments with item-scaling procedures to assess the degree to which people score on both the domination and mutualism dimensions. Following upon developmental work in Colorado (Fulton et al. 1996) and a demonstration project involving six states in the USA (Manfredo et al. 2003), a broad-based application of this approach occurred with residents in nineteen western states in 2004 in cooperation with the Western Association of Fish and Wildlife Agencies (Manfredo et al. 2009; Teel & Manfredo 2010). Survey items used as part of this effort (see Table 2.1) have also been tested in multiple studies and across various countries with some degree of consistency, particularly in Europe (e.g. Teel et al. 2010; Sijtsma et al. 2012; Hermann et al. 2013; Riepe & Arlinghaus 2014; Gamborg & Jensen 2016). However, not all survey items hold the same meaning across cultures, and countries with vastly different social structures are likely to have different values than those found in the USA.

For the sake of summarizing study results in a way useful to managers: US residents were categorized in a four-group typology based on the extent to which respondents emphasized a domination and/or mutualism orientation. (For further description of concepts, see Teel & Manfredo, 2010; McCoy et al. 2016.) The four types of people include: (1) Traditionalists, who scored above the scale mid-point on domination and at or below the scale mid-point on mutualism; (2) Mutualists, who scored above the scale mid-point on mutualism and at or below the scale mid-point on domination; (3) Pluralists, who scored above the mid-point on both domination and mutualism scales; and (4) Distanced, who scored at or below the mid-point on both scales. Consistent with others' work on values (Kluckholn & Strodtbeck 1961; Schwartz 2006), results of this study confirm that domination was commonly found across the western USA. A replication of this investigation and expansion to include all fifty states was completed in 2018 with the aim of assessing possible change in values over time (see www.wildlifevalues.org).

2.4 WILDLIFE VALUES PREDICT RESPONSE TO WILDLIFE-RELATED ISSUES

Measuring wildlife value orientations can be useful in anticipating and understanding the basis for individual thought and action. Prior research testing this notion has demonstrated a connection between

Table 2.1 *Survey items for wildlife value orientation scales[1]*

Wildlife value orientation
Basic belief dimension
  Items comprising the basic belief dimension[2]

Domination value orientation
Appropriate use belief dimension
  Humans should manage fish and wildlife populations so that humans benefit.
  The needs of humans should take priority over fish and wildlife protection.
  It is acceptable for people to kill wildlife if they think it poses a threat to their life.
  It is acceptable for people to kill wildlife if they think it poses a threat to their property.
  It is acceptable to use fish and wildlife in research even if it may harm or kill some animals.
  Fish and wildlife are on earth primarily for people to use.

Hunting belief dimension
  We should strive for a world where there's an abundance of fish and wildlife for hunting and fishing.
  Hunting is cruel and inhumane to the animals.[3]
  Hunting does not respect the lives of animals.[3]
  People who want to hunt should be provided the opportunity to do so.

Mutualism value orientation
Social affiliation belief dimension
  We should strive for a world where humans and fish and wildlife can live side by side without fear.
  I view all living things as part of one big family.
  Animals should have rights similar to the rights of humans.
  Wildlife are like my family and I want to protect them.

Caring belief dimension
  I care about animals as much as I do other people.
  It would be more rewarding to me to help animals rather than people.
  I take great comfort in the relationships I have with animals.
  I feel a strong emotional bond with animals.
  I value the sense of companionship I receive from animals.

Notes:
[1] Table adapted from Teel and Manfredo (2010).
[2] Items were measured on a scale ranging from 1 (strongly disagree) to 7 (strongly agree).
[3] Item reverse-coded prior to analysis.

wildlife value orientations and attitudes towards wildlife-related issues or acceptability of different management actions, including lethal control and carnivore conservation (e.g. Fulton et al. 1996; Whittaker et al. 2006; Dietsch et al. 2016; Manfredo et al. 2016). Other studies have also explored the relationship between value orientations and participation in wildlife-related recreation (e.g. Manfredo & Zinn 1996; Manfredo et al. 1999; Bright et al. 2000; Zinn et al. 2002). Recent applications in western Europe offer additional evidence across cultures that value orientations influence a wide range of wildlife-related attitudes and behaviours, including acceptability of lethal control (Sijtsma et al. 2012), support for the return of native land mammals (Hermann et al. 2013) and protected area management (Teel et al. 2010).

Attitudinal comparisons conducted as part of the nineteen-state investigation in the western USA allowed for a more in-depth examination of the predictive validity of the concept, showing how value orientation differences form the foundation for conflicting positions on a host of wildlife-related issues (Manfredo 2008). Out of more than 450 attitudinal measures included on state-specific versions of the survey, 71 per cent were statistically correlated with the domination scale, and 59 per cent were correlated with mutualism (Dietsch et al. 2017). Of those that were statistically significant, over 30 per cent of both domination and mutualism correlations were at or above 0.30, indicating a moderate to large effect (Cohen 1988). Findings as a whole indicated that value orientations were predictive of attitudes across a diverse array of issues and were particularly useful in explaining variability on issues involving harm to wildlife and trade-offs between human interests and wildlife protection (Teel & Manfredo 2010; Manfredo et al. 2016). Specifically, people with higher domination scores exhibit greater concern for human interests including private property rights, a healthy economy and public access to recreation areas (Figure 2.1); those with higher mutualism scores were instead more likely to prioritize wildlife-focused concerns, including habitat protection and support for threatened or endangered species (Dietsch et al. 2016; Manfredo et al. 2016).

2.5 VALUE CHANGE AS PART OF A BROADER SOCIETAL SHIFT

Although values tend to be stable within individuals due to their formation at an early age as part of a cultural transmission process, systemic value change can occur at a societal level as new generations age under different social conditions. Inglehart (1997) argues that societal change in the decades following the Second World War contributed to a substantial shift in values worldwide. He describes a process of *modernization*, indicated by increased economic well-being, urbanization and technological sophistication, which has been a driver of value shift due

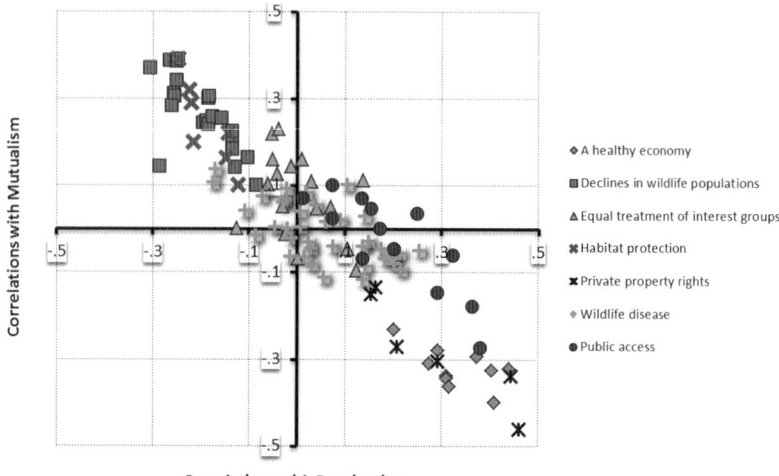

Figure 2.1 Correlations between Domination and Mutualism wildlife value orientations and attitudes towards wildlife-related issues, from a 2004 survey of western USA residents. Figure adapted from Manfredo et al. (2016). Correlations denoted as 0.1 are weak, 0.3 are moderate and 0.5 are strong relationships. (Cohen 1988)

to dramatic alterations in the affordances and challenges of people's surroundings. Specifically, Inglehart argues that modernization's impact spawned a shift from emphasizing subsistence needs to a greater focus on self-enhancement needs. This, in turn, has been accompanied by a shift from materialist values (rooted in basic needs such as safety, survival and sustenance) to post-materialist values (centred on belongingness, self-fulfilment and quality-of-life concerns). Longitudinal findings from the World Values Survey (Inglehart & Welzel 2005) offer empirical support for Inglehart's proposal, documenting a wide array of implications for such a shift, including a loss of faith in government institutions, declines in willingness to sacrifice for the greater good and increased concern for civil and environmental rights.

We further propose that this process of modernization as described by Inglehart has contributed to a shift from domination to mutualism wildlife value orientations in the USA. Results from our nineteen-state study offer support for this proposal in relation to several key areas. First, we found post-materialist values at the individual level to be positively associated with mutualism and negatively associated with domination (Manfredo et al. 2009). This relationship was also readily observable at the state level (Figure 2.2), where we found a strong

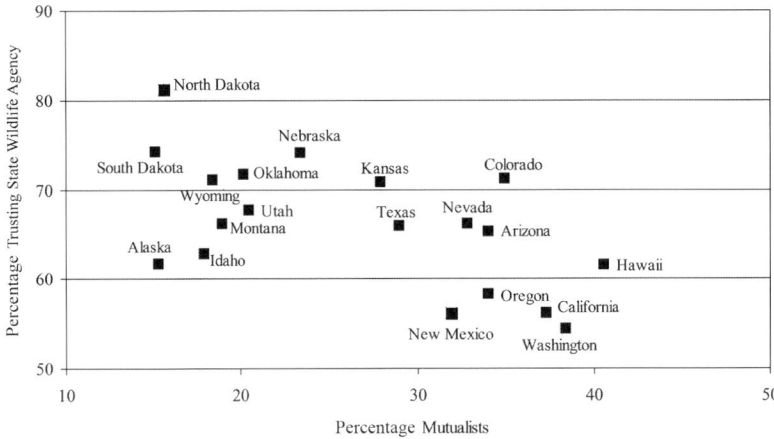

Figure 2.2 Percentage of Post-Materialists by percent of Mutualists across states, from a 2004 survey of western USA residents. (Pearson's $r = 0.62$, $p < 0.05$)

positive relationship between the percentage of residents with post-materialist values in a state and the percentage of people classified as Mutualists (those only holding a Mutualist orientation). Consistent with our micro (individual) level model, wildlife value orientations were also found to mediate the relationship between materialist/post-materialist values and wildlife-related attitudes and behaviours (Manfredo et al. 2009).

We also explored whether indicators of modernization could be used to explain variability in wildlife value orientations, and particularly whether modernization might be leading to a rise in mutualism. Indeed, Manfredo (2008) shows differences in value orientations across the western states in the USA, suggesting that the expected relationship between modernization and mutualism is plausible (Figure 2.1). Furthermore, results of multilevel modelling procedures confirmed that state-level measures of education, urbanization and income were significantly related to the composition of wildlife value orientations within states (Manfredo et al. 2009); higher scores on these measures were associated with a higher prevalence of mutualism. While limited to cross-sectional data in this first phase of our research, the findings provide evidence consistent with our proposal that macro (societal) level modernization forces are contributing to a shift towards mutualism. The America's Wildlife Values project (www.wildlifevalues.org) is aimed at assessing the possibility of such a shift over time.

A change of societal values would create significant opportunities as well as challenges for reducing conflicts and fostering coexistence with wildlife. In fact, many current challenges facing wildlife management authorities are likely evidence of a value shift. For example, the funding base for most state wildlife agencies is declining because fewer people are participating in activities, such as hunting (United States Fish and Wildlife Service (USFWS), 2007), that generate revenue by way of licence sales. This decline in hunting participation also means fewer people on-the-ground helping to control certain populations of game species. Fewer hunters, along with a history of predator removal in the USA and habitat fragmentation and loss, has led to localized surges in ungulate populations, which is associated with a greater number of automobile–wildlife collisions and increased amount of wildlife damage to agricultural crops (Conover 2001; Walter et al. 2010). Additionally, more lawsuits, ballot initiatives and other types of public protest against wildlife governing authorities have occurred (Jacobson & Decker 2008), challenging the traditional role that agencies have played in the wildlife decision arena. However, these *challenges* also indicate a growth in public participation and interest in wildlife-related decisions that could help foster increased opportunities for coexistence with a diversity of wildlife that do not necessarily directly benefit humans through *take* activities. For example, mutualism has been positively correlated with support for the reintroduction of previously extirpated carnivores and large land-based herbivores (e.g. Hermann et al. 2013; Dietsch et al. 2016). Thus, a more systems-based understanding of values – including how values operate and are shaped across time and place – can help to inform wildlife management, as well as opportunities for coexistence with wildlife.

2.6 THE ADAPTIVE NATURE OF WILDLIFE VALUES HAS LED TO VARIABILITY IN THEIR GEOGRAPHIC DISTRIBUTION

Values and their associated belief systems are considered adaptive because they can help facilitate behavioural responses that assist individuals, groups, agencies or societies in coping with and adjusting to change (Denevan 1983; Smit & Wandel 2006; Oishi 2014). Values guide humans in meeting their own biologically based needs and social affiliation needs, and help to improve the survival and welfare of groups that assist in meeting those same needs (Schwartz & Bilsky 1987). For example, research in cross-cultural psychology suggests that land use and occupational cooperation are associated with values of collectivism or independence (Talhelm et al. 2014), and the degree to which people are holistic or analytic thinkers (Uskul et al. 2008). Kitayama et al. (2006) also proposed that human migration and adaptation to the novel socio-ecological conditions (e.g. harsh environments, lack of community support) presented by the frontiers of Hokkaido, Japan and the western USA led to the rise of independence values, which were subsequently emulated elsewhere in the respective countries.

This process of adaptation to unique social-ecological conditions can lead to distinct geographic patterns associated with values. For example, research shows that psychological traits, such as personality characteristics, are not randomly distributed across the landscape (Rentfrow et al. 2008). In the USA, in particular, the emergence of domination- or mastery-oriented values was associated with eighteenth- and nineteenth-century immigration of people with distinct cultural origins, and traces of this values heritage can still be found (Manfredo et al. 2016). The persistence of values is shaped by the need for residents to fit within the prevailing social environment. People choose to live in areas where other people have traits similar to their own, or people must adapt to the norms and values of the new location in order to affiliate with others. Although the geographic pattern of values may change, it is unlikely that the drivers of change are geographically random. In our research, for example, we find that modernization in the USA is tied to urban areas where more people have post-materialist values and mutualism wildlife value orientations (Manfredo et al. 2009).

For wildlife managers, understanding the geographic distribution of human values (similar to understanding the distribution of wildlife species) is critical, because it can help determine patterns of public response to wildlife management actions across the landscape (Dietsch et al. 2016). We first illustrate this through an assessment of wildlife value orientations conducted at the county level within Washington state (Dietsch et al. 2011). As shown in Figure 2.3a, people classified as Traditionalists (holding only a domination orientation) generally out-numbered Mutualists in the eastern and some of the central parts of the state – areas that are much more rural. The reverse of this pattern was true for the more urbanized north-western counties where we found more Mutualists. Indeed, Figure 2.3b demonstrates how that distribution may relate to residents' attitudes in the form of support for lethal control of black bears involved in nuisance scenarios (e.g. getting into rubbish or pet food containers). As we would predict, the domination-oriented counties showed greater support for killing nuisance bears compared to the more Mutualist counties. This relationship was confirmed through correlational analysis at the county level using measures reported in the two figures.

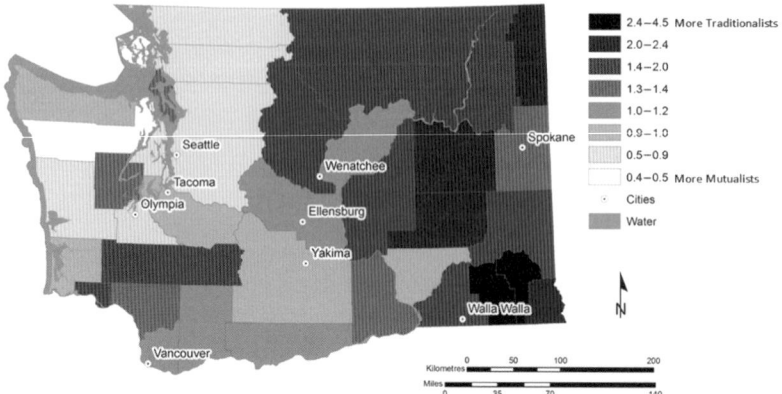

Figure 2.3a Ratio of Traditionalists to Mutualists, from a 2009 survey of Washington residents by county (Dietsch et al. 2011). Ratio represents the number of Traditionalists for every one Mutualist, whereby a number greater than one signifies that there are more Traditionalists than Mutualists in the county, and a number less than one indicates that there are more Mutualists than Traditionalists. *A black and white version of this figure will appear in some formats. For the colour version, please refer to the plate section.*

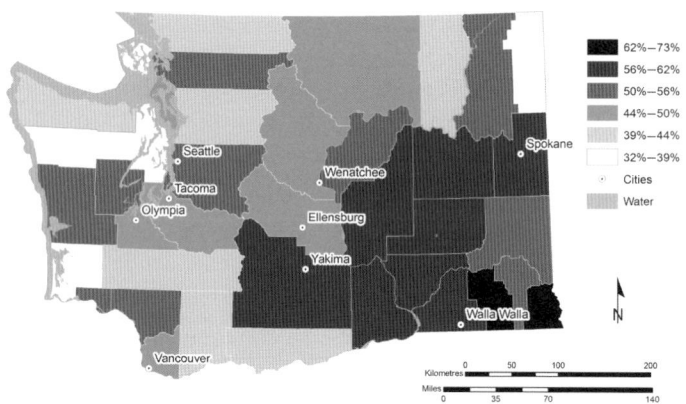

Figure 2.3b Percentage of residents accepting of lethal removal by the state wildlife agency of a nuisance black bear, from a 2009 survey of Washington residents by county (Dietsch et al. 2011). The nuisance situation was described as one where, for example, the bear is getting into rubbish or pet food containers near the resident's home. Darker colours denote a higher percentage of people who find lethal removal of a nuisance black bear to be acceptable. *A black and white version of this figure will appear in some formats. For the colour version, please refer to the plate section.*

However, the distribution of values that we observed in Washington and elsewhere in our work is not merely a matter of a rural–urban divide. Values instead appear to be the result of complex social and ecological factors (Dietsch et al. 2016). We illustrate this complexity using data collected at the census block group level through a door-to-door, bilingual survey in the urban area of Tucson, Arizona (see Dietsch et al. 2012 for a complete description of methods). We would typically expect a mutualism orientation to translate into greater opposition to lethal control of wildlife. Results here, however, show somewhat of a reverse pattern: the central part of Figure 2.4a (also the central area of the city) appeared to have a higher proportion of Mutualists. The north-eastern part of the city (on the urban fringe) appears to consist of a more even mix of Mutualists and Traditionalists. In contrast to what we would expect, we found some of the more mutualism-oriented census block groups to have *lower* levels of opposition to lethal removal compared to places with a more equal distribution of value types (Figure 2.4b). One reason for this disparity might be that wildlife is more common on the urban fringe and residents may have learned to expect and tolerate them in these places, or perhaps people have moved to the fringe to see wildlife more frequently and perhaps coexist with them. In either scenario, residents living in the urban core (despite their value orientations) may be more supportive of lethal control because they believe that wildlife *should* not be found in the middle of the city, and certainly should not be a nuisance there.

A culturally based explanation for our Tucson results is also informative. Certain areas in the urban core (e.g. in the southern portion of the city) with high proportions of Mutualists also had high proportions of Hispanic residents (United States Census Bureau, 2010). Additionally, Hispanic residents in the study area reported lower income and education levels than residents of non-Hispanic ethnicity, and had lived in Arizona, on average, for a longer time. We might hypothesize, then, that the mutualism values of these residents are rooted in traditions of Hispanic culture as opposed to being borne largely of modernization processes. If modernization did not drive mutualism in the Hispanic population, then responses towards management actions such as lethal control may not work as expected. Specifically, we found that US residents who reported 'Mexican' as their primary ethnic origin tended to have lower domination scores compared to other ancestry groups, a trend that was consistent with Schwartz's World Values data showing that residents of Mexico had lower mastery scores (Manfredo et al. 2016). Other research supports the

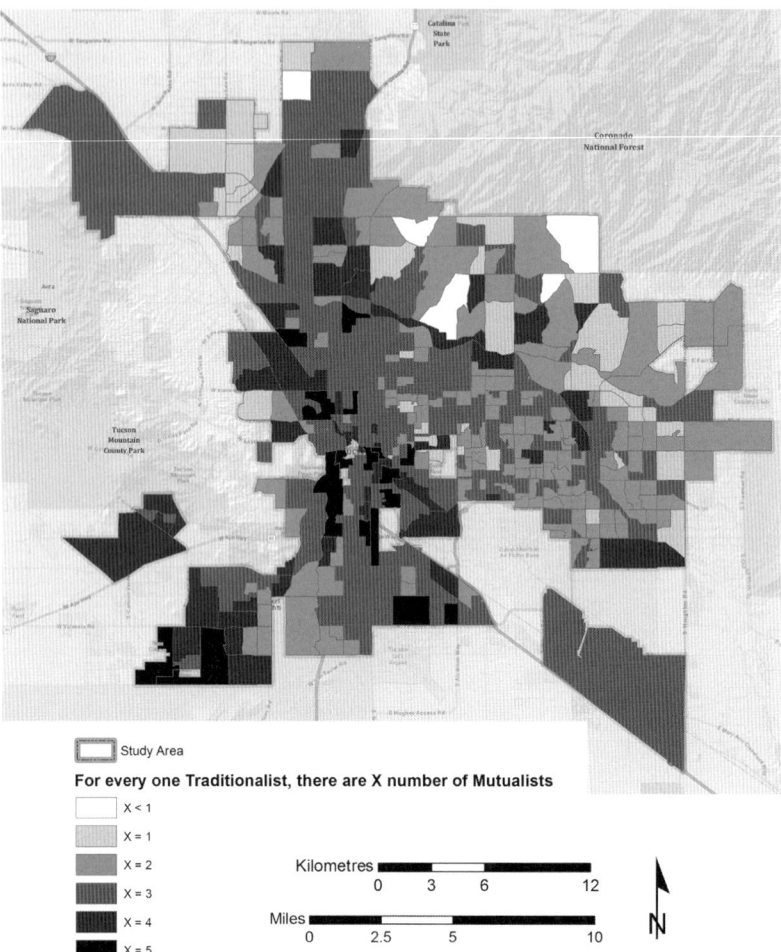

Figure 2.4a Ratio of Mutualists to Traditionalists, from a 2008–9 survey of Tucson, Arizona, residents by census block group (Dietsch et al. 2012). Ratio represents the number of Mutualists for every one Traditionalist, whereby a number greater than one signifies that there are more Mutualists than Traditionalists.

idea that Hispanics in Arizona (the vast majority of whom claim Mexican heritage) may be more mutualism-oriented than non-Hispanic whites, but exhibit stronger domination values as they become more acculturated over generations (Chase 2016). Thus, greater support for lethal control of wildlife found in areas with a higher percentage of Hispanic residents (what we found) may be tied to values that emerged from a different

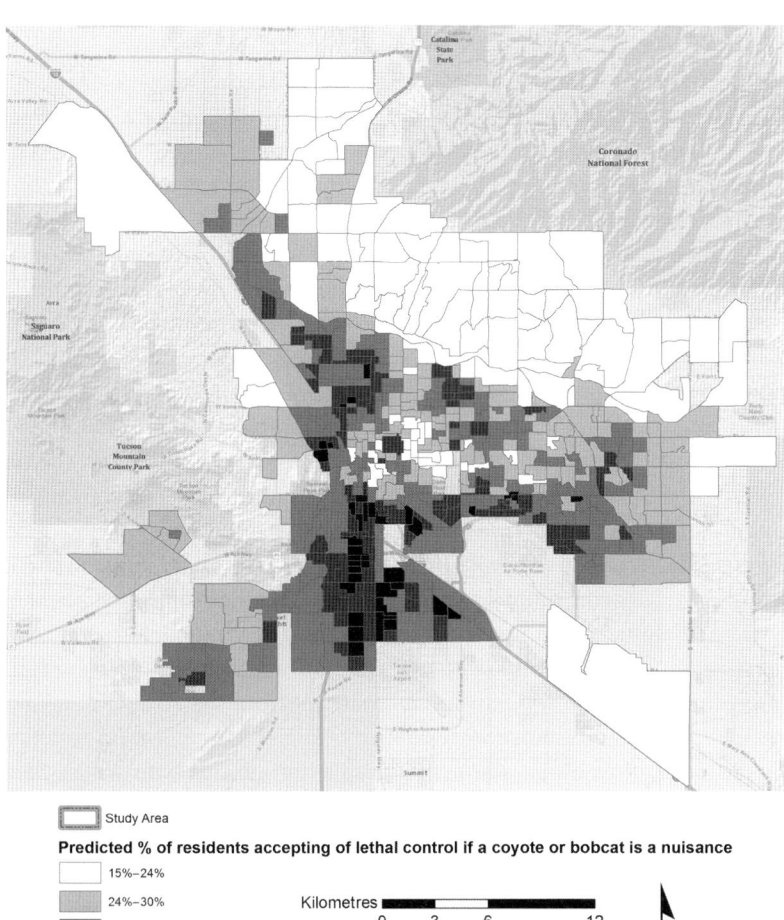

Figure 2.4b Percentage of residents accepting of lethal removal by the state wildlife agency of a nuisance bobcat or coyote, from a 2008–9 survey of Tucson, Arizona residents by census block group (Dietsch et al. 2012). The nuisance situation was described as one where, for example, the animal is getting into rubbish or damaging landscaping near the resident's home.

cultural heritage altogether – and not from modernization. Taken as a whole, these investigations highlight the importance of depicting the complex interplay of social, ecological and place-based factors that shape patterns of human thought and behaviour observed on the landscape, which will ultimately influence where people fall along the conflict-to-coexistence continuum.

2.7 THE WILDLIFE VALUES GAP IN A MULTILEVEL CONTEXT

Our systems approach to values further proposes that values permeate all levels of human existence from individual thought to cultural and political institutions (Manfredo et al. 2017a). Thus, an exploration of the nature of values at multiple levels and how values influence human behaviour within and across those levels is critically needed. Research must go beyond focusing on individual values only, and acknowledge the strong influence of groups (e.g. cultural affiliation, formal organizations, social classifications, norms) on how people think and behave. As an example, individuals' values guide their actions (purchase/own guns); give them an identity in interpersonal dealings (gun rights advocates who value their right to independence); and provide a motivational basis for group membership and socialization (belonging to groups such as the National Rifle Association (NRA) and some hunter organizations that may share individualism, power and domination values). At an organizational level, a group such as the NRA, by way of the people who act on its behalf, exerts influence back on individuals as the group articulates and demonstrates *appropriate* behaviours and attitudes that members *should* take. The group thus sustains an ideology (and group norms) through its values, and acts as its own entity to exert the power and influence that emanates from the collective, taking action at an organizational level (lobbying to protect the Second Amendment, or right to own a gun). Individuals and groups with which they affiliate are in a dynamic interchange: as new issues arise (mass shootings in schools), individuals respond (advocates argue that guns should be allowed in schools for self-defence); group emergence shapes normative positions (teachers should carry guns in schools); and those positions immerge down to individuals within the broader group when they *replicate* this new position through their own words and actions. (For more on emergence and immergence, see Conte et al. 2007.) Of course, each person identifies with or belongs to many formal and informal groups, resulting in multiple influences on one individual, and the interactions and feedbacks among these groups create a complex system of social influence that significantly complicates the decision-making arena.

Fostering tolerance for wildlife and striving for coexistence will require a multilevel approach. Taking a broad-based view, we propose that wildlife value shift occurs in the context of a societal-individual model. Nested within that model is an organizational-individual model of value stability and change, comprised of both emergent properties of wildlife management organizations (e.g. state fish and wildlife management agencies in

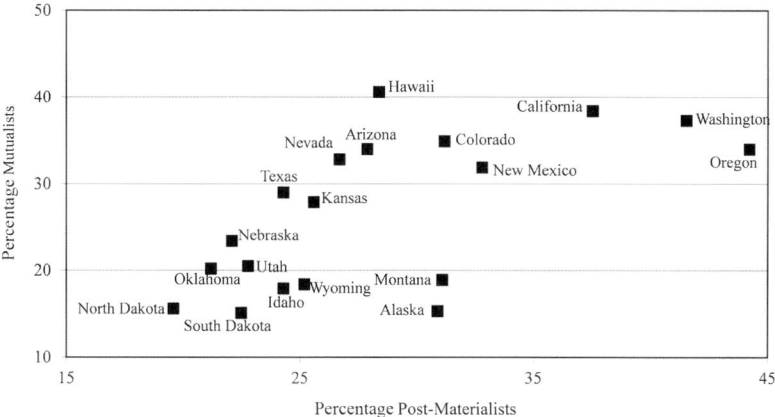

Figure 2.5 Percentage of Mutualists across states by percentage of residents indicating they trust the state wildlife agency, from a 2004 survey of western USA residents (Pearson's $r = -0.60$, $p < 0.05$). Trust was defined as the percentage of people selecting a 3 or 4 on the following response scale: 1 = almost never, 2 = only some of the time, 3 = most of the time and 4 = almost always.

the USA) and individual employees within those organizations. Organizational action, such as public outreach and decision-making, as well as challenges to management authority, for example, through citizen ballot initiatives and court cases in the USA can affect adaptive responses at every level. Our nineteen-state study suggested that a broader cultural shift stimulated by modernization may indeed be occurring, and, in turn, will lead to calls for greater public inclusion in decision-making. If these diverse public interests are not included, public trust in managing authorities is likely to decline (Decker et al. 2016). In support of this assertion, we found fewer people trusting of the state wildlife management agency in states with more people classified as Mutualists (Figure 2.5).

Our results support Inglehart's research (Inglehart 1997; Inglehart & Welzel 2005) describing a growing disconnect between emerging publics and the agencies charged with representing them in decision-making. Interestingly, however, we did not find a strong relationship between the mutualism value orientation and trust at the individual level (Manfredo et al. 2017b) across the entire nineteen-state sample, indicating that the relationship between values and trust of governing agencies is complex and demands further exploration. For example, we had hypothesized that the state relationship we found would be synonymous with less trust of a state managing agency for all individuals. However, some state managing agencies may have adjusted (or are starting to adjust) to the needs of more Mutualist-oriented constituents, and trust levels for these individuals could possibly reflect a positive response to their agency's efforts to include them. If this were the case, we might also expect individuals who traditionally hold power in decision-making (i.e. those with domination values) to be outwardly frustrated about these changes (Inglehart & Norris 2016). Such a *cultural backlash* would be spurred on by the presence of more people classified as Mutualists in a state, fostering less trust of an agency among traditional constituents as the agency attempts to become more inclusive of different audiences (Manfredo et al. 2017a).

This process of transition for agencies towards inclusivity is particularly challenging when the values of the organization differ from those of the individuals the agency represents. Results from a study in North Dakota (Gigliotti & Harmoning 2004) hint at the possibility of such disparity between public values and the values of the state wildlife management agency. For example, agency employees had stronger domination orientations than their public, and were also inaccurate in estimating public attitudes on a variety of wildlife-related issues. The value transition occurring in the public is likely to be reflected in long-term changes occurring within the agencies (e.g. increased proportion of staff with mutualism values), but emergent properties of the agency may perpetuate and reinforce values at the group level above and beyond individual values (Elder-Vass 2010). For example, Cramer et al. (1993) found that the United States Forest Service (USFS) employees felt that the agency prioritized use of National Forests for timber harvest over management for recreation and wildlife. Many of these employees, following organizational norms and in accordance with their job responsibilities at USFS, took action consistent with that prioritization even though it conflicted with their own personal beliefs about what the agency's priorities should be (i.e. managing forests for recreation and wildlife over timber). Thus, we expect that a similar values gap between the state fish and wildlife management agency (organizational level) and its employees (individual level) may exist, particularly in states where the percentage of Mutualists is relatively high and agency culture does not adapt.

2.8 CONCLUSIONS

The key to effective conservation of wildlife lies in understanding the human phenomena that influence the resilience of social-ecological systems (Martin et al. 2016). These phenomena are complex and challenging to comprehend, and the static, reductionist and individual-focused theoretical approaches applied from social psychology to date have largely failed to capture this complexity. New approaches (e.g. socio-ecological psychology; Oishi 2014) are breaking from that tradition and, in the process, opening new avenues for understanding human–environment relationships. We similarly advocate for the use of a social-ecological systems approach to understanding values and their role in directing responses to human–wildlife interactions. Such an approach can ultimately provide guidance on how to foster tolerance of wildlife involved in conflicts, and ultimately seek overall coexistence with wildlife. Fundamental to that approach is the recognition that values exist at multiple levels and research will be more productive and useful if values are studied as a multilevel concept. Our ongoing programme of research aims to provide wildlife managers and conservationists with a better understanding of the social changes occurring at these different levels and how those changes are likely to affect our relationships with and management of wildlife. It is also our hope that this understanding will help the profession cope with changes at the societal level and identify new ways of adapting to contemporary challenges that can sustain the unique biodiversity of this planet.

2.9 RECOMMENDATIONS AND FUTURE DIRECTIONS

- Relying solely on individual-level models is problematic, because such models ignore the important interplay between society, culture, institutions, groups and individuals.
- Efforts to model dynamics between two entities – whether they are values and attitudes, societal and individual values, or institutions and their constituents, etc. – must consider that such dynamics are likely reciprocal, requiring feedback loops between those entities.
- A critical next step is to investigate (among many things) how contrasting values among groups, organizations and societies affect the ability to collaborate and ultimately facilitate human–wildlife coexistence.
- Get to know your constituents – particularly their fundamental values – and learn more about how values can be affected by broader societal trends. This is not an easy task, but is crucial to being relevant in a changing world.

- Determine your own organization's values; often these values are more deeply held than you realize, which can greatly affect how the organization interacts with its constituents.
- Consider a variety of way to engage different publics. *Conflicts* over natural resources, such as wildlife, are often realized as displays of social power amongst groups of people; thus, traditional ways of gathering public input (e.g. public forums where only a handful of people can speak) may alienate some groups, creating additional conflicts and leading to a loss of trust.
- Recognize that value change is slow and often only in response to large-scale social-ecological changes. Efforts that focus on behaviour or attitude change may be more effective at increasing opportunities for human–wildlife coexistence.

2.10 References

Bright, A. D., Manfredo, M. J. & Fulton, D. C. (2000). Segmenting the public: An application of value orientations to wildlife planning in Colorado. *Wildlife Society Bulletin*, 28, 218–26.

Buttel, F. H. & Humphrey, C. R. (2002). Sociological theory and the natural environment. In R. E. Dunlap & W. Michelson, eds., *Handbook of Environmental Sociology*. Westport, CT: Greenwood Press, pp. 33–69.

Chase, L., Teel, T. L., Thornton-Chase, M. R. & Manfredo, M. J. (2015). A comparison of quantitative and qualitative methods to measure wildlife value orientations among diverse audiences: A case study of Latinos in the American Southwest. *Society & Natural Resources*, 29(5), 572–87.

Chase, L. D. (2016). Measurement of wildlife value orientations among diverse audiences: A multigroup confirmatory factor analysis among Hispanic and non-Hispanic White communities. *Human Dimensions of Wildlife*, 21(2), 127–43.

Cohen, J. (1988). *Statistical Power Analysis for the Behavioral Sciences*. 2nd edn, Hillsdale, NJ: Lawrence Erlbaum Associates.

Conover, M. R. (2001). *Resolving Human–Wildlife Conflicts: The Science of Wildlife Damage Management*. Boca Raton, FL: CRC Press.

Conte, R., Andrighetto, G., Campennì, M., Paolucci, M., Istituto, L. & Martino, S. (2007). *Emergent and Immergent Effects in Complex Social Systems*. Emergent Agents and Socialities: Social and Organizational Aspects of Intelligence. Papers from the 2007 AAAI Fall Symposium, pp. 42–7.

Cramer, L. A., Kennedy, J. J., Krannich, R. S. & Quigley, T. M. (1993). Changing forest service values and their implications for land management decisions affecting resource-dependent communities. *Rural Sociology*, 58(3), 475–91.

Dayer, A. A., Stinchfield, H. M. & Manfredo, M. J. (2007). Stories about wildlife: Developing an instrument for identifying wildlife value orientations cross-culturally. *Human Dimensions of Wildlife*, 12(5), 307–15.

Decker, D., Smith, C., Forstchen, A., Hare, D., Pomeranz, E., Doyle-Capitman, C., Schuler, K. & Organ, J. (2016). Governance principles for wildlife conservation in the 21st century. *Conservation Letters*, 9(4), 290–5.

Denevan, W. M. (1983). Adaptation, variation, and cultural geography. *The Professional Geographer*, 35(4), 399–407.

Dietsch, A. M., Manfredo, M. J. & Teel, T. L. (2017). Wildlife value orientations as an approach to understanding the social context of human-wildlife conflict. In C. M. Hill, A. D. Webber & N. E. C. Priston, eds., *Understanding Conflicts about Wildlife: A Biosocial Approach*. New York: Berghahn, pp. 107–26.

Dietsch, A. M., Teel T. L. & Manfredo, M. J. (2016). Social values and biodiversity conservation in a dynamic world. *Conservation Biology*, 30(6), 1212–21.

Dietsch, A. M., Teel, T. L., Manfredo, M. J. & Chase, L. (2012). *State Report for Arizona from the Research Project Entitled Understanding People in Places. Project Report for the Arizona Game and Fish Department*. Colorado State University, Fort Collins, CO, USA.

Dietsch, A. M., Teel, T. L., Manfredo, M. J., Jonker, S. A. & Pozzanghera, S. (2011). *State Report for Washington from the Research Project Entitled Understanding People in Places. Project Report for the Washington Department of Fish and Wildlife*. Colorado State University, Fort Collins, CO, USA.

Elder-Vass, D. (2010). *The Causal Power of Social Structures: Emergence, Structure and Agency*. Cambridge: Cambridge University Press.

Epley, N., Waytz, A. & Cacioppo, J. T. (2007). On seeing human: A three-factor theory of anthropomorphism. *Psychological Review*, 114(4), 864–86.

Folke, C., Carpenter, S. R., Walker, B., Scheffer, M., Chapin, T. & Rockstrom, J. (2010). Resilience thinking: Integrating resilience, adaptability and transformability. *Ecology and Society*, 15(4), art. 20. Available from www.ecologyandsociety.org/vol15/iss4/art20/ (accessed November 2018).

Fulton, D. C., Manfredo, M. J. & Lipscomb, J. (1996). Wildlife value orientations: A conceptual and measurement approach. *Human Dimensions of Wildlife*, 1 (2), 24–47.

Gamborg, C. & Jensen, F. S. (2016). Wildlife value orientations: A quantitative study of the general public in Denmark. *Human Dimensions of Wildlife*, 21(1), 34–46.

Gigliotti, L. M. & Harmoning, A. K. (2004). Evaluation of the North Dakota Game and Fish Department using a communications assessment model. *Human Dimensions of Wildlife*, 9(1), 79–81.

Hermann, N., Voß, C. & Menzel, S. (2013). Wildlife value orientations as predicting factors in support of reintroducing bison and of wolves migrating to Germany. *Journal of Nature Conservation*, 21, 125–32.

Homer, P. M. & Kahle, L. R. (1988). A structural equation test of the value-attitude-behavior hierarchy. *Journal of Personality and Social Psychology*, 54(4), 638.

Inglehart, R. (1997). *Modernization and Postmodernization: Cultural, Economic, and Political Change in 43 Societies*. Princeton, NJ: Princeton University Press.

Inglehart, R. & Norris, P. (2016). *Trump, Brexit, and the Rise of Populism: Economic Have-Nots and Cultural Backlash*. Social Science Research Network HKS Working Paper No. RWP16–026.

Inglehart, R. & Welzel, C. (2005). Exploring the unknown: Predicting the responses of publics not yet surveyed. *International Review of Sociology*, 151, 173–201.

Jacobson, C. A. & Decker, D. J. (2008). Governance of state wildlife management: Reform and revive or resist and retrench? *Society & Natural Resources*, 21, 441–8.

Kitayama, S. (2002). Culture and basic psychological processes: Toward a system view of culture – comment on Oyserman et al. 2002. *Psychological Bulletin*, 128(1), 89–96.

Kitayama, S., Ishii, K., Imada, T., Takemura, K. & Ramaswamy, J. (2006). Voluntary settlement and the spirit of independence: Evidence from Japan's 'Northern Frontier'. *Journal of Personality and Social Psychology*, 91(3), 369–84.

Kluckholn, F. R. & Strodtbeck, F. L. (1961). *Variations in Value Orientations*. Oxford: American Psychological Association.

Manfredo, M. J. (2008). *Who Cares about Wildlife? Social Science Concepts for Exploring Human–Wildlife Relationships and Conservation Issues*. New York: Springer-Verlag.

Manfredo, M. J., Bruskotter, J. T., Teel, T. L., Fulton, D., Schwartz, S. H., Arlighaus, R., Oishi, S., Uskul, A. K., Redford, K., Kitayama, S. & Sullivan, L. (2017b). Why social values cannot be changed for the sake of conservation. *Conservation Biology*, 31(4), 772–80.

Manfredo, M. J., Pierce, C. L., Fulton, D., Pate, J. & Gill, B. R. (1999). Public acceptance of wildlife trapping in Colorado. *Wildlife Society Bulletin*, 27(2), 499–508.

Manfredo, M. J., Teel, T. L. & Bright A. D. (2003). Why are public values toward wildlife changing? *Human Dimensions of Wildlife*, 8, 287–306.

Manfredo, M. J., Teel, T. L. & Dietsch, A. M. (2016). Implications of human value shift and persistence for biodiversity conservation. *Conservation Biology*, 30(2), 287–96.

Manfredo, M. J., Teel, T. L. & Henry, K. L. (2009). Linking society and environment: A multilevel model of shifting wildlife value orientations in the western United States. *Social Science Quarterly*, 90, 407–27.

Manfredo, M. J., Teel, T. L., Sullivan, L. & Dietsch, A. M. (2017a). Values, trust, and cultural backlash in conservation governance: The case of wildlife management in the United States. *Biological Conservation*, 214, 303–11.

Manfredo, M. J. & Zinn, H. C. (1996). Population change and its implications for wildlife management in the new west: A case study of Colorado. *Human Dimensions of Wildlife*, 1(3), 62–74.

Martin, J. L., Maris, V. & Simberloff, D. S. (2016). The need to respect nature and its limits challenges society and conservation science. *Proceedings of the National Academy of Sciences*, 113(22), 6105–12.

McCoy, C., Bruyere, B. L. & Teel, T. L. (2016). Qualitative measures of wildlife value orientations with a diverse population in New York City. *Human Dimensions of Wildlife*, 21(3), 223–39.

Mithen, S. (1996). *The Prehistory of the Mind*. London: Thames and Hudson.

Oishi, S. (2014). Socioecological psychology. *Annual Review of Psychology*, 65, 581–609.

Pattberg, P. (2007). Conquest, domination and control: Europe's mastery of nature in historical perspective. *Journal of Political Ecology*, 14, 1–9.

Pratto, F. (1999). The puzzle of continuing group inequality: Piecing together psychological, social and cultural forces in social dominance theory. *Advances in Experimental Social Psychology*, 31, 191–263.

Rentfrow, P. J., Gosling, S. D. & Potter, J. (2008). A theory of the emergence, persistence, and expression of geographic variation in psychological characteristics. *Perspectives on Psychological Science*, 3(5), 339–69.

Riepe, C. & Arlinghaus, R. (2014). Explaining anti-angling sentiments in the general population of Germany: An application of the cognitive hierarchy model. *Human Dimensions of Wildlife*, 19, 371–90.

Schwartz, S. H. (1992). Universals in the content and structure of values: Theoretical advances and empirical tests in 20 countries. *Advances in Experimental Social Psychology*, 25, 1–65.

Schwartz, S. H. (2006). A theory of cultural value orientations: Explication and applications. *Comparative Sociology*, 5, 136–82.

Schwartz, S. H. & Bilsky, W. (1987). Toward a universal psychological structure of human values. *Journal of Personality and Social Psychology*, 53 (3), 550–62.

Sijtsma, M. T., Vaske, J. J. & Jacobs, M. H. (2012). Acceptability of lethal control of wildlife that damage agriculture in the Netherlands. *Society & Natural Resources*, 25, 1308–23.

Smit, B. & Wandel, J. (2006). Adaptation, adaptive capacity and vulnerability. *Global Environmental Change*, 16, 282–92.

Stern, P. C. & Dietz, T. (1994). The value basis of environmental concern. *Journal of Social Issues*, 50(3), 65–84.

Talhelm, T., Zhang, X., Oishi, S., Shimin, C., Duan, D., Lan, X. & Kitayama, S. (2014). Large-scale psychological differences within China explained by rice versus wheat agriculture. *Science*, 344, 603–8.

Teel, T. L. & Manfredo, M. J. (2010). Understanding the diversity of public interests in wildlife conservation. *Conservation Biology*, 24, 128–39.

Teel, T. L., Manfredo, M. J., Jensen, F. S., Buijs, A. E., Fischer, A., Riepe, C., Arlinghaus, R. & Jacobs, M. H. (2010). Understanding the cognitive basis for human–wildlife relationships as a key to successful protected-area management. *International Journal of Sociology*, 40(3), 104–23.

United States Census Bureau. (2010). *American FactFinder fact sheet*. US Census Bureau, Washington, DC. Available from http://factfinder2.census.gov/ (accessed May 2014).

United States Fish and Wildlife Service (USFWS). (2007). *2006 National Survey of Fishing, Hunting, and Wildlife-Associated Recreation: National Overview*. Washington, DC: US Department of the Interior.

Uskul, A. K., Kitayama, S. & Nisbett, R. E. (2008). Ecocultural basis of cognition: Farmers and fishermen are more holistic than herders. *Proceedings of the National Academy of Sciences*, 105(25), 8552–6.

Walter, W. D., Lavelle, M. J., Fischer, J. W., Johnson, T. L., Hygnstrom, S. E. & VerCauteren, K. C. (2010). Management of damage by Elk (*Cervus elaphus*) in North America: A review. *Wildlife Research*, 37, 630–46.

Whittaker, D., Vaske, J. J. & Manfredo, M. J. (2006). Specificity and the cognitive hierarchy: Value orientations and the acceptability of urban wildlife management actions. *Society & Natural Resources*, 19(6), 515–30.

Wildavsky, A. B. (1991). *The Rise of Radical Egalitarianism*. Washington, DC: American University Press.

Zinn, H. C., Manfredo, M. J. & Barro, S. C. (2002). Patterns of wildlife value orientations in hunters' families. *Human Dimensions of Wildlife*, 7(3), 147–62.

3

Broadening the Aperture on Coexistence with Wildlife through the Lenses of Identity, Risk and Morals

MICHELLE L. LUTE AND MEREDITH L. GORE

Shifting conservation narratives from a conflict-based frame to one of coexistence minimally requires addressing human–human conflict (HHC) over how to manage wildlife in addition to direct human–wildlife conflict. HHC undermines efforts to achieve human–wildlife coexistence because social conflict focuses on the symbolic nature of a species and contests over property rights, urban–rural divisions, government authority, cultural hegemony, livelihood preservation and other often historically rooted identity clashes. HHC can prevent decision-makers and stakeholders from directly addressing the root causes of negative human–wildlife interactions and cooperating on shared goals and impacts. Like many communities across the globe, conservation communities are diversifying. Although diversity can be and is celebrated in wildlife conservation, it also can and is introducing new challenges for democratic decision-making. A common thread linking many of these challenges is the need for decision-makers to sufficiently consider the perspectives of diverse stakeholders. When stakeholder perspectives are rooted in differing values, for wildlife in our case, policies that find a satisfactory middle ground or compromise between values are often difficult if not potentially impossible to find (Nie 2003). Consider for instance the value-driven conflict over whether or not to hunt a particular species. Stakeholders who believe that someone's desire for a fur, trophy or unique hunting experience does not outweigh the intrinsic value of life will likely not change their position against hunting. Other stakeholders may contend that maintaining their hunting traditions outweighs the value of an individual animal's continued existence. What policy could possibly address both stakeholders' values and how would such a policy be developed?

In response to impossibly opposing public policy preferences, decision-makers, wildlife managers, human dimensions scholars and facilitators often turn to process, especially in the United States and other developed nations (Madden & McQuinn 2014; Lute & Axelrod 2015). In theory, if decision processes are considered broadly inclusive, fair and transparent by those holding a stake in the decision, decisions will last and not be overturned by a judge, voters or executive order by a new administration (Bruskotter et al. 2014; Manfredo et al. 2017a). Importantly, according to process proponents, the stakeholders in the aforementioned example do not need to find a true compromise. Instead, they need to work together in equitable decision-making roles or as advisors to whomever the ultimate decision-makers are. Involvement and contribution to the decision can produce stakeholder buy-in for the final decision (Wilson 2008; Wilson & Bruskotter 2009).

Although support for engagement-based proper process in stable decision-making has been documented in practical settings (e.g. De Vente et al. 2016), contentious wildlife-related policy, for instance, often defies what process proponents might expect. In the United States, in states such as Wisconsin, New Mexico and Washington, hunters, ranchers and wolf advocates, for example, have not accepted rulemaking regarding where and how wolves may be hunted and killed (e.g. Associated Press 2013; Oosting 2013). These case studies exemplify how extreme HHCs are often well described as value-based conflicts, seeming to suggest there is little to no room for common ground, or coexistence. In extreme HHC cases where process is limited in its ability to alleviate fundamental value-based differences, alternative and interdisciplinary approaches may provide a path forward because value change is rarely achievable (Manfredo et al. 2017b). For example, risk and decision science introduces mechanisms to promote risk-mitigating behaviours. Coupled human and natural systems inquiry focuses on negative interactions between humans and nature (c.g. climate change). Regardless of the informing discipline, interventions that focus not on values but perceptions (e.g. related to risk), behavioural intentions or behaviours are more likely to succeed in moving along the continuum from HHC to coexistence.

This chapter explores HHC using principles from morals, risk and human–nature interactions. We explore identity (i.e. affiliation with groups of like-minded individuals), risk perception (i.e. judgements related to harm) and moral judgements (i.e. intuitions about right and wrong) in an effort to advance understanding about the psychology of morally relevant behaviour as well as mechanisms for encouraging positive relationships between humans and their environment, from restoration of habitat and wildlife populations to reducing greenhouse gas emissions. Can societies find compromise in collectively deciding what reasons justify killing particular animals in particular contexts? What level of risk is as low as reasonably acceptable? Herein, we argue for further exploration of the three considerations of identity, risk perceptions and moral judgements and an integrated understanding of how they interact in regards to HHC over wildlife conservation. We discuss each of these considerations separately and then how they interact to influence coexistence-based conservation behaviours, which can range from tacit tolerance of the presence of carnivores to active support in the recovery and reintroduction of species to historic and appropriate habitats (for various definitions of coexistence, see e.g. Carter & Linnell 2016; Frank 2016). We operate under the assumption that encouraging coexistence behaviours facilitates stakeholder cooperation with policies and processes that move HHC along the conflict-to-coexistence continuum in the contexts where and when it is appropriate. The appropriateness of a particular property for human-wildlife coexistence where humans and wildlife share space, as opposed to land sparing where humans and wildlife are kept separate, is a question unto itself and a topic for another chapter (see Chapter 1). After we discuss the three separate considerations of identity, risk perceptions and moral judgements, we discuss how each consideration potentially interacts to influence coexistence behaviours. We conclude with recommendations for next steps in research and application.

3.1 COEXISTENCE CONSIDERATION 1: IDENTITY

By distilling patterns in stakeholder diversity of opinion, social identity helps researchers categorize, predict and understand differing perspectives and policy preferences (Tajfel & Turner 1979). Identity starts at the individual level where a person self-categorizes with a group that holds the person's same values and beliefs. This group affiliation in turn influences the individual's perceptions and behaviours. When identification with a particular group is strong, group norms strongly influence the individual's world-views, policy preferences and behaviours (Giannakakis & Fritsche 2011). Group norms also contribute to what identity researchers call in-group bias, whereby individuals emphasize positive in-group (i.e. those with which the individual identifies and affiliates) characteristics and negative aspects of out-groups (i.e. those perceived as different from the individual) in such a way as to result in a preconceived judgement about out-group members (Sherif 1967; Labianca et al. 1998). It is through group membership and in-group bias that we can start to understand certain elements of the tribal loyalty that defines many intergroup conflicts and cultures clashes. Social identity theory has been used to understand intergroup conflict in many contexts, from racial and gender bias to power dynamics between pilots and flight attendants (Navarrete et al. 2010; Ford et al. 2012). In the case of HHC about wildlife, relevant identities could include hunters, nature lovers, outdoor enthusiasts and property rights advocates (Lute & Gore 2014; Lute et al. 2014). Social identity theory suggests that in-group members share salient values, which have been shown to influence attitudes, preferences and behavioural intentions related to wildlife and wildlife management (Sponarski et al. 2015). Social trust of risk managers can mediate the relationship between salient values and attitudes (Sponarski et al. 2015). Therefore, measuring identity can enhance understanding of the salient values at play and predict policy support in the context of HHC.

3.2 COEXISTENCE CONSIDERATION 2: RISK PERCEPTIONS

Differing perspectives among identity groups may be influenced by risk perception. Risk perceptions are value-laden judgements about one's likelihood of harm and include both affective (i.e. related to intuitive feeling states; see Chapter 4) and cognitive (i.e. related to thinking through information) dimensions (Sjöberg 1998; Lazo et al. 2000; Lindquist et al. 2006). Researchers employing the psychometric paradigm to investigate cognitive elements that affect individuals' risk perception (Slovic 1987) utilize seven factors that influence people's risk judgements about nature: certainty, control, frequency, naturalness, seriousness, responsiveness and trust (Rogers 1975; Slovic 1987; Sjöberg 1998; Gore et al. 2007). The latter two factors, responsiveness and trust, refer to the managers that help mitigate or address risk and decision-makers that craft policy related to risk. Including affective risk perception (i.e. feelings about a source of risk, which can be described as dread, fear or worry) may enhance understanding of the risk perception–behaviour relationship because cognitive components only provide partial explanations (Rivers & Arvai 2007; Wilson & Arvai 2010). Risk perceptions are not only emotional but also experiential. The role of experience in wildlife-related risk can augment HHC through experiences that induce fear (e.g. lost a companion animal to a carnivore) or, for example, frustration from past risk management (e.g. wildlife managers were considered unresponsive).

Risk perceptions are important for understanding human interactions with nature, particularly wildlife that may transmit disease or carnivores that may pose threats to the health and safety of livestock and companion animals (Riley & Decker 2000; Gore et al. 2009; Johansson & Karlsson 2011). The environmental justice literature suggests that moral indignation over the asymmetrical nature of many if not most risks is in part due to issues of fairness, or a lack thereof (Earle & Siegrist 2008). The risk literature has contributed important knowledge about risk-related decision-making, politics, communication and pro-environmental or risk-reducing behaviours to cases such as pollution near low-income neighbourhoods or in common pool resources (Hatcher et al. 2000; Jurin et al. 2010). Whether the risk is contaminants in drinking water or rabid raccoons in the backyard, risk perceptions and other perceptions that relate to risk (e.g. moral judgements such as fairness) can influence HHC more than technical risk assessments and are therefore critical to address when trying to move from conflict to coexistence.

3.3 COEXISTENCE CONSIDERATION 3: MORAL JUDGEMENTS

Differing perspectives among identity groups include judgements of right and wrong. Moral judgements assess right or wrong and, similar to risk perceptions, include intuition, involve uncertainty and vary by individual (Schwartz 1968; Amit & Greene 2012). Moral Foundations Theory (MFT) was developed to explain how people come to hold their intuitions about right and wrong, and seeks to explain both the diversity and unity of moral judgements that can exist between individuals and among cultures (Haidt 2007; Graham et al. 2011, 2013). The theory posits the existence of at least five innate, universal moral categories: authority, harm, fairness, loyalty and purity (Haidt & Joseph 2004, 2007). These categories are then elaborated or attenuated based on one's experiences and culture, thereby creating the unique moralities we see within and between groups and societies (e.g. generally that conservatives emphasize respect for authority; liberals emphasize fairness).

Importantly for HHC, moral foundations researchers emphasize the primacy of gut-level moral intuitions over conscious declarative moral reasoning in how people come to their decisions about policy, politics and morality. Here, moral judgements are made as a result of pre-existing intuitions (reviewed in Haidt 2012). Researchers have demonstrated a compelling empirical case for the usefulness of conceptualizing moral judgement as composed of basic, intuitive foundations that predict a wide range of political concerns relevant to wildlife management (e.g. treatment of animals and appropriate behaviour within social groups; Haidt 2007; Graham et al. 2011, 2013).

3.4 INFLUENCES OF IDENTITY, RISK PERCEPTIONS AND MORAL JUDGEMENTS ON BEHAVIOUR

All of these differing perspectives based on identities, varying risk perceptions and individual moral judgements result in different behaviours, ranging from intolerance for the presence of wildlife to active protection of species and their habitats (Lute & Gore 2014; Lute et al. 2016). These myriad considerations may complicate efforts to predict coexistence behaviours but can help researchers understand motivations for actions that can have significant effects on wildlife populations, such as environmental crimes and intentional non-compliance with policies (Gibbs et al. 2009; Gore et al. 2013). Greater understanding of these considerations and their combined effect on behaviour can help enable coexistence among diverse stakeholders, thereby assuage HHC by moving people towards the same conservation goals, and further towards the coexistence side of the continuum.

Although the relationship between risk perception and behaviour is well-studied in different contexts (Liao et al. 2009; Dohmen et al. 2011), empirical knowledge of how risk perception and behaviour are influenced by moral judgements is virtually non-existent (Sjöberg & Winroth 1986; Sjöberg 2000). The moral aspects of risk (e.g. whether asymmetrical exposure to risk disadvantages a group of people) may be important in judgements of whether a risk is acceptable and moral values related to risk are the subject of public debate and political action (Sjöberg & Winroth 1986). Some researchers have expanded the psychometric paradigm of risk perception to include moral aspects of risk, which highly correlates with acceptance of risk among diverse individuals and cultures (Sjöberg & Winroth 1986). Moral aspects of risk may apply to justice or fairness about risk, which can also relate to risk managers. For example, stakeholders who live near and fear wolves consider reintroduction efforts unfair (a moral judgement that reintroduction actions are wrong) because they are exposed to a risk that others are not and the risk is sometimes considered out of their control and unnatural (brought by people). The asymmetry and emotional dimensions of a risk combined with the psychometric factors of control and naturalness can result in mistrust (another factor in the psychometric paradigm of risk) of decision-makers and managers that created and implemented policies to reintroduce the large carnivore species.

Specific moral intuitions may also relate to risk perceptions that influence behaviour. Moral concerns about harm/care in relation to humans may lead to heighted perceptions of risk posed by an element in nature (e.g. hurricanes, tigers) to those humans. On the other hand, if nature is seen as something requiring protection (e.g. an endangered or rare species), concerns about decreasing harm of nature may be more salient (Lute et al. 2014). Other studies have used disgust (i.e. the opposite of purity) to explain fear of various animals (Johansson & Karlsson 2011). Intuitions about authority and in-group loyalty may influence risk perceptions via social norms (Lute & Gore 2014). For example, persons may judge their own level of risk related to natural disasters based in part on risk perceptions of respected authorities or other identity group members and their ability to help respond to risks (e.g. government assistance in the form of food supplies and temporary shelter). Lastly, if what is considered fair is more acceptable to persons, they may view natural risks as more acceptable and less threatening than unnatural risks that are man-made. Consider, for instance, a situation in which farmers seem to accept the natural, albeit unpredictable and potentially significant, risks posed by weather but strongly object to exposure to less likely risks posed by federally protected carnivores (Nie 2002, 2003a). Fairness may explain – in concert with certainty, control and other psychometric factors – such differences in risk perception because the presence of carnivores is seen as an unfair situation created by centralized governments more concerned with other interests than those of the local farmer (Skogen & Krange 2003; Naughton-Treves et al. 2003).

Because moral intuitions and risk perception may relate to each other and are important for predicting myriad human behaviours (Slovic 1987; O'Connor et al. 1999), exploring the relationships between moral intuitions, risk perception and behaviour in a single causal model may provide a more comprehensive account of human judgements according to context and natural resource-related behaviour (see Text Box 3.1). Accordingly, we propose a conceptual model, consistent with the general MFT approach, suggesting a step-wise psychological process in which basic moral intuitions influence more specific judgements about risk and finally specific coexistence actions (Baron & Kenny 1986). We sought to validate this conceptual model in the Text Box 3.1 case study below.

Text Box 3.1 Conceptual Model for Coexistence with Wolves in Michigan

In 2013, we conducted an online survey of active and aware Michigan wolf stakeholders in Qualtrics (qualtrics.com) and distributed via snowball sampling (for detailed methodology, see related study Lute et al. 2016). The survey included items measuring moral foundations, affective and cognitive risk perceptions, coexistence behaviours and socio-demographic characteristics through multiple choice-type questions using binary response options and five-point Likert-type scales (reported results collapse responses into agree and disagree). Scale items were evaluated for internal consistency using Cronbach's alpha (all scales were ≥0.8; Table 3.1). Means were used to create indices of items with continuous response options: affective and cognitive risk perceptions and the five moral foundations. One summated index was created for coexistence behaviours because it was measured with binary response options. Through zero-order correlations and mediation (i.e. path analyses; Baron & Kenny 1986), we explored relationships in our conceptual model which posited that moralities are foundational intuitions about right and wrong that can directly influence coexistence behaviours (Figure 3.2; Schwartz 1968; Holsman 2000), defined as direct or indirect actions taken to benefit nature or some component of nature (in this case, wolves and wolf habitat). We also posited that the influence of moral intuitions on behaviour may be filtered by both affective and cognitive risk perceptions (Sjöberg 1998) as intermediate steps between a person's foundational moral intuitions and their behaviour or policy positions.

Table 3.1 *Descriptive statistics: mean, standard deviation (SD), Cronbach's alpha and n*

Concept	Item	Mean	SD	Alpha	n
Socio-demographics	Age	53.80	13.64	N/A	855
	Education	5.18	1.94		
	Gender	1.52	0.88		
	Income	6.37	1.93		
	Political Party	3.57	1.54		
	Political Orientation	4.41	1.57		

(*continued*)

Table 3.1 (*continued*)

Concept	Item	Mean	SD	Alpha	n
Moral Foundations	Authority	3.03	0.94	N/A	972
	Fairness	3.55	0.86		
	Harm/Care	2.98	1.05		
	In-group	2.90	0.97		
	Purity	2.97	1.14		
Cognitive Risk Perception	*Frequency*: Evidence of human–wolf conflict is rare in my community.	4.11	1.22	0.92	960
	Control: I believe that I have control over risks posed by wolves.	3.53	1.18		
	Certainty: If the wolf population increases, human–wolf interactions will increase.	4.13	0.89		
	Trust: I trust wildlife managers to manage wolves appropriately.	3.68	1.30		
	Naturalness: Problems involving wolves are increased by environmental factors.	3.78	0.93		
	Seriousness: The risks posed by wolves are acceptably low.	3.24	1.27		
	Responsiveness: Wildlife managers are responsive to wolf problems.	3.49	1.10		
Affective Risk Perception	I worry about risks posed by wolves to... Children	3.34	1.31	0.95	895
	Game species	3.22	1.37		
	Hunting dogs	3.49	1.32		
	Livestock	3.80	1.13		
	My health	2.09	1.10		
	My hunting traditions	2.70	1.42		
	My livelihood	1.89	1.06		
	My personal safety	2.30	1.23		
	Pets	3.61	1.22		

(continued)

Table 3.1 (*continued*)

Concept	Item	Mean	SD	Alpha	n
Coexistence Behaviours	Attended a legislative hearing or organizational meeting	0.16	0.37	0.82	855
	Boycotted or avoided buying the products of a company because of their stance on wolf management	0.09	0.29		
	Donated money to a group	0.17	0.38		
	Called or wrote a letter to a legislator	0.24	0.43		
	Educated others	0.37	0.48		
	Managed land to create or conserve wolf habitat	0.05	0.21		
	Read newsletters, magazines or other publications	0.57	0.50		
	Signed a petition	0.29	0.46		
	Volunteered with a group	0.11	0.32		
	Voted for a candidate in an election based at least in part because of his/her stance on wolf management	0.13	0.33		
	Wrote a letter to a newspaper or called in to a news programme	0.07	0.25		

Results

Of the final sample of 1,239 Michigan residents, respondents skewed white (68%) and male (76%). Hunters (32%) were overrepresented in our sample compared to published recreational participation records in Michigan (n=795,535/8% for hunters and n=10,241/0.1% for trappers; Frawley 2013). Conservationists (19%) and environmentalists (10%) were the second- and third-largest identity groups. Animal rights or welfare advocates (6%), farmers (3%), gun rights advocates (8%), property rights advocates (4%) and wildlife advocates (4%) made up smaller proportions of respondents.

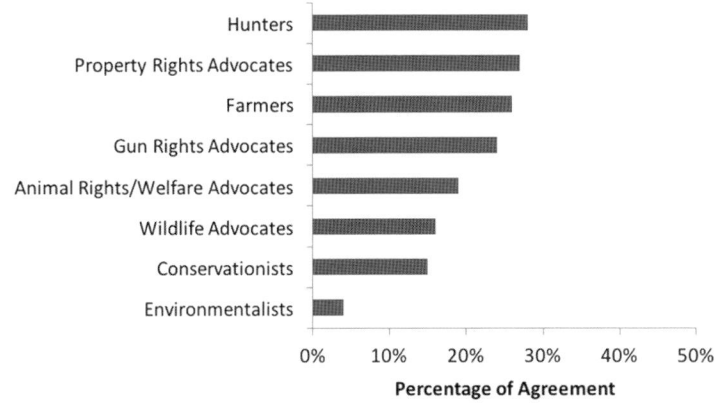

Figure 3.1 Affective risk perceptions by identity. When grouped by self-described identity, percentage of respondents vary in whether they report greater worry (i.e. agreement) that wolves pose risks to nine targets (livestock, companion animals, hunting dogs, children, game species, hunting traditions, personal safety, human health, livelihoods).

Affective Risk Perception: Respondents were more likely to report worry associated with risks posed by wolves to livestock (73%), companion animals (65%), hunting dogs (57%), children (55%) and game species (51%) than hunting traditions (32%), personal safety (18%), human health (9%) or livelihoods (6%). When grouped by identity, environmentalists reported the lowest affective risk perceptions and property rights advocates and hunters showed the highest (Figure 3.1).

Cognitive Risk Perception: Majorities of the total sample agreed that risks were controllable (risk perception factor *control*; 59%), acceptably low (*seriousness*; 48%) and rare (*frequency*; 79%); that wildlife managers were responsive (*responsiveness*; 52%) and trusted (*trust*; 64%); and 'problems involving wolves are increased by environmental factors' (*naturalness*; 70%). High agreement with these measures may indicate low cognitive risk perception. Majorities also agreed 'if the wolf population increases, human–wolf interactions will increase' (*certainty*; 84%). When grouped by identity, property rights advocates show the highest disagreement followed by hunters, wildlife and gun rights advocates.

Testing the Influences of Identity, Risk Perceptions and Moral Judgements on Behaviour

Evidence for mediation is supported when three conditions are met: the relationship between (1) mediator and independent variable is significant, (2) mediator and dependent variable is significant and (3) independent and dependent variable is significantly smaller when the effect of the mediator is controlled (Baron & Kenny 1986). Three moral foundations (i.e. authority,

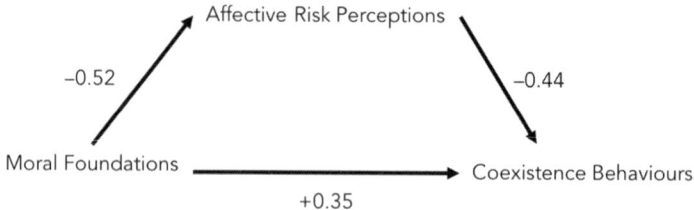

Figure 3.2 Mediation to test moral and risk-related influences on coexistence behaviours. Our affective risk perception index mediated 45% of the total effect of moral foundations on behaviour (β = 0.15, S.E. = 0.02, p $\leq$ 0.001; 95% CI = [0.11, 0.20]).

in-group loyalty, harm/care) and the affective risk perception index met the initial criteria for mediation (the zero-order correlation between cognitive risk perception and behaviour was not significant). Therefore, in order to test whether affective risk perception mediated the link between each moral intuition and coexistence behaviour, we conducted a series of regression analyses to infer a causal psychological process occurring in steps: (1) moral foundation→ (2) affective risk perception→ (3) behavioural intentions, where affective risk perception mediates or *carries* the effect between the moral foundation in question and coexistence behaviour (Baron & Kenny 1986). A mediating variable explains a significant proportion of the relationship between independent and dependent variables.

Affective risk perception mediated the relationship between three moral foundations and coexistence behaviour. A negative link between *authority* and coexistence behaviour was found in the first step (β = −0.20). However, when affective risk perception was added, the effect of authority was reduced 59% (β = −0.12, S.E. = 0.02, p $\leq$ 0.001; 95% CI = [−0.09, −0.15]). A negative link between *in-group loyalty* and coexistence behaviour was also found in the first step (β = −0.15). When affective risk perception was added, the effect of in-group loyalty was reduced by 85% (β = −0.12, S.E. = 0.02, p $\leq$ 0.001; 95% CI = [−0.09, −0.16]). We found a significant positive relationship between *harm/care* and coexistence behaviour (β= 0.27), which was reduced by 26% (β = 0.06, S.E. = 0.01, p $\leq$ 0.001; 95% CI = [0.04, 0.09]) when affective risk perception was added to the model. Each analysis verified that affective risk perception carries a significant portion of the relationship between moral intuitions about (1a) authority, (1b) in-group loyalty and (1c) harm/care and (2) coexistence behaviours.

In this study of active wolf stakeholders, we found support for our proposed model that moral intuitions predict behavioural intentions, and that affective risk perceptions filter these relationships. Affective risk perception was a particularly salient intermediate in the process of translating moral consideration of loyalty to a social group to behaviour. Results indicate that intuitions about both risk and morality are important drivers of behaviour.

3.5 ADDRESSING HUMAN–HUMAN CONFLICT AND ENCOURAGING COEXISTENCE

By recognizing common moral judgements during decision-making and identifying the level of perceived risk that the higher number of stakeholders can accept, decision-making about wildlife may be able to move towards coexistence and reduced HHC (i.e. encourage cooperation) between groups. Without understanding the ways in which moral and risk considerations compete and complement judgements that eventually influence behaviour, wildlife conservation policy working to reduce the scope and magnitude of HHC has lacked potentially useful tools in addressing disagreement over contentious decisions. Identity helps identify patterns in how moral and risk-related judgements may predict behaviours.

The role of science in evidence-based decision-making is central; however, as with other sources of information, stakeholders intentionally introduce specific scientific studies to support their arguments. This motivated reasoning can be particularly influential when identity cues are strong (Hart & Nisbet 2012). Stakeholders involved in controversial wildlife policy or practice (i.e. HHC) may accuse others of allowing emotions to drive judgement when in fact emotions play a role in most human judgements (Lute & Gore 2014). The combination of motivated reasoning with a lack of awareness about the role of emotions can complicate HHC mitigation (Smith & DeCoster 2000; Paxton & Greene 2010). Nevertheless, understanding the science-based arguments of involved identity groups and how they relate to the group's accepted level of risk can shepherd decision processes aimed at addressing HHC and create long-term options for wildlife conservation. For example, wildlife advocates often emphasize studies on the trophic cascades and ecosystem benefits of large carnivores (e.g. Ripple & Beschta 2012; Wallach et al. 2015). They consider those benefits as outweighing risks carnivores pose to domestic livestock. On the other hand, ranchers focus on science detailing negative impacts of carnivores on livestock (e.g. Treves et al. 2011), which may serve to reinforce worry and perhaps decrease acceptance of any level of risk. The information deficit model would suggest that providing all best available science to all stakeholders should address motivated reasoning. But the deficit model is limited in assuaging identity-driven conflict over controversial issues (Hart & Nisbet 2012). In such contexts, emotions can be recognized, not vilified and addressed to mitigate HHC and encourage coexistence.

Understanding which moral intuitions influence stakeholders involved in particular HHCs can inform efforts to enhance resolution and focus the picture of a moral emotional landscape, which in turn can inform understanding of behavioural motivations and thereby how best to address HHC and encourage cooperation. In Text Box 3.1 highlighting aspects of Michigan wolf management, it is evident that moral intuitions clearly shape feelings about risks posed by wolves, which in turn motivate conservation-related behaviour. Moral foundations of in-group loyalty and authority worked in concert with affective risk perception to result in decreased willingness to conserve. The exact mechanism by which authority and in-group loyalty influence affective risk perception is still unclear. For example, the relationship between respect for authority and affective risk perception might indicate that people believe humans should have authority over nature, which correlates with worry about risks (even unlikely ones) posed by wolves and not engaging in activities to benefit wolves. Among those who prioritize moral concerns of authority and in-group loyalty, personal worry about wolves may also be influenced by the worry of respected authorities and identity group members. Just as HHC is often about the symbolic nature of wildlife, HHC can also be about norms and expectations within an identity group (and contests between groups) more than the actual risks wildlife pose to people (hence the worry about low likelihood risks).

In contrast to authority and in-group loyalty, moral concerns about harm/care may encourage coexistence behaviours if the target of care is nature or wildlife. Harm/care concerns may encourage cooperation among differing identity groups if the target of care is other people. However, as the study in Text Box 3.1 suggests, affective risk perceptions may filter the effect of harm/care concerns on behaviour. The direct positive relationship between intuitions about harm/care and coexistence behaviour suggests that when reducing harm to and increasing care of wildlife is salient, conservation-related behaviours are increased. However, reducing harm to (and increasing care of) people may result in decreased coexistence behaviours when worry about wildlife is considered. To avoid decreasing the coexistence behaviours, risk perceptions need to be addressed and mitigated in appropriate ways.

Efforts to reduce HHC by leveraging insight about risk perceptions require careful consideration of the fact that affective risk perceptions, more so than cognitive risk perceptions and purely rational risk assessments, may play an important role in influencing coexistence behaviours. In other words, disagreement over wildlife management may not be about uncertainty surrounding the likelihood of risks but about worry associated with risks and how they are managed regardless of the frequency of exposure. In the Text Box 3.1 study, risks were considered rare and acceptably low. Yet affective risk perception levels in relation to certain targets (e.g. children, livestock) may be high among certain stakeholders in this and other contexts (Sponarski et al. 2015). Wildlife management strategies might aim to mitigate human–wildlife conflict as well as HHC by identifying what level of perceived risk is acceptable to the greatest number of stakeholders while also addressing legitimate concerns of minority groups (to avoid both tyrannies of the majority and minority where possible). Results from the study in Text Box 3.1 suggest that the degree of perceived risk may be irrelevant or that stakeholders may be aiming for very low perceived risk in relation to wolves (perhaps because of ideas about human authority over nature as hypothesized above). Although the assessed risk from wildlife to humans may be low, it will likely never be reduced to zero. Thus, a more pragmatic objective may be to attenuate perceived risk to accurately reflect the low level of assessed risk. Risk messaging that successfully reduces stakeholder worry about wildlife-related threats to vulnerable others may be more effective at mitigating risk-related disagreement (rather than aiming to reduce the likelihood of already low-level risks; Gore & Knuth 2009; Muter et al. 2009).

Affective risk perceptions may have the potential to decrease coexistence behaviours associated with HHC but concerns about reducing harm to and caring for nature may increase participation in positive human–nature interactions. One way to promote positive behaviours or conversely to discourage negative ones is to frame communications and outreach to stakeholders appropriately and specifically to different identity groups. Communication that only addresses cognitive aspects of risk perception (e.g. judgements about probabilities that a carnivore will attack livestock) may fall short of objectives without also addressing emotional aspects and moral judgements, such as harm to and care for nature. In order to address emotional aspects of risk, effective and targeted engagement of identity groups with high affective risk perceptions (in the study in Text Box 3.1, hunters, gun and property rights advocates) may offer: (1) support for worries as valid (regardless of likelihood), and (2) ways to reinforce residents' sense of control to protect vulnerable others (Keller et al. 2006). With effective risk communication, wildlife conservation can transcend HCC and wildlife conservation can progress along the conflict-to-coexistence continuum.

3.6 RECOMMENDATIONS AND FUTURE DIRECTIONS

This chapter sought to explore some ways to promote coexistence and influence HHC through the lenses of social identity, risk perception and moral foundations theories. Key points for takeaway and recommendations for meeting the challenges of value-based conflicts over wildlife include:

- Affective risk perceptions related to worry about vulnerable others, particularly game species and domestic animals, can decrease positive coexistence behaviours more so than cognitive risk perceptions related to certainty, control, frequency, naturalness, responsiveness, seriousness and trust.
- Moral concerns related to authority and in-group loyalty can reduce coexistence behaviours among certain stakeholders, especially when those stakeholders worry about the risks posed by carnivores to vulnerable others.
- Conversely, moral concerns of reducing harm and maximizing care can encourage coexistence among diverse stakeholders.
- Recognizing and understanding the kinds of moral judgements that are most salient for a person or identity group (e.g. reducing harm to carnivores, caring for people, being loyal to a group) can help stakeholders feel heard, which may alleviate conflict and encourage engagement in a decision process that will lead to a decision that diverse stakeholders can support.
- HHC may be reduced if perceived and assessed levels of risk are closer to matching among all involved identity groups. To assuage high levels of perceived risk among some stakeholders, risk communications should address worry related to vulnerable others rather than, or in addition to, appealing to cognitive risk perceptions.
- An overemphasis on science that ignores the role of emotions in judgements with moral and risk dimensions will likely hamstring efforts to effectively address HHC.
- Experiments with risk messaging and other interventions to encourage positive human–nature interactions.
- Confirmation of Text Box 3.1 study findings in other wildlife contexts and among broader stakeholder groups.
- Further exploration of the relationships between moral intuitions about authority/in-group loyalty and affective risk perception.
- Direct measures of psychological processes (e.g. implicit association tests) related to judgements about wildlife, other elements in nature and fellow wildlife stakeholders.

3.7 References

Amit, E. & Greene, J. D. (2012). You see, the ends don't justify the means: Visual imagery and moral judgment. *Psychological Science*, 23, 861–8.

Associated Press. (2013, 22 May). Wolf hunt referendum to go on 2014 Michigan ballot. *Detroit Free Press*. Lansing, Michigan.

Baron, R. & Kenny, D. (1986). The moderator–mediator distinction in social psychological research: Conceptual, strategic and statistical considerations. *Journal of Personality & Social Psychology*, 51, 1173–82.

Bruskotter, J. T., Vucetich, J. A., Enzler, S., Treves, A. & Nelson, M. P. (2014). Removing protections for wolves and the future of the U.S. Endangered Species Act (1973). *Conservation Letters*, 7, 401–7.

Carter, N. H. & Linnell, J. D. C. (2016). Co-adaptation is key to coexisting with large carnivores. *Trends in Ecology & Evolution*, 31, 575–8.

De Vente, J., Reed, M. S., Stringer, L., Valente, S. & Newig, J. (2016). How does the context and design of participatory decision-making processes affect their outcomes? Evidence from sustainable land management in global drylands. *Ecology & Society*, 21, art. 24.

Dohmen, T., Falk, A. & Huffman, D. (2011). Individual risk attitudes: Measurement, determinants and behavioral consequences. *Journal of the European Economic Association*, 9, 522–50.

Earle, T. C. & Siegrist, M. (2008). On the relation between trust and fairness in environmental risk management. *Risk Analysis*, 28, 1395–414.

Ford, J., O'Hare, D. & Henderson, R. (2012). Putting the 'we' into teamwork: Effects of priming personal or social identity on flight attendants' perceptions of teamwork and communication. *Human Factors: The Journal of the Human Factors & Ergonomics Society*, 55, 499–508.

Frank, B. (2016). Human–wildlife conflicts and the need to include tolerance and coexistence: An introductory comment. *Society & Natural Resources*, 1920, 1–6.

Frawley, B. J. (2013). *2012 Michigan Furbearer Harvest Survey*. Lansing, Michigan.

Giannakakis, A. E. & Fritsche, I. (2011). Social identities, group norms, and threat: On the malleability of ingroup bias. *Personality & Social Psychology Bulletin*, 37, 82–93.

Gibbs, C., Gore, M. L., McGarrell, E. F. & Rivers, L. (2009). Introducing conservation criminology: Towards interdisciplinary scholarship on environmental crimes and risks. *British Journal of Criminology*, 50, 124–44.

Gore, M. L. & Knuth, B. A. (2009). Mass media effect on the operating environment of a wildlife-related risk-communication campaign. *Journal of Wildlife Management*, 73, 1407–13.

Gore, M. L, Knuth, B. A., Curtis, P. D. & Shanahan, J. E. (2007). Factors influencing risk perception associated with human–black bear conflict. *Human Dimensions of Wildlife*, 12, 133–6.

Gore, M. L., Ratsimbazafy, J. & Lute, M. L. (2013). Rethinking corruption in conservation crime: Insights from Madagascar. *Conservation Letters*, 6, 430–8.

Gore, M. L., Wilson, R. S., Siemer, W. F., Hudenko, H. W., Clarke, C. E., Hart, P. S., Maguire, L. A. & Muter, B. A. (2009). Application of risk concepts to wildlife management: Special issue introduction. *Human Dimensions of Wildlife*, 14, 301–13.

Graham, J., Haidt, J., Koleva, S., Motyl, M., Iyer, R., Wojcik, S. & Ditto, P. H. (2013). Moral Foundations Theory: The pragmatic validity of moral pluralism. *Advances in Experimental Social Psychology*, 47, 55–130.

Graham, J., Nosek, B. A., Haidt, J., Iyer, R., Koleva, S. & Ditto, P. H. (2011). Mapping the moral domain. *Journal of Personality & Social Psychology*, 101, 366–85.

Haidt, J. (2007). The new synthesis in moral psychology. *Science*, 316, 998–1002.

Haidt, J. (2012). *The Righteous Mind: Why Good People Are Divided by Politics and Religion*. New York: Pantheon Books.

Haidt, J. & Joseph, C. (2004). Intuitive ethics: How innately prepared intuitions generate culturally variable virtues. *Dædalus*, 133, 55–66.

Haidt, J. & Joseph, C. (2007). The moral mind: How five sets of innate intuitions guide the development of many culture-specific virtues, and perhaps even modules. In P. Carruthers, S. Laurence & S. Stich, eds., *The Innate Mind*. New York: Oxford University Press, pp. 367–91.

Hart, P. S. & Nisbet, E. C. (2012). Boomerang effects in science communication: How motivated reasoning and identity cues amplify opinion polarization about climate mitigation policies. *Communication Research*, 39, 701–23.

Hatcher, A., Jaffry, S., Thébaud, O. & Bennett, E. (2000). Normative and social influences affecting compliance with fishery regulations. *Land Economics*, 76, 448–61.

Holsman, R. H. (2000). Goodwill hunting? Exploring the role of hunters as ecosystem stewards. *Wildlife Society Bulletin*, 28, 808–16.

Johansson, M. & Karlsson, J. (2011). Subjective experience of fear and the cognitive interpretation of large carnivores. *Human Dimensions of Wildlife*, 16, 15–29.

Jurin, R. R., Roush, D. & Danter, J. (2010). *Environmental Communication*, 2nd edn. Dordrecht: Springer.

Keller, C., Siegrist, M. & Gutscher, H. (2006). The role of the affect and availability heuristics in risk communication. *Risk Analysis*, 26, 631–9.

Labianca, G., Brass, D. J. & Gray, B. (1998). Social networks and perceptions of intergroup conflict: The role of negative relationships and third parties. *The Academy of Management Journal*, 41, 55–67.

Lazo, J. K, Kinnell, J. C. & Fisher, A. (2000). Expert and layperson perceptions of ecosystem risk. *Risk Analysis*, 20, 179–93.

Liao, C., Lin, H-N. & Liu, Y-P. (2009). Predicting the use of pirated software: A contingency model integrating perceived risk with the theory of planned behavior. *Journal of Business Ethics*, 91, 237–52.

Lindquist, K. A., Barrett, L. F., Bliss-Moreau, E. & Russell, J. A. (2006). Language and the perception of emotion. *Emotion*, 6, 125–38.

Lute, M., Bump, A. & Gore, M. L. (2014). Identity-driven differences in stakeholder concerns about hunting wolves. *PLoS ONE*, 9, e114460.

Lute, M., Navarrete, C. D., Nelson, M. P. & Gore, M. L. (2016). Assessing morals in conservation: The case of human–wolf conflict. *Conservation Biology*, 30, 1200–11.

Lute, M. L. & Axelrod, M. (2015). Public preferences for wolf management processes in Michigan. *Human Dimensions of Wildlife*, 20, 95–7.

Lute, M. L. & Gore, M. L. (2014). Stewardship as a path to cooperation? Exploring the role of identity in intergroup conflict among Michigan wolf stakeholders. *Human Dimensions of Wildlife*, 19, 267–79.

Madden, F. & McQuinn, B. (2014). Conservation's blind spot: The case for conflict transformation in wildlife conservation. *Biological Conservation*, 178, 97–106.

Manfredo, M. J., Bruskotter, J. T., Teel, T. L., Fulton, D., Schwartz, S. H., Arlinghaus, R., Oishi, S., Uskul, A. K., Redford, K., Kitayama, S. & Sullivan, L. (2017a). Why we can't change social values for the sake of conservation. *Conservation Biology*, 31, 772–80.

Manfredo, M. J., Teel, T. L., Sullivan, L. & Dietsch, A. M. (2017b). Values, trust, and cultural backlash in conservation governance: The case of wildlife management in the United States. *Biological Conservation*, 214, 303–11.

Muter, B. A., Gore, M. L. & Riley, S. J. (2009). From victim to perpetrator: Evolution of risk frames related to human–cormorant conflict in the Great Lakes. *Human Dimensions of Wildlife*, 14, 366–79.

Naughton-Treves, L., Grossberg, R. & Treves, A. (2003). Paying for tolerance: Rural citizens' attitudes toward wolf depredation and compensation. *Conservation Biology*, 17, 1500–11.

Navarrete, C. D., McDonald, M. M., Molina, L. E. & Sidanius, J. (2010). Prejudice at the nexus of race and gender: An outgroup male target hypothesis. *Journal of Personality and Social Psychology*, 98, 933–45.

Nie, M. (2003). Drivers of natural resource-based political conflict. *Policy Sciences*, 36, 307–41.

Nie, M. A. (2002). Wolf recovery and management as value-based political conflict. *Ethics, Place & Environment*, 5, 65–71.

O'Connor, R. E., Bord, R. J. & Fisher, A. (1999). Risk perceptions, general environmental beliefs, and willingness to address climate change. *Risk Analysis*, 19, 461–71.

Oosting, J. (2013, 26 November). New Michigan group seeks to protect future wolf hunts with citizen-initiated legislation. *MLive*. Lansing, Michigan.

Paxton, J. M. & Greene, J. D. (2010). Moral reasoning: Hints and allegations. *Topics in Cognitive Science*, 2, 511–27.

Riley, S. J. & Decker, D. J. (2000). Risk perception as a factor in wildlife stakeholder acceptance capacity for cougars in Montana. *Human Dimensions of Wildlife Management*, 4, 50–62.

Ripple, W. J. & Beschta, R. L. (2012). Trophic cascades in Yellowstone: The first 15 years after wolf reintroduction. *Biological Conservation*, 145, 205–13.

Rivers, L. & Arvai, J. (2007). Win some, lose some: The effect of chronic losses on decision making under risk. *Journal of Risk Research*, 10, 1085–99.

Rogers, E. M. (1975). *Diffusion of Innovations*, 4th edn. New York: The Free Press.

Schwartz, S. H. (1968). Awareness of consequences and the influence of moral norms on interpersonal behavior. *Sociometry*, 31, 355–69.

Sherif, M. (1967). *Group Conflict and Cooperation: Their Social Psychology*. London: Routledge & Kegan Paul Ltd.

Sjöberg, L. (1998). Worry and risk perception. *Risk Analysis*, 18, 85–93.

Sjöberg, L. (2000). Factors in risk perception. *Risk Analysis*, 20, 1–12.

Sjöberg, L. & Winroth, E. (1986). Risk, moral value of actions, and mood. *Scandinavian Journal of Psychology*, 27, 191–208.

Skogen, K. & Krange, O. (2003). A wolf at the gate: The anti-carnivore alliance and the symbolic construction of community. *Sociologia Ruralis*, 43, 309–25.

Slovic, P. (1987). Perception of risk. *Science*, 236, 280–5.

Smith, E. R. & DeCoster, J. (2000). Dual-process models in social and cognitive psychology: Conceptual integration and links to underlying memory systems. *Personality & Social Psychology Review*, 4, 108–31.

Sponarski, C. C., Vaske, J. J. & Bath, A. J. (2015). Attitudinal differences among residents, park staff, and visitors toward coyotes in Cape Breton Highlands National Park of Canada. *Society & Natural Resources*, 28, 720–32.

Tajfel, H. & Turner, J. C. (1979). An integrative theory of intergroup conflict. In W. G. Austin & S. Worchel, eds. *The Social Psychology of Intergroup Relations*. Monterey, CA: Brooks/Cole, pp. 33–47.

Treves, A., Martin, K. A., Wydeven, A. P. & Wiedenhoeft, J. E. (2011). Forecasting environmental hazards and the application of risk maps to predator attacks on livestock. *BioScience*, 61, 451–8.

Wallach, A. D., Ripple, W. J. & Carroll, S. P. (2015). Novel trophic cascades: Apex predators enable coexistence. *Trends in Ecology & Evolution*, 30, 146–53.

Wilson, R. S. (2008). Balancing emotion and cognition: A case for decision aiding in conservation efforts. *Conservation Biology*, 22, 1452–60.

Wilson, R. S. & Arvai, J. L. (2010). Why less is more: Exploring affect-based value neglect. *Journal of Risk Research*, 13, 399–409.

Wilson, R. S. & Bruskotter, J. T. (2009). Assessing the impact of decision frame and existing attitudes on support for wolf restoration in the United States. *Human Dimensions of Wildlife*, 14, 353–65.

4

Understanding Emotions As Opportunities for and Barriers to Coexistence with Wildlife

MAARTEN JACOBS AND JERRY J. VASKE

Scholars have made a distinction between cognition, affection and volition as the three basic mental functions of humans (Ajzen 2005). Research into human–wildlife relationships has largely focused on cognitions, such as values, value orientations, beliefs, attitudes, norms or risk perceptions. Empirical research on emotions is relatively scarce in social scientific research pertaining to human–wildlife relationships (Manfredo 2008; Jacobs et al. 2012). Some work has explored generic emotion theory (Manfredo 2008) to unravel the psychological mechanisms that underlie emotional responses to wildlife (Jacobs 2009). Other studies have used self-reported measures to study specific emotions towards particular animals, most notably fear (e.g. Davey et al. 1998; Johansson & Karlsson 2011). Other research has used animals (e.g. snakes and spiders) as stimuli to understand fear reactions (Öhman & Mineka 2001; Öhman et al. 2001; Öhman 2009). Overall, however, the study of emotions towards wildlife is fragmentary.

Arguably, emotions are fundamental in understanding human–wildlife relationships (Manfredo 2008; Jacobs et al. 2012). In the course of biological evolution, emotions have emerged earlier than *higher-order* mental capacities such as using complex language and abstract thinking (Jacobs 2009). Emotions are powerful and essential to humans. Once activated, emotions often control mind and behaviour (LeDoux 1998). Emotions are central to everyday human experience (Dolan 2002), and are essential for assigning value to objects and events (Jacobs 2012). Emotions also influence other mental processes and dispositions (Jacobs et al. 2012), such as memories (Talarico & Rubin 2007), motivation (Frijda 1986; Izard 2009), decision-making (Damasio 1999; Loewenstein & Lerner 2003) and perception (Dolan 2002).

Imagine walking in the woods and suddenly encountering a deer. For many individuals, this event would evoke a positive emotion that is strongly felt, central to the experience and interrupting other thought processes. Such an event will likely be remembered more vividly than less emotional events that may have occurred during the walk (Jacobs 2009). Emotions such as these constitute important internal forces that drive our attraction to wildlife (Manfredo 2008) and our motivation to view wildlife (Jacobs 2009). Research suggests that emotion-driven dispositions inform decisions about wildlife-related behaviours (Slagle et al. 2012), that individuals with strong emotions towards wildlife are more likely to identify wildlife in a complex natural scene (Öhman et al. 2001), and that neurons in the amygdala (a brain structure involved in emotional processing) respond preferentially to pictures of animals relative to other pictures (Mormann et al. 2011). Emotions can therefore play a role in determining the direction (negative to positive) and intensity (weak to strong) of dispositions towards species, thus influencing where a person stands along the conflict-to-coexistence continuum.

In this chapter, we consider human emotions towards wildlife in the context of human–wildlife coexistence. Human emotions can present opportunities for and challenges to coexistence. Positive emotions towards deer, for example, can serve as an avenue for peaceful coexistence. Fear of large carnivores, on the other hand, might be an obstacle for coexistence. This chapter uses emotion theory to create an understanding of the general working of emotions and the mechanisms that explain wildlife-related emotions. We also provide an overview of research into human emotions towards wildlife. This understanding provides a starting point for exploring and solving knowledge gaps in research about emotions towards wildlife, and offers opportunities for breaking down barriers to human–wildlife coexistence if this understanding would identify suggestions to influence emotions. We illustrate human–wildlife coexistence with an example that attempts to mitigate human fear towards wolves in public communication.

4.1 EXAMPLE: LIVING WITH WOLVES IN THE NETHERLANDS

The *Wolven in Nederland* (wolves in the Netherlands) project is a cooperation of various Dutch nature and wildlife agencies that strives for peaceful coexistence with wolves by increasing support for wolves. A wolf sighting was reported in the Netherlands in 2011 (Jacobs et al. 2014). The DNA analyses related to later sightings have confirmed that

---

Text Box 4.1 Wolven in Nederland about the Wolf and Humans

Wolf and Humans

If the wolf appears in news coverage, many people respond fiercely and emotionally. Some mothers do not dare to let their children play outdoors, farmers fear for their stock, and hunters for their prey. Others appraise this animal as the perfect crown on nature and would like nothing so much as seeing a live wolf. Responses are often fuelled by what we know about the wolf. After all, we know the wolf mainly from fairy tales and the zoo.

Do We Need to Be Afraid?

For centuries, humans and wolves lived side by side, and rarely, people are attacked or killed. Yet many stories exist in Europe about wolves leaving behind a trail of death and destruction.

(www.wolveninnederland.nl/wolf-en-mens, accessed 28 November 2016, translation into English by first author)

---

wolves have entered the Netherlands for short periods of time. By providing information, the *Wolven in Nederland* cooperation hopes that the wolf is welcome in the minds and hearts of inhabitants of the Netherlands (www.wolveninnederland.nl, accessed 28 November 2016). Text Box 4.1 above presents the page dealing with wolf–human relationships on this website.

Four possible causes of fear were mentioned on the webpage: (1) rabies, (2) humans as prey for wolves, (3) defence by wolves if being attacked by humans and (4) feeding wolves. The webpage explains that each potential cause is unlikely in contemporary Europe. As an example, rabies has disappeared in Europe, and hence there is no reason to fear attacks by wolves that have rabies.

The webpage presents the following reasoning. First, human emotions towards wolves are a potential obstacle for coexistence. Second, fear is the most relevant emotion towards wolves (note, however, that the third sentence in the quotation hints at positive emotions – yet these are not referenced in the extensive text on the webpage; only fear is addressed). Third, fear is based on beliefs about wolves that are basically false. Finally, dismantling these false beliefs could increase support for the existence of wolves in the Netherlands. This model, or its variants, are often used by wildlife managers and policy-makers who interact with the public. For instance, the concept of the *knowledge deficit model* expresses the idea that conservationists frequently believe that public responses can be changed by education, thus fixing a lack of knowledge that is assumed to be the cause of undesired responses (Heberlein 2012).

The first three steps in this reasoning raise questions that merit attention. Are human emotions towards wolves (or other species) a threat to coexistence? Is fear the most relevant emotion? Is fear (or any other emotion) based on beliefs? In this chapter, we review literature to reflect on these questions. We also widen the scope to include other species and other emotions. The aim is not to criticize *Wolven in Nederland* (or any organization that used similar models). On the contrary, we believe that the cooperation is sincere about intentions and goes at length to present accurate and up-to-date factual information about wolves, based on scientific literature and expert consultation. The quoted webpage illustrates the utility of knowledge on emotions in the quest for coexistence. We hope the chapter helps readers to understand how emotions play a role in influencing conflict-to-coexistence dispositions and increase organizations' effectiveness in information provision. We first describe the working of emotion.

4.2 NATURE AND OPERATION OF EMOTIONS

A commonly shared definition of emotion does not exist (Izard 2007). Scholars, however, agree that emotional responses consist of: (1) physiological reactions (e.g. increased heartbeat), (2) expressive reactions (e.g. smiling), (3) behavioural tendencies (e.g. approaching) and (4) emotional experiences (e.g. interpreting the situation, feeling happy) (Kleinginna & Kleinginna 1981; Izard 2007; Jacobs et al. 2012).

Mounting empirical evidence suggests that emotions emerged in the course of biological evolution as adaptive responses that foster survival and well-being (LeDoux 1998). The evolutionary background of emotions is evident from the existence of automatic and non-learned reactions as components of emotional reactions. For instance, the increase of adrenaline in the blood system as part of a fear response enhances the bodily conditions for fight-or-flight behaviour (LeDoux 1998). People do not learn to release adrenaline, yet every healthy human does so, automatically and involuntarily. Physiological responses are often similar across species. Erection of body hair during a fear response occurs in every organism with hair (Darwin 1865/1972). The adaptive nature of emotions – their function to promote survival and well-being – can be illustrated by the fear response of prey towards a predator. The erection of body hair, for example, makes the prey look bigger and may prevent an attack by the predator. Specific responses probably originated in biological evolution before different hairy creatures emerged. Cross-cultural research in facial expressions suggests that emotional facial displays are universally recognized and labelled (Ekman 1992). This research identified fear, disgust, joy, sadness, surprise and anger as universal basic emotions. While different research methods and contexts have elicited slightly different lists of basic emotions, most researchers arrive at the convergent conclusion that at least fear, joy, sadness and anger universally exist (LeDoux 1998).

In a basic model of the working of emotion, automatic appraisal makes us respond to specific stimuli. Some structure(s) in the sub-cortical human brain (the *automatic appraisal mechanism*) have the capacity to detect the emotional relevance of stimuli and initiate bodily and physiological emotional responses. Explicit knowledge is not needed for this type of emotional response to occur: detection of emotionally relevant stimuli and the emotional response occur automatically and without conscious deliberation (LeDoux 1998; Damasio 1999). 'Because of the time constraints of predator–prey encounters, the more rapid the defence recruitment, the more likely the potential prey is to survive the encounter. Thus, the fear module's judgment of the fear relevance of stimuli is likely to rely on a quick and dirty process that rather risks false positives than false negatives' (Öhman & Mineka 2001, p. 487).

Emotional experiences are interpretations of bodily and physiological responses by cortical brain structures constituting conscious experiences, as well as feedback from the sub-cortical automatic appraisal mechanism to the same cortical structures. Apart from feelings (e.g. happiness when seeing wildlife), emotional experiences (i.e. the content of consciousness during an emotional response) consist of other components: (1) valence, which is the pleasure–displeasure dimension of emotional states (e.g. liking or disliking encountering an animal), (2) a level of arousal (e.g. feeling activated by the sight of an animal), (3) situational appraisals (e.g. judging how conducive encountering an animal is to one's goals) and (4) emotionally laden thoughts (e.g. retrieved memories of past sights of animals) (Barrett et al. 2007; Jacobs et al. 2012).

The sequence of events in this basic model of the operation of emotion is as follows: stimulus onset – automatic emotional appraisal – physical emotional response – emotional experience. The emotional experience follows from the interpretation of emotional bodily and physiological responses (James 1894). This is contrary to the frequently held belief that one must first have a conscious experience of emotion towards an object before bodily responses set in (LeDoux 1998).

The automatic subconscious appraisal mechanism contains innate tendencies to respond to some classes of stimuli. Newborn babies tend to cry after hearing a loud unexpected sound. Without having seen a cat previously, laboratory-raised rats freeze or try to escape when exposed to a cat (Blanchard & Blanchard 1988). Yet the automatic appraisal mechanism is modifiable. Conditioning is a psychological mechanism that can affect the appraisal mechanism (LeDoux 1998; Jacobs 2009). If a neutral stimulus gets associated with an emotional stimulus upon repeated simultaneous presence, the neutral stimulus becomes an emotional stimulus. Through conditioning, for instance, scavengers might become emotional stimuli if associated with dead animals.

The model of the operation of emotions offers a potential explanation for why seeing a wolf can evoke a fear response, and why in general some people might fear wolves. The fear response has been adaptive for our ancestors. The automatic appraisal mechanism has the tendency to respond with fear to wolves; people become aware of this tendency and thus also consciously fear wolves in general. The model does not explain, however, why people who are fearful of wolves in the wild do not fear the presence of wolves in a zoo. The model also does not explain why a dedicated birdwatcher would have a positive emotion to seeing a small grey rare bird that is not yet on his or her list. The basic model needs expansion to account for these events. The cognitive system plays a role here. While the cognitive and the emotional systems are largely different systems in the human brain, these systems interact (LeDoux 1998). Explicit learning through personal experiences and cultural transmission affect cognitive systems. The learned knowledge can subsequently affect the operation of emotions in various ways. Text Box 4.2 below summarizes the operation of emotions.

As the model expresses, activation of cognitive dispositions can evoke feedback to the automatic appraisal mechanism. Knowledge that a wolf in an enclosed environment in a zoo does not present any danger suppresses an initial fear response of the automatic appraisal mechanism. Importantly, cognitive appraisal takes more time than automatic appraisal does (Jacobs et al. 2012). Thus, while the automatic mechanism does still respond immediately, the cognitive system presents a second opinion, overrules the automatic mechanism and restrains the bodily fear response (Jacobs 2009). By the same token, the cognitive

Text Box 4.2 Operation of Emotion

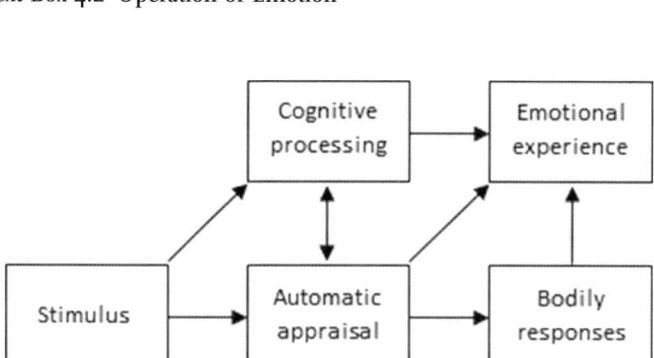

Figure 4.1 A model for the operation of emotion.

A stimulus causes a flow of information to the automatic appraisal mechanism. If this mechanism evaluates the stimulus as emotionally relevant, a bodily emotional response follows. Feedback from the body is interpreted into an emotional experience. Simultaneously, the stimulus causes a flow of information to the cognitive system that processes the information. Output of this processing is sent to the automatic appraisal mechanism, where it can cause a change in emotional appraisal (suppression of initial appraisal or initiation of appraisal), and indirectly influence bodily responses. Output of the appraisal mechanism and cognitive processing can also influence the way bodily emotional responses are interpreted into an emotional experience. Output of the automatic appraisal mechanism, in turn, can influence cognitive processing, such as the tendency to focus on the emotional stimulus.

system can initiate an emotional response even if the automatic appraisal system does not on its own (Jacobs 2009). Feedback from the cognitive system to the appraisal mechanism tells the mechanism to respond to the grey bird as a second opinion. The cognitive system also influences the way feedback of emotional bodily responses is interpreted into an emotional experience (Jacobs 2009). An automatic fear response to a wolf in a zoo might be interpreted into a positive sensation.

Returning to the living with wolves example, the model for the working of emotions reflects on the assumption that fear of wolves is based on beliefs about wolves. Beliefs in the cognitive system might initiate a fear response to wolves. Once activated, these beliefs feed the automatic appraisal mechanism. The model, however, also presents another potential explanation. The disposition to fear wolves might be part of the automatic appraisal mechanism, and thus not a consequence of beliefs about wolves. The fear can also be based on a mixture of beliefs and the automatic appraisal mechanism. Only empirical research can be conclusive at this point; we review the relevant (but not yet conclusive) research currently available on emotions towards animals. Understanding human emotions towards wildlife necessitates an initial understanding of emotional dispositions towards wildlife; the principal internal causes of emotional responses.

4.3 EMOTIONAL DISPOSITIONS

The model of emotions presented in the previous section implies the existence of emotional dispositions. For emotional appraisal to work, criteria in the brain, against which emotional relevance of stimuli are judged, must exist (Frijda 1986; Jacobs et al. 2012). These criteria are labelled *emotional dispositions*. Similar to all mental dispositions, emotional dispositions are traits. While states reflect *how* you are, traits reflect *who* you are (Hamaker et al. 2007). As opposed to states, traits are always there, even if they are not active. For example, personality characteristics are properties of an individual, even if they do not always guide current behaviour, thought or experience.

The literature has defined *emotional dispositions* using two fundamentally different concepts. First, the term can reflect emotionally laden *personality traits* (Watson & Clark 1984; Digman 1990; Hamid & Cheng 1996). With this definition, emotional dispositions refer to a general tendency to be happy or sad, or to be emotionally stable or unstable. Second, the term can denote criteria against which the emotional relevance of stimuli is appraised (Frijda 1986; Lerner & Keltner 2000). We use the concept of emotional dispositions in the second sense.

People do not exhibit emotional reactions randomly, but rather in response to specific objects, events or situations. The objective nature of a stimulus does not determine the emotional response (Scherer 1999). If this was the case, we could not explain why two people respond differently to spiders. Rather, a process of emotional appraisal occurs. Even 'a completely automatic, reflexive defence reaction of the organism constitutes an intrinsic assessment, a valuation, of the noxiousness of the stimulus' (Scherer 1999, p. 647). The evaluation of the stimulus, the appraisal, leads to an emotional response. Emotional dispositions guide the appraisal process.

Emotional dispositions as traits are relatively stable as compared to states. Being scared by a bear is a temporary state that can switch on and off and vary in intensity depending on the situation. A dispositional fear for bears, on the other hand, is usually stable. The stability of emotional dispositions is illustrated by the fact that it is hard to overcome phobias.

Like all concepts denoting mental dispositions, emotional dispositions are characterized by a specific level of abstraction. A dispositional fear of snakes, for example, is abstract because *snakes* is an abstract category that denotes a collection of species. In sum, emotional dispositions are criteria against which emotional relevance is judged, must exist, are always there, are relatively stable and have a certain level of abstraction.

Two types of emotional dispositions can be distinguished: (1) general emotional dispositions that guide the unfolding of emotional responses, and (2) specific emotional dispositions towards wildlife. Scholars of emotional appraisal list general criteria (labelled *appraisal dimensions, appraisal criteria, stimulus evaluation checks* or simply *appraisals*) that are employed to evaluate the emotional relevance of situations and to guide emotional responses (Roseman et al. 1996; Scherer 1999; Ellsworth & Scherer 2003; Sander et al. 2005).While these lists differ in the number and kinds of appraisal dimensions (e.g. Smith & Ellsworth 1985; Frijda 1986; Frijda et al. 1989; Roseman et al. 1996; Scherer 1999), considerable consensus exists about a limited set of primary dimensions. Theory and research suggest that humans evaluate the emotional relevance of stimuli in terms of five dimensions: (1) novelty (has anything changed?), (2) valence (is it good or bad?), (3) goals (is it obstructive or conducive to current goals?), (4) agency (what is the cause and can it be controlled or predicted?) and (5) norms (is it compatible with standards?) (Smith & Ellsworth 1985; Scherer 1999). Phrased as questions, these appraisals evaluate situations as follows: Is there anything new (novelty), is it relevant (valence), are there consequences (goals), can I cope (agency) and is it normatively significant (norms) (Sander et al. 2005)? Theorists contend that these appraisals usually occur sequentially (e.g. relevance needs to be detected before goal conduciveness can be detected) (Ellsworth & Scherer 2003), although people do not necessarily notice the sequence because emotional appraisal occurs quickly and may not enter consciousness (Ellsworth & Scherer 2003; Sander et al. 2005).

As an illustration of how these general appraisal criteria work, imagine that a person sees a moose. The appearance of the moose is appraised as novel, drawing attention and interest. If the individual generally likes moose, the appearance is rated as positive or pleasant. The appearance is then evaluated against current goals. If the moose is blocking the road, for instance, the appraisal might depend on whether the person is wildlife viewing or driving to work. For the latter situation, the emotional response will vary depending on the person's perceived control and prediction regarding events in the near future (e.g. thinking one can easily pass the moose versus thinking the moose will cause an accident). While the appearance of the moose would normally not be appraised in terms of being compatible with norms, another situation, such as a person feeding the moose, could lead to this kind of appraisal.

In the context of human–wildlife coexistence, these general emotional dispositions are potentially relevant. As an illustration, the migration of wolves to regions where they have been absent for a long time (such as the Netherlands) might evoke strong emotional responses because the situation is new (novelty), might be seen as relevant (valence), might be seen as having consequences for what people normally want to do (goals), might be judged as an event that can be hard to cope with (agency) and might be seen as normatively significant if people find it not compatible with existing norms (norms). For agencies aiming to foster coexistence, the general appraisal criteria present questions to consider, and subjects for conversations with the public. In the living with wolves example the presence of wolves could be discussed in the light of these appraisal criteria with stakeholders. If the criterion of goals presents a problem to stakeholders, information about exactly where wolves are present and absent could assist stakeholders in finding wolf-free places for their recreation activities. If the criterion of agency is relevant, information about coping with the presence of wolves could be useful to stakeholders.

People also have emotional dispositions towards specific objects such as wildlife (Ellsworth & Scherer 2003). Snake and spider phobias, for instance, are common (Cook & Mineka 1989). Appraisal theorists have not focused on these object-related emotional dispositions, perhaps a consequence of their focus on generic principles that work in every situation.

Emotional dispositions can be innate (i.e. a consequence of biological evolution) or learned (Jacobs 2009). As wildlife were crucial to early hominids, humans inherited emotional responses to wildlife (Öhman & Mineka 2001; Manfredo 2008; Jacobs 2009). Experiments with newborn infants have revealed an innate attraction to biological movement, as babies tend to attend for a longer period of time to a movie representing a walking hen represented by dots than to a similar movie not representing biological movement (Simion et al. 2008). Other dispositions are learned. The delight of a dedicated birdwatcher who sees a rare bird after a long search is a learned disposition in which the knowledge that the bird is seldom seen plays a role. While this example illustrates a consciously learned disposition, emotional dispositions can also be learned unconsciously via conditioning (Öhman & Mineka 2001; Jacobs 2009). Emotional dispositions do not have to be either exclusively innate or learned. For instance, an innate tendency to fear large carnivores might be amplified by cultural discourse depicting these as harmful. A study of 2,000 twins revealed that animal phobias arise from both genetic vulnerability and traumatic childhood experiences (Kendler et al. 1992). Animal phobias had both a genetic component, as suggested by commonalities between two individuals comprising a twin, and as a learned component, suggested by differences between these individuals.

Emotional dispositions towards species of wildlife are relevant as well. Coexistence is easier with species for which humans have positive emotional dispositions, and more difficult with species for which humans have negative emotional dispositions. The next section summarizes research findings on human emotional dispositions towards wildlife.

4.4 EMOTIONAL DISPOSITIONS TOWARDS WILDLIFE: RESEARCH FINDINGS

While emotions towards wildlife have not been extensively studied to date, different strands of empirical research have emerged. First, descriptive research has identified emotional dispositions towards specific species. For example, a Norwegian study found that 57 per cent of respondents were very much afraid of brown bears, 48 per cent of wolves, 27 per cent of lynx and 20 per cent of wolverines (Røskaft et al. 2003). In Finland, fear towards wolves varied from 24 per cent of the respondents in regions with stable wolf populations, to 47 per cent of those living in areas with dispersing wolves (Bisi et al. 2007). Such descriptive research is rare and confined to fear dispositions (that is, overall feelings of fear were assessed, and not specific fear responses during an encounter) towards large carnivores in specific human populations. Additional research is necessary to obtain a representative picture of other (not only fear) emotional dispositions towards a variety of species (not only large carnivores) in various nations, and among different segments of populations. This research can facilitate identifying avenues for and challenges to coexistence from the perspective of human emotions.

Second, some research has addressed patterns of human emotions across species. A few studies have identified fear dimensions by factor-analysing self-reported fear items towards a variety of animals (Davey 1994; Ware et al. 1994; Tucker & Bond 1997; Davey et al. 1998; Arrindell 2000). Different lists of species were used in these studies. For instance, large predators were included in the most comprehensive study that included fifty-one species fairly reflecting the animal kingdom (Davey et al. 1998), but not in all studies (e.g. Arrindell 2000). Naturally, different lists produce different factor solutions. Convergence, however, has occurred in at least three categories: fear-relevant animals, disgust-relevant animals and fear-irrelevant animals. The fear-relevant category consists of large carnivores that can attack humans, such as lions, bears, crocodiles and tigers. The disgust-relevant category contains smaller species that do normally not attack humans, but are feared nevertheless (e.g. cockroaches, spiders, worms, rats). The fear-irrelevant category comprises herbivores (mammals, birds), such as chickens, hamsters, pigs and rabbits. The wolf resides in the fear-relevant category (Ware et al. 1994; Tucker & Bond 1997; Davey et al. 1998), co-varying with fear towards other large carnivores.

A recent study aimed to extend knowledge on patterns of emotional dispositions towards wildlife by investigating whether factor analyses upon valence towards species presents the same categorization (Jacobs unpublished data). Valence is the positive–negative, or pleasure–displeasure dimension of emotions. Theorists have argued that valence is the most fundamental aspect of emotion (Russell & Barrett 1999; Russell 2009). Empirical research has confirmed that measures of valence explain a large portion of the total variability in emotions (Mehrabian & Russell 1974; Bradley & Lang 2000), larger than any other measure (e.g. fear). Factor analyses over valence ratings revealed the same pattern as previously found in fear studies (Jacobs unpublished data). Moreover, valence on the level of these groups of species (as represented by indices calculated as mean valence of all species in the fear-relevant, disgust-relevant and fear-irrelevant categories) predicted perceived existence value for specific species as well as did valence towards those specific species. For example, the correlation between valence towards lions and perceived existence value of lions was 0.35, while the correlation between valence towards the fear-relevant group of species (to which lions belonged) and perceived existence value of lions was 0.36 (Jacobs unpublished data). These findings support the assertion that emotional dispositions exist and are functional (i.e. consequential) on the level of groups of species, because species belonging to one group presented similar challenges and opportunities to our ancestors. Simple emotional dispositions on the level of groups would make the automatic appraisal fast (no need to check for a large number of different species), fostering a very quick adaptive emotional response and thus survival. While the pattern of emotions across species seems robust – as it is similar in samples from different nations (Davey et al. 1998) and between fear and valence (Davey et al. 1998; Jacobs unpublished data) – further research is needed to substantiate the interpretation that humans have innate emotional dispositions on the level of groups of species. For example, neuro-imaging studies could examine the hypothesis that neural responses to pictures representing species belonging to the same functional group are relatively similar, and likewise, neural responses to species from different groups are relatively distinctive. Regardless of the question whether dispositions on the level of groups are innate (for which ultimate evidence is hard to conceive), the findings of these factor-analytical studies are relevant for identifying opportunities and obstacles to human–wildlife coexistence. On the basis of human emotions, coexistence is likely to be much easier with species belonging to the fear-irrelevant group than for species belonging to the fear-relevant and disgust-relevant groups. Understanding which of those categories a species falls into can help in predicting if a human–wildlife interaction will be perceived as negative or positive, thus clarifying where a person is likely to position himself along the conflict-to-coexistence continuum.

Third, some studies have addressed the relationship between emotions towards wolves and specific cognitions that are relevant to wildlife policy and management. Fear towards large carnivores was negatively associated with willingness to pay for large carnivore policy (Johansson et al. 2012). Positive or negative feelings towards wolves predicted beliefs about the outcomes of wolf recovery (Slagle et al. 2012). Sympathy for wolves, sympathy for ranchers and anger about wolves (all assessed with items using these labels) explained a substantial portion of the variance of acceptability of lethal wolf management actions (Vaske et al. 2013). Another study has addressed the predictive potential of emotional dispositions towards wolves for acceptability of lethal wolf control, in three different scenarios: wolves are present, wolves attack sheep and wolves kill a hiker (Jacobs et al. 2014). The discrete emotions of joy and disgust were significant predictors in all scenarios. The predictive potential of fear was lower, and limited to two scenarios. The dimension of valence was a better predictor than any other emotion measure. These findings suggest that fear towards wolves has limited relevance for understanding acceptability of lethal control. These results do not suggest that people do not feel fear towards wolves: the average level of fear was approximately 3 on a scale from 0 – no fear – to 6 – strong fear. Yet, importantly, the average level of joy towards wolves was slightly higher.

Overall, this set of studies has immediate relevance in the context of human–wildlife coexistence. Emotions predict other mental dispositions that are likely to flag willingness to coexist (i.e. willingness to pay, acceptability of lethal control). Further research is needed to include more emotions, more species and more dependent variables relevant to coexistence (e.g. acceptability of the presence of species in one's living area).

Fourth, studies have examined mental processes that are associated with snake and spider phobias (see Öhman 2009 for an overview). People fearful of snakes or spiders demonstrated fear responses (measured as skin conductance responses) to pictures of snakes and spiders that were presented for only 30 ms, too short for the pictures to be consciously perceived (Öhman & Soares 1994). Fear of snakes and spiders is mediated by quick and automatic processes that bypass the cognitive system. Whether the mechanisms that underpin snake and spider phobias apply to large carnivores is yet to be determined. A study into the constitution of fear towards bears and wolves suggests that different mechanisms play a role (Flykt et al. 2013). Studies on the nature of emotional dispositions – whether they are innate, conditioned or consciously learned – are relevant for designing strategies to influence emotional dispositions to foster coexistence. If emotional dispositions are learned and based on beliefs, communication aimed to change these beliefs might have an intended effect. However, this communication strategy might have no effect if dispositions are innate, as the automatic appraisal mechanism containing these dispositions is virtually immune to cognitive information. Since extensive research into the nature of emotional dispositions is confined to snakes and spiders, similar research on other species would expand our understanding.

Fifth, some papers have addressed the cognitive antecedents of fear towards large carnivores. Perceived danger and perceived uncontrollability of one's own reactions when encountering a species were positively associated with fear towards wolves among a small sample of Swedish stakeholders (Johansson & Karlsson 2011). This finding was replicated in a larger sample of Swedes living in areas where bears and wolves were present (Johansson et al. 2012). This study also revealed that trust in authorities responsible for carnivore management predicted fear towards wolves. These studies suggest that beliefs might influence emotional dispositions, in line with the assertion in the living with wolves example. Because this research is correlational, however, it is inconclusive about cause–effect relationships. The beliefs could be outcomes, rather than causes, of the emotional dispositions. Future research should address this interpretation problem by manipulating beliefs and measuring the effects on emotions towards species in an experimental design. If beliefs can explain emotional dispositions, such research could inform communication strategies aimed to mitigate negative emotions as a means to foster coexistence.

4.5 COEXISTENCE: WHAT CAN WE LEARN FROM EMOTION THEORY AND RESEARCH?

In the living with wolves example, we posed three questions. Are human emotions towards wolves (or other species) a threat to coexistence? Is fear the most relevant emotion? Is fear (or any other emotion) based on beliefs? Existing research does not provide conclusive answers to these questions, yet presents findings that allow reflection. We do not know if human emotions towards wolves are a threat to coexistence. In Finland and Norway, a considerable share of respondents were fearful of wolves (Røskaft et al. 2003; Bisi et al. 2007). Fear towards wolves was on average moderate in Dutch and Canadian student samples; joy levels were slightly higher (Jacobs et al. 2014). Another study suggested that fear towards wolves was relevant for coexistence, as fear was negatively associated with willingness to pay for large carnivore policy (Johansson et al. 2012). However, other research found that fear had limited influence on the acceptability of lethal wolf control, and that joy and disgust towards wolves were much stronger predictors of acceptability (Jacobs et al. 2014). Human emotions towards wolves might not be a threat to coexistence if positive emotions are equally as strong as negative emotions.

Research does not substantiate the assertion that beliefs are the cause of fear towards wolves, or other fear-relevant animals. Beliefs about danger posed by wolves and uncontrollability of one's own behaviour if encountered by a wolf were associated with fear for wolves (Johansson et al. 2012). Cause–effect relationships, however, cannot be inferred from this correlational study. Strong evidence that fear towards wolves is genetically constituted or otherwise originating in the automatic appraisal mechanism does not exist. The factor-analytical studies suggest that humans have innate fear dispositions towards large carnivores in general (Jacobs unpublished data) – dispositions that apply to wolves.

Overall, the communication strategy of living with wolves might be ineffective. If joy towards wolves is more relevant than fear, the almost exclusive focus on fear might miss an opportunity to foster human–wolf coexistence. If fear towards wolves is innate and constituted in the automatic appraisal mechanism, attempts to reduce fear by presenting factual information will likely be ineffective. If fear towards wolves is immune to communication, the text might make readers feel unintelligent if they still fear wolves after considering the information offered by *Wolven in Nederland*. Still, it is probably useful to explain why wolves hardly pose a threat to humans these days. Acknowledgement that humans simply might have an innate fear towards wolves that is not likely to change with factual information could prevent negative feelings towards the organization.

Wildlife managers and policy-makers who promote coexistence via human emotions should be careful about making strong assertions. First, research on human emotions towards wildlife is fragmentary and primarily focused on fear towards problem species. Hence, the research represents a selective piece of a pie. Practitioners, however, should consider the whole pie by looking broader than the existing research. Positive emotions should not be overlooked as coexistence promotion opportunities. An experimental study among whale watchers (Jacobs & Harms 2014) presents encouraging findings. Stories told by on-board guides to whale watchers were manipulated to constitute four conditions: (1) a factual story about whales, (2) a normative story about how the behaviours of individuals affect whales, (3) a story that promotes positive emotions towards whales and (4) no story. Whale conservation intentions were measured before and after the trip. In the emotion condition, increases in these intentions were larger and more widespread than in the other conditions (Jacobs & Harms 2014). As whales are charismatic species, and whale watchers arguably have an interest in seeing whales, these results cannot be blindly generalized to other species, people and contexts. This study, however, does demonstrate that at least in some circumstances, nurturing positive emotions towards wildlife can have a positive effect on intentions that are probably closely associated with coexistence intentions.

4.6 RECOMMENDATIONS AND FUTURE DIRECTIONS

Research on emotions towards wildlife is not extensive, and often not conclusive enough to anticipate clear recommendations for strategies and measures to promote human–wildlife coexistence. Research is needed to:

- Identify human emotions towards the variety of wildlife species;
- Understand both negative and positive emotions towards wildlife and how those affect people's positions along the conflict-to-coexistence continuum;
- Include both problematic and less problematic wildlife;
- Understand how emotions towards wildlife influence willingness to coexist;
- Avoid making strong assumptions about which emotions are felt towards which species, how those emotions are constituted and what the consequences of those emotions are – current research does not provide a solid knowledge base to make those kind of inferences;
- Avoid negativity bias by only focusing on negative emotions and problem species – positive emotions and less problematic wildlife are probably an avenue to influence those who hold negative-to-neutral perceptions towards species and foster stronger coexistence values in those positively inclined towards wildlife;
- Think beyond current scientific knowledge as existing research does not reflect the range of emotions and species;
- Use the general appraisal criteria of novelty, valence, goals, agency and norms as prompts to discuss the presence of wildlife with stakeholders. This approach will provide a foundation for understanding how stakeholders appraise the presence of wildlife and what measures will contribute to move people to the coexistence side of the continuum.

4.7 References

Ajzen, I. (2005). *Attitudes, Personality, and Behavior.* London: McGraw-Hill Education.

Arrindell, W. A. (2000). Phobic dimensions: IV. The structure of animal fears. *Behaviour Research & Therapy*, 38(5), 509–30.

Barrett, L. F., Mesquita, B., Ochsner, K. N. & Gross, J. J. (2007). The experience of emotion. *Annual Review of Psychology*, 58, 373–403.

Bisi, J., Kurki, S., Svensberg, M. & Liukkonen, T. (2007). Human dimensions of wolf (*Canis lupus*) conflicts in Finland. *European Journal of Wildlife Research*, 53(4), 304–14.

Blanchard, D. C. & Blanchard, R. J. (1988). Ethoexperimental approaches to the biology of emotion. *Annual Review of Psychology*, 39, 43–68.

Bradley, M. M. & Lang, P. J. (2000). Measuring emotion: Behavior, feeling, and physiology. *Cognitive Neuroscience of Emotion*, 25, 49–59.

Cook, M. & Mineka, S. (1989). Observational conditioning of fear to fear-relevant versus fear-irrelevant stimuli in rhesus monkeys. *Journal of Abnormal Psychology*, 98(4), 448–59.

Damasio, A. R. (1999). *The Feeling of What Happens: Body and Emotions in the Making of Consciousness*. New York: Harcourt Brace.

Darwin, C. (1865/1972). *The Expression of Emotions in Man and Animals*. Chicago: Chicago University Press.

Davey, G. C. (1994). Self-reported fears to common indigenous animals in an adult UK population: The role of disgust sensitivity. *The British Journal of Psychology*, 85, 541–54.

Davey, G. C. L., McDonald, A. S., Hirisave, U., Prabhu, G. G., Iwawaki, S., Jim, C. I. C. & Reimann, B. (1998). A cross-cultural study of animal fears. *Behaviour Research & Therapy*, 36(7–8), 735–50.

Digman, J. M. (1990). Personality structure: Emergence of the five-factor model. *Annual Review of Psychology*, 41(1), 417–40.

Dolan, R. J. (2002). Emotion, cognition, and behavior. *Science*, 298(5596), 1191–4.

Ekman, P. (1992). An argument for basic emotions. *Cognition & Emotion*, 6(3–4), 169–200.

Ellsworth, P. C. & Scherer, K. R. (2003). Appraisal processes in emotion. In R. J. Davidson, K. R. Scherer & H. H. Goldsmith, eds., *Handbook of Affective Sciences*. Oxford: Oxford University Press, pp. 572–95.

Flykt, A., Johansson, M., Karlsson, J., Lindeberg, S. & Lipp, O. V. (2013). Fear of wolves and bears: Physiological responses and negative associations in a Swedish sample. *Human Dimensions of Wildlife*, 18(6), 416–34.

Frijda, N. H. (1986). *The Emotions*. Cambridge: Cambridge University Press.

Frijda, N. H., Kuipers, P. & Ter Schure, E. (1989). Relations among emotion, appraisal, and emotional action readiness. *Journal of Personality and Social Psychology*, 57(2), 212–28.

Hamaker, E. L., Nesselroade, J. R. & Molenaar, P. C. M. (2007). The integrated trait-state model. *Journal of Research in Personality*, 41(2), 295–315.

Hamid, P. N. & Cheng, S. T. (1996). The development and validation of an index of emotional disposition and mood state: The Chinese affect scale. *Educational & Psychological Measurement*, 56(6), 995–1014.

Heberlein, T. A. (2012). Navigating environmental attitudes. *Conservation Biology*, 26(4), 583–5.

Izard, C. E. (2007). Basic emotions, natural kinds, emotion schemas, and a new paradigm. *Perspectives on Psychological Science*, 2(3), 260–80.

Izard, C. E. (2009). Emotion theory and research: Highlights, unanswered questions, and emerging issues. *Annual Review of Psychology*, 60, 1–25.

Jacobs, M. H. (2009). Why do we like or dislike animals? *Human Dimensions of Wildlife*, 14(1), 1–11.

Jacobs, M. H. (2012). Human emotions toward wildlife. *Human Dimensions of Wildlife*, 17(1), 1–3.

Jacobs, M. H. & Harms, M. (2014). Influence of interpretation on conservation intentions of whale tourists. *Tourism Management*, 42, 123–31.

Jacobs, M. H., Vaske, J. J., Dubois, S. & Fehres, P. (2014). More than fear: Role of emotions in acceptability of lethal control of wolves. *European Journal of Wildlife Research*, 60, 589–98.

Jacobs, M. H., Vaske, J. J. & Roemer, J. M. (2012). Toward a mental systems approach to human relationships with wildlife: The role of emotional dispositions. *Human Dimensions of Wildlife*, 17(1), 4–15.

James, W. (1894). The physical basis of emotion. *Psychological Review*, 1(5), 516–29.

Johansson, M. & Karlsson, J. (2011). Subjective experience of fear and the cognitive interpretation of large carnivores. *Human Dimensions of Wildlife*, 16(1), 15–29.

Johansson, M., Karlsson, J., Pedersen, E. & Flykt, A. (2012). Factors governing human fear of brown bear and wolf. *Human Dimensions of Wildlife*, 17(1), 58–74.

Johansson, M., Sjöström, M., Karlsson, J. & Brännlund, R. (2012). Is human fear affecting public willingness to pay for the management and conservation of large carnivores? *Society & Natural Resources*, 25(6), 610–20.

Kendler, K. S., Neale, M. C., Kessler, R. C., Heath, A. C. & Eaves, L. J. (1992). The genetic epidemiology of phobias in women: The interrelationship of agoraphobia, social phobia, situational phobia, and simple phobia. *Archives of General Psychiatry*, 49(4), 273–81.

Kleinginna Jr, P. R. & Kleinginna, A. M. (1981). A categorized list of emotion definitions, with suggestions for a consensual definition. *Motivation & Emotion*, 5(4), 345–79.

LeDoux, J. (1998). *The Emotional Brain: The Mysterious Underpinnings of Emotional Life*. New York: Simon and Schuster.

Lerner, J. S. & Keltner, D. (2000). Beyond valence: Toward a model of emotion-specific influences on judgement and choice. *Cognition & Emotion*, 14(4), 473–93.

Loewenstein, G. & Lerner, J. S. (2003). The role of affect in decision making. In R. J. Davidson, K. R. Scherer & H. H. Goldsmith, eds., *Handbook of Affective Sciences*. Oxford: Oxford University Press, pp. 619–42.

Manfredo, M. J. (2008). *Who Cares about Wildlife? Social Science Concepts for Exploring Human–Wildlife Relationships and Conservation Issues*. New York: Springer US.

Mehrabian, A. & Russell, J. (1974). *An Approach to Environmental Psychology*. Cambridge, MA: MIT Press.

Mormann, F., Dubois, J., Kornblith, S., Milosavljevic, M., Cerf, M., Ison, M. & Adolphs, R. (2011). A category-specific response to animals in the right human amygdala. *Nature Neuroscience*, 14(10), 1247–9.

Öhman, A. (2009). Of snakes and faces: An evolutionary perspective on the psychology of fear. *Scandinavian Journal of Psychology*, 50(6), 543–52.

Öhman, A., Flykt, A. & Esteves, F. (2001). Emotion drives attention: Detecting the snake in the grass. *Journal of Experimental Psychology: General*, 130(3), 466–78.

Öhman, A. & Mineka, S. (2001). Fears, phobias, and preparedness: Toward an evolved module of fear and fear learning. *Psychological Review*, 108(3), 483–522.

Öhman, A. & Soares, J. J. F. (1994). 'Unconscious anxiety': Phobic responses to masked stimuli. *Journal of Abnormal Psychology*, 103(2), 231–40.

Roseman, I. J., Antoniou, A. A. & Jose, P. E. (1996). Appraisal determinants of emotions: Constructing a more accurate and comprehensive theory. *Cognition & Emotion*, 10(3), 241–77.

Røskaft, E., Bjerke, T., Kaltenborn, B., Linnell, J. D. C. & Andersen, R. (2003). Patterns of self-reported fear towards large carnivores among the Norwegian public. *Evolution & Human Behavior*, 24(3), 184–98.

Russell, J. A. (2009). Emotion, core affect, and psychological construction. *Cognition & Emotion*, 23(7), 1259–83.

Russell, J. A. & Barrett, L. F. (1999). Core affect, prototypical emotional episodes, and other things called emotion: Dissecting the elephant. *Journal of Personality & Social Psychology*, 76(5), 805–19.

Sander, D., Grandjean, D. & Scherer, K. R. (2005). A systems approach to appraisal mechanisms in emotion. *Neural Networks*, 18(4), 317–52.

Scherer, K. R. (1999). Appraisal theory. In T. Dalgleish & M. Power, eds., *Handbook of Cognition and Emotion*. London: John Wiley & Sons, pp. 637–63.

Simion, F., Regolin, L. & Bulf, H. (2008). A predisposition for biological motion in the newborn baby. *Proceedings of the National Academy of Sciences of the United States of America*, 105(2), 809–13.

Slagle, K. M., Bruskotter, J. T. & Wilson, R. S. (2012). The role of affect in public support and opposition to wolf management. *Human Dimensions of Wildlife*, 17(1), 44–57.

Smith, C. A. & Ellsworth, P. C. (1985). Patterns of cognitive appraisal in emotion. *Journal of Personality & Social Psychology*, 48(4), 813–38.

Talarico, J. M. & Rubin, D. C. (2007). Flashbulb memories are special after all; in phenomenology, not accuracy. *Applied Cognitive Psychology*, 21(5), 557–78.

Tucker, M. & Bond, N. W. (1997). The roles of gender, sex role, and disgust in fear of animals. *Personality & Individual Differences*, 22(1), 135–8.

Vaske, J. J., Roemer, J. M. & Taylor, J. G. (2013). Situational and emotional influences on the acceptability of wolf management actions in the greater Yellowstone ecosystem. *Wildlife Society Bulletin*, 37(1), 122–8.

Ware, J., Jain, K., Burgess, I. & Davey, G. (1994). Disease-avoidance model: Factor analysis of common animal fears. *Behaviour Research & Therapy*, 32(1), 57–63.

Watson, D. & Clark, L. A. (1984). Negative affectivity: The disposition to experience aversive emotional states. *Psychological Bulletin*, 96(3), 465–90.

5

Tolerance for Wildlife

A Psychological Perspective

KRISTINA SLAGLE AND JEREMY T. BRUSKOTTER

In the field of conservation, the term *coexistence* is generally employed as the antonym of *conflict* – as in Woodroofe et al. (2005)'s book, *People and Wildlife, Conflict or Coexistence?*. Indeed, research on human–wildlife conflict (hereafter, HWC) is generally motivated by the desire to provide humans with a means of avoiding conflicts, thereby paving the way for coexistence. The logic here is simple: humans will be less likely to kill animals that do not impinge upon human interests and activities (i.e. animals perceived to pose unacceptable risks, or cause unacceptable conflicts). Although this simple framework has both intuitive and practical appeal, it both oversimplifies how human beings respond to their social and biophysical environment, and limits the focus to conflict, thereby reducing the potential set of solutions (Frank 2016). In fact, a substantial proportion of research in HWC is not about human conflicts with wildlife at all, but rather, human conflicts *over* wildlife conservation issues (Redpath et al. 2015). Results from recent studies suggest that human intolerance of wildlife can be rooted in conflicts among human groups; our intolerance for one another plays out as intolerance for a species (Krange & Skogen 2011; Rust et al. 2016), even in places where levels of HWC are demonstrably low (Krange & Skogen 2011). Moreover, these acts occur within a broader socio-cultural context that shapes both cultural norms and values, as well as the incentives of individual actors (Skogen & Thrane 2007; Manfredo 2008). Starting with the premise that human behaviour, and therefore coexistence, is rooted partly in psychology (i.e. an individual's thoughts and emotions), we first discuss the concept of tolerance for wildlife drawing upon theory and empirical research from psychology, and second, detail potential internal (psychological) and interpersonal mechanisms that foster tolerant attitudes and behaviours, focusing on two theories that have been broadly applied to human–wildlife conflicts.

5.1 TOWARDS A SHARED UNDERSTANDING OF TOLERANCE FOR WILDLIFE

Coexistence is a slippery concept. Merriam-Webster provides two definitions of *coexist*: (1) to exist together or at the same time, and (2) to live in peace with each other especially as a matter of policy (Merriam-Webster Dictionary, accessed online 25 January 2017). The conceptual space occupied by these two definitions is vast. Under the first, coexistence simply refers to the sharing of space (a synonym for *sympatric*), with no formal rules for how the entities that are coexisting should relate to one another. Under such a definition, one could assert that grey wolves (*Canis lupus*), grizzly bears (*Ursus arctos horribilis*) and cougars (*Puma concolor*) *coexist* with elk (*Cervus canadensis*), mule deer (*Odocoileus hemionus*) and pronghorn (*Antilocapra americana*) in Yellowstone National Park, or that various species of the order rodentia coexist with human populations all over the world, despite continuous human efforts to eradicate them. Under the second definition, however, coexistence means much more than sharing space. Living peaceably with wildlife as a matter of policy requires some human action. At the societal level, this action may include laws that are meant to conserve wildlife or preserve them through prohibitions against killing. At the personal level, it might require individuals to refrain from killing or harming individual animals or their habitat (peaceful coexistence or tolerance); and in some cases, it might even require individuals to take actions designed to benefit wildlife (active coexistence or stewardship; Bruskotter & Fulton 2012; Frank 2016). The remainder of this chapter is dedicated to explaining how individual human actors might come to engage in actions that promote or inhibit coexistence with wildlife through the lens of psychology.

Researchers in the field of wildlife conservation and management have long been interested in factors that motivate behaviour aimed at coexisting with, or eliminating the threats posed by wildlife. This interest gave rise to parallel lines of inquiry on *acceptance of* and *tolerance for* wildlife (for review, see: Bruskotter & Fulton 2012; Bruskotter et al. 2015; Kansky et al. 2016). Researchers examining *acceptance* of wildlife have primarily been interested in the extent to which individuals or stakeholder groups are willing to accept (or tolerate) local wildlife populations (Decker & Purdy 1988; Carpenter et al. 2000; Riley & Decker 2000; Lischka et al. 2008). Decker and Purdy (1988) conceptualized *acceptance capacity* as reflecting the *maximum wildlife population level in an area that is acceptable* and offered this as a means of explaining how low levels of human acceptance might limit the size and distribution of a particular population of wild animals. Thus, for example, if a particular population is perceived by an individual to have exceeded some internal threshold, the population becomes unacceptable, which may motivate the individual to take action to reduce that population. Similarly, a variety of researchers have explored the concept of *tolerance* for a species or population, often as a means of evaluating the effectiveness of policies or interventions designed to increase tolerance, or promote more positive attitudes (Naughton-Treves et al. 2003; Treves et al. 2009; Agarwala et al. 2010).

Although the measures used to assess tolerance and acceptance vary greatly, Bruskotter and Fulton (2012) argued that researchers following both lines were interested in the same general phenomena – both sets of researchers sought to understand what motivated individuals and societies to take actions that impact wildlife populations. They pointed out that acceptance and tolerance share *at least* two important characteristics (p. 99):

> For both, passive restraint or inaction on the part of affected individuals or societies is the default or *normal* state – most people, most of the time tolerate wildlife –, and . . . both concepts posit (either explicitly or implicitly) that there is some point at which individuals' or societies' inaction ceases, and actions designed to negatively impact species/populations are undertaken (i.e. intolerance, or unacceptability).

Further, they proposed that wildlife conservation behaviours could be arrayed along a continuum ranging from actions intended to harm wildlife, to inaction, to actions intended to benefit wildlife, which were labelled *intolerance, acceptance/tolerance* and *stewardship*, respectively. From this perspective, *intolerance* and *stewardship* represent opposing behavioural poles on a single continuum. However, they acknowledged the impracticality of collecting this type of behavioural data. After all, intolerant behaviours were often illegal, and anyway, in most human populations, most of the time, passive inaction (i.e. tolerance) was likely to be the behavioural default, and therefore, the far most common state. Given these limitations, they suggested that *tolerance* might also be assessed by measuring individuals' *intentions* to engage relevant behaviours. Treves (2012) countered that attitudes towards a particular species might also be assessed to indicate intolerance. In response, Bruskotter et al. (2015) correlated two attitudinal measures of attitudes towards wolves (i.e. Wildlife Acceptance Capacity and a semantic differential scale) with measures of self-reported policy-relevant behaviour directed at wolves, as well as intentions to engage in the same behaviours. They found both attitudinal measures were strongly (r > 0.60) related with self-reports of prior behaviour, as well as reports of behavioural intentions; however, they cautioned that their sample was comprised of people who were both interested in and knowledgeable about wolves, which likely inflated the correlation between attitudinal and behavioural measures. While attitudes and behaviour are typically linked (Eagly & Chaiken 1993; Fishbein & Ajzen 2010), they are by no means synonymous. Keeping the two concepts separate allows us to tap into extensive literatures on the antecedents to both concepts, thereby gaining a stronger understanding of drivers for coexistence.

5.1.1 Developing a Typology of Tolerance: Thoughts, Evaluations and Actions

Grounding the concept of tolerance in the basic psychological literature allows conservationists to draw upon nearly a century of theoretical and empirical research linking belief, attitude and behaviour. This base of theory provides insights into how to connect the disparate literatures on tolerance for (and acceptance of) wildlife, which have operationalized tolerance/acceptance in a wide variety of ways. Teasing out the conceptual differences requires a bit of vocabulary, so we turn briefly to outlining some basic psychological jargon.

Psychologists use the term *attitude* to refer to the tendency for an individual to evaluate a particular entity or object with some degree of favour (or disfavour) (Eagly & Chaiken 1993). Simply put, attitudes are evaluations of things (people, policies, places, etc.) along a good–bad or positive–negative continuum. In contrast, *beliefs* are conceptualized as an individual's subjective assessment 'of the certainty that a proposition is true' (Wyer & Albarracín 2005, p. 274). Beliefs can carry evaluative meaning; for example, an individual might believe a sixth mass extinction is under way, and that this mass extinction is terrible. However, beliefs do not need to be tied to evaluations. One might believe that speciation occurs as a result of natural selection, for example, without attaching any particular evaluative meaning to that proposition.

The term *affect* refers to 'the specific quality of *goodness* or *badness*: (1) experienced as a feeling state (with or without consciousness), and (2) demarcating a positive or negative quality of a stimulus' (Slovic et al. 2007, p. 1333). Affect is inclusive of both moods and emotions (Forgas 2000). Though affect and attitude are similar in that both concepts carry an evaluative component, they differ in a couple of important respects. Attitudes refer to psychological *tendencies* that are encoded in memory (specifically, the tendency to evaluate an object favourably/unfavourably), whereas the term affect is employed to relatively transient states, that need not be connected to a particular object in a consistent manner. Put simply, attitudes are relatively stable (Ajzen & Fishbein 2000), where affect is susceptible to situational influence (Isen 2008).

Finally, although lay and academic views of *behaviour* generally align with one another, some discussion of behaviour is relevant to our understanding of tolerance. First, psychologists often differentiate between behaviours that are under conscious or volitional control (e.g. voting in an election, purchasing a new vehicle) and behaviours that are involuntary (e.g. a startle reaction) or, at the very least, not entirely under one's control (e.g. use of addictive substances). We might anticipate that volitional behaviours (especially those deemed important) will often follow some effortful processing of relevant information. In contrast, *automatic* behaviours may occur without any conscious recognition whatsoever. Second, we might also differentiate between single behavioural acts (e.g. casting one's vote for the office of President) versus classes of related acts (e.g. an individual's voting record over time). Single acts can be extremely hard to explain or predict, as any given individual at any given moment may be impacted by a variety of contextual (environmental) factors that impede his/her ability to engage in the behaviour of interest (Weigel & Newman 1976; Fishbein & Ajzen 1977). Moreover, for behaviours with few barriers, one might opt out of opportunities to engage in the behaviour without altering the underlying attitude. Under such conditions, one's lack of action, when inconsistent with one's attitude, is easily justified because it can always be performed at a later date. Thus, an individual who loves ice cream might drive by an ice cream parlour every day for a month without stopping.

Research on beliefs, attitudes and behaviour suggests that how researchers choose to conceptualize tolerance is not merely an academic question. Rather, its conceptualization has practical implications for those interested in understanding and explaining intolerance, and in turn, promoting coexistence with wildlife. If, for example, we define *intolerance* for wildlife as an attitude, we run into the problem that attitudes are not always reliably predictive of behaviour (Heberlein 2012). For example, one might have an extremely negative attitude towards wolves, but not have access to a firearm, or the proficiency to discharge it, and thereby lack the opportunity to kill a wolf. Thus, defining intolerance relative to attitudes could provide misleading information (if one was interested in illegal killing, for example). Defining intolerance relative to an overt behaviour (e.g. illegally killing wildlife), risks the opposite problem (i.e. underestimating intolerance). Again, numerous people may hold very negative attitudes towards a particular species without ever having the opportunity to engage in the behaviour of interest. These examples illustrate the need for caution in operationalizing measures of intolerance, and highlight the value of using multiple measures.

5.1.2 Understanding Human Judgements: Two Systems

Psychologists often contrast two approaches to judgement and decision-making: on one hand, humans are capable of acquiring and assimilating vast amounts of information through effortful processing of relevant information; yet we are also capable of making judgements rapidly, with very little effort or information. Chaiken et al. (1989) refer to the former approach as *systematic*, the latter as *heuristic* – their *heuristic-systematic model* is one version of what is known generally as a *dual process* model. The extent to which an individual engages in systematic processing of information is largely a function of their motivation to come to a correct or accurate judgement, and their capacity to devote meaningful effort to the task. The use of heuristic processing, though prone to lead to errors or biases in judgement, is more efficient (Kahneman et al. 1982). The differences between systematic and heuristic processing are also captured by the notion of *hot* cognition (Simon 1990; also called system 1) or *cold* cognition (system 2; Fazio 1986; Eagly & Chaiken 1993). Hot cognition is fast, intuitive and affect-laden, whereas cold cognition is slow, effortful and reasoned. Unfortunately, these depictions of human information processing perpetuate the idea that the systems are independent. In fact, the division between the systems is more a convenient metaphor than a strict, physical division. Text Box 5.1 describes how understanding information processing under these two systems shapes human judgements and decisions in different contexts, and why this is important for understanding tolerance.

Text Box 5.1 An Illustration of Systematic versus Intuitive Processing

Dual process models of judgement and decision-making have important implications for the study of tolerance. To illustrate, consider the following scenario: John farms a variety of row crops and has noticed a substantial increase in crop depredation, presumably from white-tailed deer (*Odocoileus virginianus*). At first, it was not significant, but this last year was bad enough that it is affecting his profits. John learns that the university has an expert in wildlife damage management, and makes an appointment as a first step towards learning how he can reduce or prevent future damages. Now consider a second scenario: Dick runs a hobby cattle ranch, with a few hundred animals. Late one night he hears a calf bellowing and heads outside to find three wolves attacking the calf. From the standpoint of judgement and decision-making, these two scenarios stand in stark contrast. In the first scenario, John has noticed crop damage after the fact, which motivates him to seek help. As his crops are out of the ground, there is no immediate need to figure out how to deal with the problem. In contrast, Dick stumbles into a situation that requires an immediate judgement; he needs to decide what (if anything) to do to save his calf. These situations lend themselves to different approaches to decision-making. John has the time and motivation to track down relevant information and make an informed decision. Moreover, he has a reliable (and presumably trusted) source. These factors suggest John will take the effortful, systematic route – carefully weighing the pros and cons of various approaches to dealing with his depredation problem. In contrast, Dick's problem is immediate; he lacks the time to find and acquire relevant information. The situation forces Dick to use the quick, intuitive route. The point of this vignette is to illustrate that some behaviours and decisions that implicate tolerance are products of deliberate and explicit reasoning, while others arise from automatic and intuitive reasoning; the decision-making context greatly influences which system is employed. For researchers, these examples illustrate how the model used to understand behaviour in one context might not be appropriate for other contexts.

5.2 PSYCHOLOGICAL ANTECEDENTS OF TOLERANCE

Psychological research on the acceptance of hazards provides additional insight concerning what factors promote *acceptance* of wildlife perceived as potentially hazardous. Theory and research indicate that beliefs about the risks and benefits of a hazard are important antecedents of acceptance of that hazard (Bruskotter & Wilson 2014). Assessment of risk beliefs incorporate judgements concerning the perceived *likelihood* of risk-event, and the perceived *severity* associated with the hazard: such that low-likelihood, low-severity events are deemed the least risky overall (Yates 1992). Conscious consideration of the likelihood of a particular event would seem to invoke cold cognition (i.e. slow, effortful and reasoned processing). However, an individual's affective (emotive) reaction to a stimulus (a form of hot cognition) can shape subsequent judgements, including how we judge the risks and benefits associated with a hazard. In the context of understanding tolerance, we anticipate a positive affective reaction to a particular species (or representation of that species) will be associated with greater perceived benefits, and fewer perceived risks. Indeed, such intuitive affective reactions may actually drive the consistent, inverse relationship between risk and benefit in our minds (Alhakami & Slovic 1994).

To elaborate, positive affective reactions towards potential hazards are consistently associated with the belief that the hazard is more beneficial and less risky, and this relationship becomes stronger when a time pressure is introduced (Finucane et al. 2000). This suggests that there is a robust, observed relationship between affect, risk and benefit. The inverse relationship between risk and benefit holds for a host of hazards ranging from nuclear power (de Groot et al. 2013) to genetically modified organisms (Siegrist 2000) to tiger conservation in Bangladesh (Inskip et al. 2016). The tendency to perceive a negative relationship between risk and benefit may inflate or deflate one's perception of the risk or benefit of a hazard. If perceptions of the risks related to animals are high, people may discount real potential benefits, and vice versa, thus affecting their tolerance for wildlife.

Affect impacts other cognitive processes in a way that might be thought of as motivated reasoning, such that quick affective reactions directly influence choice, bypassing any deliberative or normatively rational cognitions (Taber & Lodge 2016). Moreover, the effect of affect on choices is stronger under time pressure (Lodge & Taber 2005). The result is a systematic biasing of our judgements and decisions in a manner that is consistent with our affective reactions (psychologists refer to this as the *affect heuristic*). The affect heuristic may not be a problem if one is an expert, as experts tend to use affect with greater acuity, but for the uninterested public, biased processes often lead to political choices that do not match their core values (Lau & Redlawsk 2001). In USA presidential elections spanning from 1972 to 2004, voters with high political knowledge and strong partisan identification were far more likely to vote for candidates that represented their stated interests, a concept described as *correct voting* (Lau et al. 2008).

Because attitudes are expressed as behavioural tendencies, both hot and cold processes are involved in their formation and change (Eagly &

Chaiken 1993). One might form an attitude towards a particular group of animals based on a conscious and effortful reflection of the characteristics of those animals (e.g. charismatic, beautiful, useful, harmful; see Batt 2009); however, one's attitude towards that group of animals might also be influenced by the emotional arousal one felt when one first encountered a particular member of that group (Chapter 4). Thus, attitude formation and change involves both conscious and effortful processing as well as intuitive, emotional processing of information. In contrast, affect originates in the limbic system, a more primitive part of the brain, and therefore precedes more effortful information processing (LeDoux 1996). Indeed, Zajonc convincingly demonstrates that one might form affective evaluations of objects completely independent of conscious recognition (Zajonc 1980, 2000). Attitudes and affect differ in that attitudes require an object of evaluation – something towards which the tendency to evaluate is expressed. In contrast, affect *can* occur entirely independently of any particular object (as when one is in a bad mood). Put simply, affect is quick, intuitive *gut feeling*, with or without a specific stimulus, while an attitude is an evaluative tendency towards some object based, in part, on articulable thoughts/beliefs about that object.

5.2.1 Psychological Models of Tolerance and Their Applicability

In understanding the psychological drivers of tolerance as a behaviour, we turn to two models of human behaviour – one broad and one narrow – sharing several common factors, and a few notable exceptions. Hazard acceptance models are drawn from the literature on judgement and decision-making under uncertainty, and have recently been adapted to wildlife. Wildlife are imbued with uncertainty, particularly surrounding the potential outcomes of their conservation, making hazard approaches uniquely useful for explaining tolerance for wildlife (Bruskotter & Wilson 2014; see Text Box 5.2). While the hazard acceptance model of tolerance is narrowly focused, the theory of reasoned action/theory of planned behaviour (TRA/TPB) encompasses a range of behaviours (Figure 5.1; Fishbein & Ajzen 2010). The aim of the TRA/TPB is to explain the broadest possible set of human behaviours, and remains one of the most widely researched models of human behaviour (though not without modifications or critique; see Sniehotta et al. 2014). The TRA/TPB includes variables now familiar to the reader: attitudes and beliefs predicting behaviours, plus two additional predictor variables: norms and control. Broadly, norms are what one perceives others think they

Text Box 5.2 Perceptions of the Risks and Benefits of Wildlife

Empirical investigation of the perceived risks and benefits related to wildlife is recent, but patterns have emerged in the findings related to predators. Riley and Decker (2000) examined wildlife acceptance capacity for cougars, and its relationship to risk perceptions, and found that lower risk perceptions were related to higher support for an increase in cougar populations. Gore et al. (2006), after interviewing campers in rural New York about black bears (*Ursus americanus*), found certainty and dread, two hypothesized dimensions of risk perceptions (Slovic 1987), to influence the perceptions of the threats posed by black bears. Specifically, certainty about the known risks of black bears, and how to prevent exposure likely drove lower risk perceptions, however dread related to the risk from black bears was mixed, in that campers reported feeling highly variable levels of dread. Furthermore, the two factors suggested by the data were different from the psychometric paradigm mentioned above. A follow-up, large-sample mail survey again confirmed these findings (Gore et al. 2007). Generally, their results suggest that, psychologically, threats related to black bears were perceived similarly to other hazards, in that similar characteristics of the hazard influenced risk perceptions (volitional exposure, severity of consequences, dread, control). Also studying black bears in rural New York, Siemer et al. (2009) found that positive beliefs about black bears were negatively associated with concern about the risks from black bears. Tolerance to black bears in Ohio was well explained ($R^2 = 62\%$) almost equally by risk and benefit, which were strongly and negatively associated, and, in turn, driven by trust and control (Zajac et al. 2012). Similarly, perceptions of risk and benefit are very predictive of tolerances for tiger populations, though benefits are the more important of the two (Carter et al. 2012; Inskip et al. 2016). And indeed, tolerance for wolves among an interested and knowledgeable public was driven by perceptions of risk and benefit, and benefit again was the more important direct predictor (Slagle et al. 2012); however, the indirect effects showed affect towards wolves was the primary driver of tolerance.

should do, and control refers to the perception that some behaviour will be more or less difficult to perform due to contextual factors. To draw out the earlier example of John, facing crop damage and planning to consult an extension specialist, add to his story that his friend, another farmer, has mentioned that many people in the area consult the extension specialists for help with crop damage. John now has a norm to follow: other people in his position typically ask extension agents for help. John meets with the specialist, but on learning of the two preferred management options, he knows that other demands on the farm may prevent him from executing the most effective management tactics for

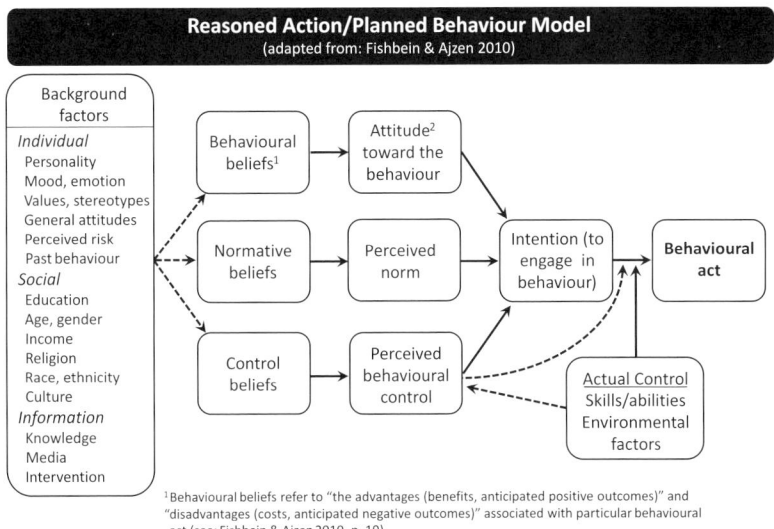

Reasoned Action/Planned Behaviour Model
(adapted from: Fishbein & Ajzen 2010)

[1] Behavioural beliefs refer to "the advantages (benefits, anticipated positive outcomes)" and "disadvantages (costs, anticipated negative outcomes)" associated with particular behavioural act (see: Fishbein & Ajzen 2010, p. 19).

[2] Attitudes refer specifically to the behavioural act, as opposed to a particular object.

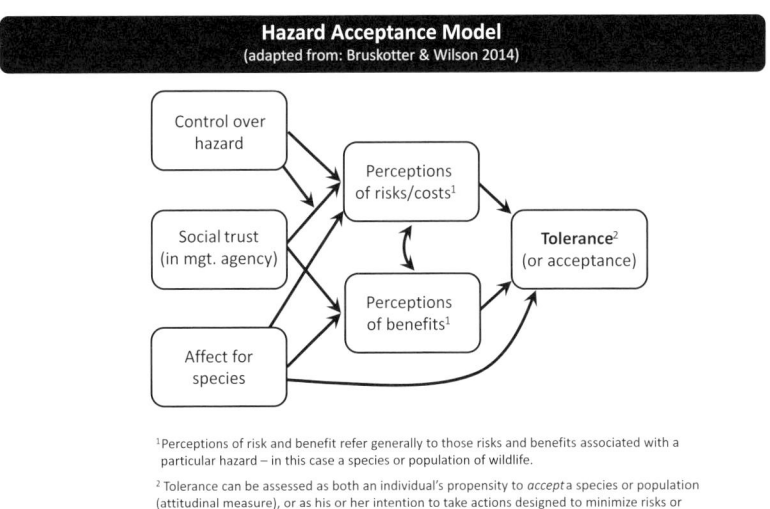

Hazard Acceptance Model
(adapted from: Bruskotter & Wilson 2014)

[1] Perceptions of risk and benefit refer generally to those risks and benefits associated with a particular hazard – in this case a species or population of wildlife.

[2] Tolerance can be assessed as both an individual's propensity to *accept* a species or population (attitudinal measure), or as his or her intention to take actions designed to minimize risks or increase benefits associated with that species or population.

Figure 5.1 Variables and pathways of the TRA/TPB and the hazard acceptance model.

reducing damage. This could result in John feeling less control over the behaviours that would fix his problem – a common perception preventing people from following through with behaviours that would seem otherwise to match their attitudes, beliefs and norms.

The TRA/TPB model is most useful for explaining variation in specific behavioural acts, and the specificity with which one measures four elements (i.e. the action or behaviour of interest, the context in which the action is performed, the target at which the action is directed and the time at which it is performed) all affect the correspondence between model components (Ajzen & Fishbein 1977). When attitudes and behaviours are measured at similar levels of specificity, the correspondence between attitudes and behaviours increases (Kraus 1995; Ajzen & Cote 2008). A meta-analysis of 185 TRA/TPB studies found that the average correlation between attitude towards a behaviour and intention in these studies was 0.49, and that attitudes and perceived behavioural control were typically better predictors of intentions than subjective norm (Armitage & Conner 2001). On average, model components explained 39 per cent of the variation in intention, and 27 per cent of the variation in behaviour.

In developing strategies to improve coexistence with wildlife, the TRA/TPB model offers insights into how one might design interventions to increase (or decrease) tolerance for a species. Specifically, the model suggests that interventions designed to modify behavioural, normative and control beliefs (the antecedents of attitudes, perceived norms and perceived behavioural control, respectively) will be most useful in modifying behaviour. For example, a case study assessing the efficacy of non-lethal predator management strategies may have modified ranchers' behavioural beliefs as well as their perceived control (though these psychological variables were not explicitly assessed by the authors; Stone et al. 2017). By supplying trainers and equipment, the funding organization may have increased a sense of control among ranchers, and the eventual results lent evidence that specific predator management strategies could reduce conflict. The exercise may have changed behavioural beliefs among some ranchers regarding the efficacy of non-lethal predator management; after initial reluctance to participate, they continued the management strategies beyond the funding period.

Drawbacks to the TRA/TPB are linked together. First, research suggests it is most useful in explaining specific behaviours (e.g. one's intention to shoot an animal caught killing livestock). However, as in the study described above, practitioners may be interested in a wide range of behaviours or repeated behaviours (e.g. use of both lethal and non-lethal control techniques, and engagement in political activities designed to impact a species or population) that may occur over an extended time period (see Bruskotter & Fulton 2012). Use of the TRA/

TPB when a diverse range of behaviours is applicable may become overly burdensome – requiring researchers to assess not only multiple behavioural criteria, but corresponding attitudes, norms and control measures for each behavioural act of interest. In other words, while specificity of measurement may increase the model's predictive ability, it also reduces the generalizability of the finding.

Relatedly, some behaviours that may be of great importance in a particular area or among a particular population may be irrelevant in others. For example, while it may be extremely relevant to assess hunters' or rural landowners' intentions to *shoot a wolf if I saw one*, this behaviour is irrelevant for a large portion of the population who either do not live or recreate where wolves exist, or own a firearm (for examples). Consequently, the TRA/TPB is most useful when practitioners are interested in single behavioural acts that are time-bounded (voting for a ballot measure, for example).

In contrast to the TRA/TPB, the hazard acceptance model is designed to deal with multiple behavioural acts related specifically to tolerance or intolerance – acts that may indicate either passive tolerance, active intolerance or active stewardship of a species (see Bruskotter & Fulton 2012; Bruskotter et al. 2015 for discussion). A primary difference between these models is the subject of interest. In the TRA/TPB, the subject of all model components is the behaviour of interest. In contrast, the subject of hazard model components is the species or population itself (sometimes called the behavioural *target* in attitude-behaviour studies). The TRA/TPB and hazard acceptance models share several parallels in their assessments of specific behaviours or species, respectively, and one important difference (Figure 5.1). Where the TRA/TPB relies on a combination of beliefs about how likely is it that a behaviour will lead to specific outcomes (coexistence with wildlife), the hazard acceptance model focuses on a range of risks and benefits of conservation. While not equivalent, these ideas are related, in that each requires at least an implicit weighing of likelihoods and outcomes. Similarly, attitudes in the TRA/TPB are replaced by measures of affect in the hazard acceptance model, and control over behaviour in the TRA/TPB is modified to include perceived control of the hazard itself in the hazard acceptance model. Within the TRA/TPB, norms serve to connect the individual to social influences; however, in the hazard acceptance model, the unique nature of wildlife and their management by unknown others requires a different approach. Instead of perceptions of how others think one should behave, the variable the variable focuses on how one thinks some other entity should (or will) behave, summed up by their trust in a management entity.

Trust is a notably slippery concept, but two perspectives critical to environmental risk have emerged: calculative trust, placing confidence elsewhere for the management of risk, and relational trust, which is placed with those in whom we perceive shared values (Earle & Cvetkovich 1995; Siegrist 2000). Trust is a necessary mechanism that confers responsibility to some outside entity for the things we ourselves cannot control or manage. Because it involves uncertainty, it is risky to choose to trust, but doing so reduces the vast complexity of our modern lives and makes them manageable (Earle & Cvetkovich 1995). Calculative trust in an organization to manage wildlife effectively can be lost with one mistake in management, while relational trust is more resilient to mistakes, as it is based in perceived shared values between the stakeholder and the organization. It acts as a heuristic in our decision-making, and specific to wildlife management; it simplifies complex, landscape-scale problems with bureaucratic and ecological dimensions. In trusting an agency to manage climate change, individuals trust those agencies they perceive to share values and agreement with on the issue (Earle 2004), and when considering endangered species management by the US Forest Service, individuals with greater trust in the agency were also more accepting of management actions (Cvetkovich & Winter 2003). Higher trust may serve to reduce perceptions of risk and raise tolerance for carnivores, particularly for the less-informed public (Bruskotter & Wilson 2014). Knowing how the public places trust in an organization can help individuals within the organization gauge the acceptability of taking certain risks in their management. Organizations with stakeholders perceiving a trust based in shared values might be empowered to take greater risks in their management, as mistakes or failures might be discounted or forgiven; although once lost, trust is difficult and costly to recover, so risky moves should not be undertaken lightly.

The difference here between the TRA/TPB and the hazard model in the inclusion of a socially focused variable (TRA/TPB-norms, hazard-trust), belies a limitation of both: neither assesses the influence of interpersonal factors on coexistence with wildlife. The omission is understandable, as these are psychological models focused on individual behaviour, but future work might consider how to better include the influence of social factors on behaviour. One useful direction may lie in considering group identities as guiding forces shaping individual thoughts and behaviours.

5.3 LIMITATIONS AND FUTURE WORK: INTERPERSONAL DRIVERS FOR TOLERANCE

Although research on tolerance for wildlife diverges both in terms of theories employed and in terms of how tolerance is operationalized, there is much greater consistency in the design of such studies. The vast majority of these studies employ cross-sectional surveys using self-administered questionnaires (or in some cases, interviews). Several issues arise from such designs. First, although cross-sectional surveys allow for a wide variety of factors to be studied, simultaneous assessment of dependent and independent variables impedes researchers' ability to assess causation.

Second, self-administered questionnaires are generally completed in environments very different to those in which people encounter wildlife – at one's desk or kitchen table, as opposed to on a forest trail, for example. Yet it is reasonable to assume that an individual's ability to predict how they might respond in a scenario is impacted both by their physical and social environment – i.e. the elements that make each interaction unique. Researchers attempt to counter this in two ways – by either *decontextualizing* measures (i.e. designing response items that are very general in nature and apply to a variety of settings) or, alternatively, attempting to create context by providing specific scenarios that define for respondents some aspect of the interaction. Both types of measures have limitations. Decontextualized response items seek to gauge general dispositions – one's inclination to believe that wolves compete with big game hunters for prime trophy animals, for example. Yet the context can significantly impact how one responds to such items. Using the prior example, the belief that wolves compete with big game hunters could depend, for example, on the density of the local wolf or deer population, or other local factors. The point, for clarity, is that while respondents might wish to provide caveats (e.g. I think wolves compete with big game hunters under some circumstances, but not others), they are forced simply to agree or disagree with a generalized response item. Alternatively, items that attempt to provide context by detailing particular scenarios are limited in that they can become overly specific; that is, they may not generalize beyond that very specific scenario, thereby limiting their applicability to management.

From a sampling perspective, the predominant approach used in nearly all surveys dealing with tolerance is to draw random samples of households (or individuals) within some geopolitical unit (e.g. country, state/province or some region there within). However, due to spatial clustering in urban areas, such sampling limits exploration of how social and ecological conditions impact tolerance; that is, it allows researchers to explain variation across individuals, not across space. Furthermore, when coupled with cross-sectional design, such studies are limited in the capacity to address how tolerance changes within individuals, or likewise, examine factors that contribute to changes over time.

Moving forward, a better understanding of the way groups influence their members' thoughts and behaviour may prove critical, as society continues to divide along tribal lines (Haidt 2013). Social or group identity broadly refers to the degree to which one identifies with a specific group within a certain context. Social identity theory was originally developed to explain the observation that people prefer in-groups to out-groups, but evolved into a perspective for understanding the psychological basis for group conflict (Tajfel & Turner 1986). It posits that individuals belong to social groups and associate some value or emotional attachment to maintaining identification with their respective groups (Hogg 2006). Identity may influence tolerance by informing an individual's beliefs or attitudes towards a species, where an individual seeks to match their beliefs to the perceived beliefs of the group in order to further confirm their sense of belonging and reinforce their identity with the group (Bruskotter et al. 2009; Lute & Gore 2014). Performing behaviours also associated with the group (i.e. purchasing a hunting tag every year and self-identifying as a hunter) can also serve to solidify identity and beliefs associated with the group (Bem 1967; Terry et al. 1999).

Alternatively, a lack of identity with any wildlife-focused group may influence people to be passively tolerant of wildlife populations. Without a guiding identity of some kind, people may simply not know what to think about a wildlife population or policy, nor perhaps has it been prioritized in their lives to acquire any kind of related identity. These people may not show an interest in wildlife, and may be difficult to engage as stakeholders, but this does not necessarily indicate to an unimportant portion of the population. Practitioners should stay prepared to engage with the disinterested public once some triggering event provides the opportunity to do so. For example, for professionals focused on climate change, guidance documents give advice on the best approaches for engaging the public about climate change after extreme weather events (Cutting 2012). Similar styles of outreach development may prove useful for wildlife professionals in speaking to the newly interested public after conflict events or other triggering event.

A consideration of group identity is but one potential path for adapting psychological models of tolerance and coexistence; others might include nesting individuals within cultural or ecological contexts that influence our continued coexistence with other species. Linking individual behaviour with socio-ecological systems is the next big challenge to understanding coexistence, and regardless of how psychological models are further refined, making these linkages will inevitably require an interdisciplinary perspective on predicting and influencing human behaviour.

5.4 RECOMMENDATIONS AND FUTURE DIRECTIONS

- The development of effective coexistence strategies requires a better understanding of why people tolerate (or fail to tolerate) wildlife. Psychological theory and research provides insights useful for both investigating human tolerance of wildlife, and developing means of increasing tolerance for wildlife (thereby, promoting coexistence).
- In the human–wildlife conflict literature, tolerance has been conceptualized and measured in a wide variety of ways. In the parlance of psychology, these can be referred to as *beliefs*, *attitudes* and *behaviours*. Because human behaviours most directly impact conservation outcomes, it is imperative that we develop and revise models that explain behaviours relevant to the conservation of wildlife. Researchers and practitioners should think critically on which conceptualization of tolerance will be most useful, and employ appropriate measures.
- Two approaches that hold promise for explaining conservation actions are the TRA (aka TPB) and the hazard acceptance model. The former is particularly well-suited for explaining specific actions that occur within fixed time periods, while the latter is adapted to understanding a wide range of human behaviours that may define stewardship of – and intolerance for wildlife. These models are generally concerned with how various thoughts (cognitions) impact behaviour, and downplay the relevance of other potentially important factors (e.g. social group affiliation or identity, trust).
- Behavioural models and existing research suggest behaviours aimed at positively impacting species are likely when individuals: (1) perceive that they (or others) are likely to incur some benefit from the species/population, (2) perceive relatively low risks from a species/population, (3) believe that those charged with managing a species are trustworthy, (4) believe that they can control risks associated with a

species through their own actions, (5) feel positive emotions (affect) towards a species, (6) have positive attitudes towards the behaviour, (7) believe that others who are important to them would likely engage in that behaviour (or when they believe they *should* engage in the behaviour) and (8) feel that it is within their ability/capacity to perform the behaviour.

- Calculative trust in an organization to manage wildlife effectively can be lost with one mistake in management, while relational trust is more resilient to mistakes, as it is based in perceived shared values between the stakeholder and the organization. Knowing how the public places trust in an organization can help individuals within the organization gauge the acceptability of taking certain risks in their management.

- Be prepared to engage with uninformed/disinterested individuals on wildlife issues when an event or issue raises their attention. Climate change literature on engaging the public after extreme weather events may provide insights.

- Existing approaches to understanding tolerance are limited in several ways that impact the generalizability and usefulness of their insights. Typical research in this field uses self-administered questionnaires, cross-sectional designs and regression methods to explain variation in tolerance (i.e. beliefs, attitudes or behaviours) across individuals. These approaches rarely address questions such as (1) how/why does tolerance change? (2) how do changing social and ecological conditions impact tolerance?

- There is a need for multilevel approaches that seek to understand simultaneously the influence of individual and group-level factors on tolerance.

5.5 References

Agarwala, M., Kumar, S., Treves. A. & Naughton-Treves, L. (2010). Paying for wolves in Solapur, India and Wisconsin, USA: Comparing compensation rules and practice to understand the goals and politics of wolf conservation. *Biological Conservation*, 143, 2945–55.

Ajzen, I. & Cote, N. G. (2008). *Attitudes and Attitude Change*. New York: Psychology Press.

Ajzen, I. & Fishbein, M. (1977). Attitude–behavior relations: A theoretical analysis and review of empirical research. *Psychological Bulletin*, 84, 888–918.

Ajzen, I. & Fishbein, M. (2000). Attitudes and the attitude–behavior relation: Reasoned and automatic processes. *European Review of Social Psychology*, 11, 1–33.

Alhakami, A. S. & Slovic, P. (1994). A psychological study of the inverse relationship between perceived risk and perceived benefit. *Risk Analysis*, 14, 1085–96.

Armitage, C. J. & Conner, M. (2001). Efficacy of the theory of planned behaviour: A meta-analytic review. *British Journal of Social Psychology*, 40(4), 471–99.

Batt, S. (2009). Human attitudes towards animals in relation to species similarity to humans: A multivariate approach. *Bioscience Horizons: The International Journal of Student Research*, 2, 180–90.

Bem, D. J. (1967). Self-perception: An alternative interpretation of cognitive dissonance phenomena. *Psychological Review*, 74, 183–200.

Bruskotter, J. T. & Fulton, D. C. (2012). Will hunters steward wolves? A comment on Treves and Martin. *Society & Natural Resources*, 25, 97–102.

Bruskotter, J. T., Singh, A., Fulton, D. C. & Slagle, K. (2015). Assessing tolerance for wildlife: Clarifying relations between concepts and measures. *Human Dimensions of Wildlife*, 20, 255–70.

Bruskotter, J. T., Vaske, J. J. & Schmidt, R. H. (2009). Social and cognitive correlates of Utah residents' acceptance of the lethal control of wolves. *Human Dimensions of Wildlife*, 14(2), 119–32.

Bruskotter, J. T. & Wilson, R. S. (2014). Determining where the wild things will be: Using psychological theory to find tolerance for large carnivores. *Conservation Letters*, 7, 158–65.

Carpenter, L. H., Decker, D. J. & Lipscomb, J. F. (2000). Stakeholder acceptance capacity in wildlife management. *Human Dimensions of Wildlife*, 5, 5–19.

Carter, N. H., Riley, S. J. & Liu, J. (2012). Utility of a psychological framework for carnivore conservation. *Oryx*, 46(4), 525–35.

Chaiken, S. & Eagly, A. H. (1989). Heuristic and systematic information processing within and beyond the persuasion context. In J. Uleman & J. Bargh, eds., *Unintended Thought*. New York: The Guilford Press, pp. 212–52.

Cutting, H. (2012). *Connecting the Dots: A Communications Guide to Climate Change and Extreme Weather*. New York: Climate Nexus.

Cvetkovich, G. & Winter, P. L. (2003). Trust and social representations of the management of threatened and endangered species. *Environment & Behavior*, 35, 286–307.

Decker, D. J. & Purdy, K. G. (1988). Toward a concept of Wildlife Acceptance Capacity in wildlife management. *Wildlife Society Bulletin*, 16, 53–7.

De Groot, J. I., Steg, L. & Poortinga, W. (2013). Values, perceived risks and benefits, and acceptability of nuclear energy. *Risk Analysis: An International Journal*, 33(2), 307–17.

Eagly, A. H. & Chaiken, S. (1993). *The Psychology of Attitudes*. Fort Worth, TX: Harcourt Brace Jovanovich College Publishers.

Earle, T. (2004). Thinking aloud about trust: A protocol analysis of trust in risk management. *Risk Analysis*, 24, 169–83.

Earle, T. C. & Cvetkovich, G. (1995). *Social Trust: Toward a Cosmopolitan Society*. Westport, CT: Praeger Publishers.

Finucane, M. L., Alhakami, A., Slovic, P. & Johnson, S. M. (2000). The affect heuristic in judgments of risks and benefits. *Journal of Behavioral Decision Making*, 13, 1–17.

Fishbein, M. & Ajzen, I. (2010). *Predicting and Changing Behavior: The Reasoned Action Approach.* New York: Psychology Press.

Forgas, J. P. (2000). Introduction: The role of affect in social cognition. In J. P. Forgas, ed., *Feeling and Thinking: The Role of Affect in Social Cognition.* New York: Cambridge University Press, pp. 1–28.

Frank, B. (2016). Human–wildlife conflicts and the need to include tolerance and coexistence: An introductory comment. *Society & Natural Resources*, 29, 738–43.

Gore, M. L., Knuth, B. A., Curtis, P. D. & Shanahan, J. E. (2006). Stakeholder perceptions of risk associated with human–black bear conflicts in New York's Adirondack Park campgrounds: Implications for theory and practice. *Wildlife Society Bulletin*, 34(1), 36–43.

Gore, M. L., Knuth, B. A., Curtis, P. D. & Shanahan, J. E. (2007). Factors influencing risk perception associated with human–black bear conflict. *Human Dimensions of Wildlife*, 12(2), 133–6.

Haidt, J. (2013). *The Righteous Mind: Why Good People Are Divided by Politics and Religion.* New York: Vintage Books.

Heberlein, T. A. (2012). *Navigating Environmental Attitudes.* Oxford: Oxford University Press.

Hogg, M. A. (2006). Social identity theory. In P. J. Burke, ed., *Contemporary Social Psychological Theories.* Stanford, CA: Stanford University Press, pp. 111–36.

Inskip, C., Carter, N., Riley, S., Roberts, T. & MacMillan, D. (2016). Toward human–carnivore coexistence: Understanding tolerance for tigers in Bangladesh. *PLoS ONE*, 11, e0145913.

Isen, A. M. (2008). Some ways in which positive affect influences decision making and problem solving. *Handbook of Emotions*, 3, 548–73.

Kahneman, D., Slovic, P. & Tversky, A. (1982). *Judgment under Uncertainty: Heuristics and Biases.* Cambridge: Cambridge University Press.

Kansky, R., Kidd, M. & Knight, A. T. (2016). A wildlife tolerance model and case study for understanding human–wildlife conflicts. *Biological Conservation*, 201, 137–45.

Krange, O. & Skogen, K. (2011). When the lads go hunting: The 'Hammertown mechanism' and the conflict over wolves in Norway. *Ethnography*, 12, 466–89.

Kraus, S. J. (1995). Attitudes and the prediction of behavior: A meta-analysis of the empirical literature. *Personality and Social Psychology Bulletin*, 21(1), 58–75.

Lau, R. R., Andersen, D. J. & Redlawsk, D. P. (2008). An exploration of correct voting in recent U.S. presidential elections. *American Journal of Political Science*, 52, 395–411.

Lau, R. R. & Redlawsk, D. P. (2001). Advantages and disadvantages of cognitive heuristics in political decision making. *American Journal of Political Science*, 45, 951–71.

LeDoux, J. E. (1996). *The Emotional Brain: The Mysterious Underpinnings of Emotional Life.* New York: Simon & Schuster.

Lischka, S. A., Riley, S. J. & Rudolph, B. A. (2008). Effects of impact perception on acceptance capacity for white-tailed deer. *Journal of Wildlife Management*, 72, 502–9.

Lodge, M. & Taber, C. S. (2005). The automaticity of affect for political leaders, groups, and issues: An experimental test of the hot cognition hypothesis. *Political Psychology*, 26, 455–82.

Lute, M. L. & Gore, M. L. (2014). Stewardship as a path to cooperation? Exploring the role of identity in intergroup conflict among Michigan wolf stakeholders. *Human Dimensions of Wildlife*, 19(3), 267–79.

Manfredo, M. J. (2008). *Who Cares about Wildlife? Social Science Concepts for Exploring Human–Wildlife Relationships and Conservation Issues.* New York: Springer Science.

Naughton-Treves, L., Grossberg, R. & Treves, A. (2003). Paying for tolerance: The impact of livestock depredation and compensation payments on rural citizens' attitudes toward wolves. *Conservation Biology*, 17, 1500–11.

Redpath, S. M., Bhatia, S. & Young, J. (2015). Tilting at wildlife: Reconsidering human–wildlife conflict. *Oryx*, 49, 222–5.

Riley, S. J. & Decker, D. J. (2000). Wildlife stakeholder acceptance capacity for cougars in Montana. *Wildlife Society Bulletin*, 28, 931–9.

Rust, N. A., Tzanopoulos, J., Humle, T. & MacMillan, D. C. (2016). Why has human–carnivore conflict not been resolved in Namibia? *Society & Natural Resources*, 29, 1079–94.

Siegrist, M. (2000). The influence of trust and perceptions of risks and benefits on the acceptance of gene technology. *Risk Analysis*, 20, 195–204.

Siemer, W. F., Hart, P. S., Decker, D. J. & Shanahan, J. E. (2009). Factors that influence concern about human–black bear interactions in residential settings. *Human Dimensions of Wildlife*, 14(3), 185–97.

Simon, H. A. (1990). Alternative visions of rationality. In P. K. Moser, ed., *Rationality in Action: Contemporary Approaches.* New York: Cambridge University Press, pp. 189–204.

Skogen, K. & Thrane, C. (2007). Wolves in context: Using survey data to situate attitudes within a wider cultural framework. *Society & Natural Resources*, 21, 17–33.

Slagle, K. M., Bruskotter, J. T. & Wilson, R. S. (2012). The role of affect in public support and opposition to wolf management. *Human Dimensions of Wildlife*, 17, 44–57.

Slovic, P. (1987). Perception of risk. *Science*, 236, 280–5.

Slovic, P., Finucane, M. L., Peters, E. & MacGregor, D. G. (2007). The affect heuristic. *European Journal of Operational Research*, 177, 1333–52.

Sniehotta, F. F., Presseau, J. & Araújo-Soares, V. (2014). Time to retire the theory of planned behaviour. *Health Psychology Review*, 8, 1–7.

Stone, S. A., Breck, S. W., Timberlake, J., Haswell, P. M., Najera, F., Bean, B. S. & Thornhill, D. J. (2017). Adaptive use of nonlethal strategies for minimizing wolf–sheep conflict in Idaho. *Journal of Mammalogy*, 98, 33–44.

Taber, C. S. & Lodge, M. (2016). The illusion of choice in democratic politics: The unconscious impact of motivated political reasoning. *Political Psychology*, 37, 61–85.

Tajfel, H. & Turner, J. C. (1986). The social identity theory of intergroup behavior. *Psychology of Intergroup Relations*, 2, 7–24.

Terry, D. J., Hogg, M. A. & White, K. M. (1999). The theory of planned behaviour: Self-identity, social identity and group norms. *British Journal of Social Psychology*, 38, 225–44.

Treves, A. (2012). Tolerant attitudes reflect an intent to steward: A reply to Bruskotter and Fulton. *Society & Natural Resources*, 25(1), 103–4.

Treves, A., Jurewicz, R. L., Naughton-Treves, L. & Wilcove, D. (2009). The price of tolerance: Wolf damage payments after recovery. *Biodiversity & Conservation*, 18, 4003–21.

Weigel, R. H. & Newman, L. (1976). Increasing attitude–behavior correspondence by broadening the scope of the behavioral measure. *Journal of Personality and Social Psychology*, 33, 793–802.

Woodroffe, R., Thirgood, S. & Rabinowitz, A. (2005). *People and Wildlife, Conflict or Coexistence?* Cambridge: Cambridge University Press.

Wyer, R. S. & Albarracín, D. (2005). Belief formation, organization, and change: Cognitive and motivational influences. In D. Albarracín, B. T. Johnson and M. P. Zanna, eds., *The Handbook of Attitudes*. London: Routledge, pp. 273–322.

Yates, J. (1992). *Risk-Taking Behavior*. Chichester: John Wiley & Sons.

Zajac, R. M., Bruskotter, J. T., Wilson, R. S. & Prange, S. (2012). Learning to live with black bears: A psychological model of acceptance. *The Journal of Wildlife Management*, 76, 1331–40.

Zajonc, R. B. (1980). Feeling and thinking: Preferences need no inferences. *American Psychologist*, 35, 151–75.

Zajonc, R. B. (2000). Feeling and thinking: Closing the debate over the independence of affect. In J. P. Forgas, ed., *Feeling and Thinking: The Role of Affect in Social Cognition*. New York: Cambridge University Press, pp. 31–58.

A Framework for Assessing and Quantifying Human–Wildlife Interactions in Urban Areas

CARL D. SOULSBURY AND PIRAN C. L. WHITE

Humans and wildlife in urban areas will inevitably interact, but the type of interactions, their frequency and severity will vary substantially. Whilst most work on human–wildlife interactions has focused on conflict, we emphasize that interactions can indeed be negative, positive and neutral, sometimes at the same time and in different directions. In this chapter, we review relevant frameworks from different disciplinary contexts that could be used to help build a better understanding of human–wildlife interactions in urban areas. We illustrate how these may be applied and discuss their usefulness and limitations through a relevant example. We conclude by providing recommendations on how human–wildlife interactions need to progress to the point of shifting from conflict to tolerance and coexistence with urban wildlife.

6.1 TYPES OF INTERACTIONS

Human–wildlife interactions are daily occurrences for people across the globe. The nature of such interactions varies substantially on different continuous scales, ranging from rare to frequent, minor to major, brief moments to long encounters (Dickman 2010; Soulsbury & White 2015; Nyhus 2016). These interactions include conflicts about wildlife impacting humans and their belongings (Ogra 2008; White et al. 2009; Decker et al. 2010), through a continuum of more positive interactions such as tolerating and coexisting with species across spaces and time (see Chapter 1). They are also typically viewed as being from one perspective, usually the human one. In fact, interactions can be viewed from human or animal perspectives and may be simultaneously positive and negative. This broadness and variedness make human–wildlife interactions hard to categorize and to study; most studies seek to address only parts of the human–wildlife interaction such as human–wildlife conflict (Barua et al. 2013) or positive outcomes (Clark et al. 2014). The problem with this is that human–wildlife interactions exist on a continuum (Soulsbury & White 2015; Frank 2016) and that viewing or even studying one part of an interaction misses other vital components or processes.

Despite representing the majority of interactions, neutral and positive ones are poorly understood (Soulsbury & White 2015). Direct human–wildlife conflict frequently involves wildlife causing damage to human property or interest, and can often be quantified as some form of financial cost. In contrast, the impacts of other forms of interaction can be harder to value. Some positive interactions can be quantified financially; for example, non-consumptive wildlife watching in the USA generated between $104 billion and $217 billion in 2006 as consumer surplus (Sun et al. 2015) and the annual spend on bird seed, dispensers and other peripherals in UK gardens approaches $300–$360 million each year (Jones & Reynolds 2008). In addition, people engaged in activities such as bird feeding show a positive willingness to pay for conservation (Clucas et al. 2015). Harder to assess are the less tangible impacts of interactions on human well-being and health that many forms of human–wildlife interaction can promote (Curtin 2009; Sandifer et al. 2015; Shanahan et al. 2016). In particular, there is evidence that encounters benefiting humans can also be beneficial indirectly to wildlife by promoting a stronger connection to nature in the humans involved and a greater likelihood of willingness to conserve the natural world (Curtin 2009; Zelenski et al. 2015; Hobbs & White 2016; Soga et al. 2016b). Promoting coexistence between humans and wildlife therefore has potential to be beneficial both to humans and to wildlife.

6.2 HUMAN–WILDLIFE INTERACTIONS IN URBAN AREAS

Today, 54 per cent of the world's population lives in urban areas, a proportion that is expected to increase to 66 per cent by 2050. Nearly 90 per cent of the increase has been concentrated in Asia and Africa (United Nations 2014). At the same time, the very nature of urban areas is changing. Historically, cities were well-defined areas that grew outwards from a dense urban core to less dense, more dispersed peri-urban[1] areas (Seto et al. 2010; Ramhalo & Hobbs 2012). Contemporary cities are no longer so sharply defined and are increasingly dispersed and expansive, especially in Europe, Australia and North America (Seto et al. 2010). This change in shape and structure has critical implications for human–urban wildlife interactions. First, there is greater opportunity of interactions in low–medium[2] density housing, because these typically have the highest species richness and reduced species extinction rates (McKinney 2008; Magle et al. 2016) and greatest areas of green space and diversity of land cover (Smith et al. 2005; Loram et al. 2007). Second, as the size of urban areas grows, so does the size of the urban–rural interface, resulting in an increase of urban green spaces under the form of house gardens and public green areas for recreation. Both factors are linked to increases in human–wildlife interactions (Kretser et al. 2008; Teixeira et al. 2015a, 2015b; Hosaka & Numata 2016; Poessel et al. 2017). In addition, the human element within these areas is critical. People living in low–medium density housing tend to have the highest per capita income which is associated with higher rates of positive interactions (Fuller et al. 2012; Davies et al. 2012), yet these people can also react most negatively to human–wildlife conflict (Teixeira et al. 2015b; Wine et al. 2015). In combination, the urban space has a critical role to play in shaping and defining a number of key features of human–wildlife interactions, including their frequency and their type.

With urban areas increasing globally, it is clear that the interactions between humans and urban wildlife are going to increase steadily. To better understand the nature of human–urban wildlife interactions and to inform decisions to reduce risks from or to maximize benefits of them, there needs to be some form of unified framework that provides a mode of assessing and evaluating human–urban wildlife interactions (Morzillo et al. 2014). Considering human–urban wildlife interactions in a more structured way should help to pull together the various factors affecting interactions and lead to better decision-making. It will also allow better characterization of the conflict-to-coexistence continuum by offering an additional tool to understand when and how a human–wildlife interaction turns into conflict or coexistence.

---

[1] Peri-urban = the interface between urban areas and surrounding ex-urban landscape.
[2] Low density housing = 6–49 houses/km²; medium density housing = 49–741 houses/km².

6.3 FRAMEWORKS TO ASSESS HUMAN–WILDLIFE INTERACTIONS

Human–wildlife interactions, whether they are related to urban areas or not, can be broken down into components that integrate both biological and human dimensions. These include human–wildlife encounter probability, consequences of the encounter and outcomes of the encounter (Gore & Knuth 2009).

We searched for frameworks using Google Scholar with the search terms *human–wildlife, interactions* and *framework*, which generated 1,120 matching articles (as of November 2016). Google Scholar searches within articles so is advantageous in finding more papers on a broader range of topics. Titles were first screened for relevance, before abstracts from 230 papers were read to assess suitability (i.e. they contained a framework or model). Based on this initial assessment, 23 papers were selected for full appraisal. Eight of these used frameworks in contexts that were not directly applicable to human–wildlife interactions (e.g. habitat suitability modelling), which left 15 papers providing suitable frameworks for analysis. The majority of studies related to human–wildlife conflict in some form and only one focused on human well-being (Table 6.1). Frameworks were generally either rural-focused or generic (i.e. they were not specifically applied to urban or rural areas). Only one had a more urban focus (Table 6.1). The focus of frameworks was strongly taxonomically biased towards mammals, and more specifically to carnivores (Table 6.1). Only one included an example of human–avian conflict.

Table 6.1 *Categorization of papers used during framework analysis*

	Topic	Number of papers
Type of study	Disease	1
	Conflict or management	10
	Tourism	3
	Well-being	1
Location of study	Rural	8
	Urban	1
	Generic	6
Wildlife involved	Carnivores	3
	Bats	2
	Primates	1
	Elephants	1
	Raptors	1

Table 6.2 *List of categories and elements within analysed papers.*

Numbers of papers refers to those papers containing those elements.

Category	Element	Number of papers
Socio-environment	Co-occurrence	3
	Direct encounter risk (human factors)	5
	Direct encounter risk (animal factors)	5
	Perceived risk/attitudes	6
Outcomes	Direct effect	6
	Indirect effect	6
	Short term	2
	Long term	2
Impacts	Animal-focused	3
	Human-focused	3
Impact direction	Animal-negative	4
	Animal-positive	3
	Human-negative	2
	Human-positive	4

To critically assess existing frameworks from across disciplines, we categorized the (1) type of article based on the focus, (2) location (urban, rural or generic), (3) wildlife involved and (4) what were the elements within the framework. As outcomes of this assessment, we (1) report key elements of frameworks, (2) identify gaps in frameworks and (3) construct a new framework. We assessed frameworks on three basic categories of elements: (1) environmental factors, including risks and attitudes; (2) outcomes of interactions, which includes direct or indirect effects and short- or long-term effects; and (3) impacts of interactions (Table 6.2, Decker et al. 2010). We additionally include the direction of impacts, because we recognize that interactions can be both positive and negative. In the following section, we discuss these three elements in more detail and how frameworks use them.

6.3.1 Socio-Environmental Factors

Socio-environmental factors place humans and wildlife in situations where they can interact. The majority of frameworks dealt with the direct (or perceived) encounter risk of this happening and what may cause this (Table 6.2). Human-related factors that drive encounter risk include economic and social drivers and human behaviours (White et al. 2009; Dickman 2010). For example, probability of interaction with coyotes (*Canis latrans*) was highest in areas where more residents were employed in industries that require outdoor work, which may simply reflect that these residents spend more time outdoors (Wine et al. 2015). Specific human behaviours can inherently increase probabilities of encounters; for example, most snakebite victims are trying to remove the animals from somewhere they are not wanted (Morandi & Williams 1997). In human–urban macaque (*Macaca fascicularis*) interactions, approximately 25 per cent of interactions are provoked by a human who initiated action towards a macaque (e.g. chasing, pointing at close range, approaching closely; Sha et al. 2009). Encounters will also be driven by both animal and environmental factors (White et al. 2009). Encounters with venomous Eastern brown snakes (*Pseudonaja textilis*) depended on season, time of day, habitat type and weather conditions (wind and air temperature) (Whitaker & Shine 1999), whereas animal–vehicle collisions were dependent on season and time of day (Baker et al. 2007; Chen & Wu 2014). The emphasis in frameworks on environmental factors reflects frequent targets for management. For example, changing human behaviour that may increase conflict (i.e. wildlife feeding, securing waste) is a key target for media and risk communication campaigns (Gore & Knuth 2009). As targets for management, environmental factors are also critical for promoting coexistence. By reducing or focusing on the source of conflict (e.g. food sources; Kurosawa et al. 2003; Kaplan et al. 2011), this has the spin-off benefit of increased wildlife tolerance and acceptance of wildlife in urban areas (crows–humans in Tokyo: Kurosawa et al. 2003).

6.3.2 Risks and Attitudes

A key part of the socio-environment within the analysed papers are the perceived risks and attitudes – the so-called *human dimension*. These are a critical component of most frameworks (Table 6.2), because it is well established that attitudes and risks are an important modifier of all components of the human–wildlife interaction and have important roles in shaping management options and pathways (Kansky et al. 2016). Attitudes are typically rooted in a combination of personal, historical and cultural contexts, alongside current societal information sources including mass media, the internet and education (Manfredo & Dayer 2004; Decker et al. 2010). Attitudes are often the best predictors of response to interactions and to management (McCleery 2009). In urban areas, people often lack historical and cultural contexts for wildlife (Leong 2009), either because they lack direct experience themselves or in their social community (Manfredo et al. 2009; Soga et al. 2016a). Within these contexts, there is the opportunity for information from the media or education programmes to strongly drive attitudes, which in turn can amplify or reduce people's perception of risk (Gore et al. 2005; Dudo et al. 2007; Gore & Knuth 2009; Manfredo et al. 2009).

How people react or respond to encounters is often dependent on their prior expectation or perceptions (Kansky et al. 2016). If an action or encounter is *unexpected*, then reactions can be stronger than if the encounter was *expected*. Indeed, the type of encounter is often critical in shaping attitudes. Nuisance behaviours may be tolerated, but when there is financial damage, this tolerance may be reduced (Hill et al. 2007; Kansky et al. 2014). Where people have *loss of control* of a situation, this often leads to negative attitudes (Thomas & Jones 1997). Stronger reactions to rare (or even perceived) negative encounters or threats are often more important than neutral to positive interactions as they can lead to shifts in attitude. For example, coyote attacks and resulting media coverage led to increased concern about possible threats (Siemer et al. 2014), whilst tolerance to grey wolves declined because of perceived threats to livestock and competition for deer (Treves et al. 2013). Even things such as a new disease entering a location would change how people view and interact with wildlife (Sparkes et al. 2017). Shifts in perceptions and attitudes shape the nature of the response, including changes to management such as increased lethal control (Treves et al. 2013; Sponarksi et al. 2016). Similarly, residents who have a lower positive attitude towards conservation, in turn are less likely to change their behaviour to the benefit of conservation (koala conservation: Shumway et al. 2014). Together, this highlights the importance of attitudes in shaping not only the nature of the interaction but also further behaviours in the future.

6.3.3 Outcomes of Interactions

Many frameworks dealt with the direct and indirect outcomes of wildlife interactions (Table 6.2), i.e. the consequences of the interaction happening. Direct outcomes are typically *visible*, e.g. property damage (e.g. Ogra 2008). Instead, indirect costs are less visible and usually harder to quantify, e.g. increased workloads or lost opportunities (Ogra

2008; Barua et al. 2013) or increased mental health problems (Chowdhury et al. 2016). Categorization into direct and indirect outcomes works equally well for understanding positive interactions (Clark et al. 2014). Direct outcomes may include de-stressing and relaxation (Cox & Gaston 2016) and indirect outcomes for example may include changing appreciation of the local environment. Children who listen to mixes of bird song were found to be appreciative of urban settings (Hedblom et al. 2014) and children given educational material about conservation increased the entire household's knowledge (Damerell et al. 2013). Linked to both direct and indirect interactions is the time frame of such outcomes (Table 6.2). There is an increasing difficulty in assessing outcomes over longer time frames, especially for indirect interactions. For example, recent evidence suggests that nature connectedness in children will determine their later emotional connection to nature as adults (Soga et al. 2016a) and their tolerance to human–wildlife conflict (Hosaka et al. 2017).

6.3.4 Impacts of Interactions

Understanding the ultimate impacts (or consequences) of interactions is typically challenging because the focus tends to be on the specific interaction. Ultimately, the impact represents the broad effect and gives a wider context to the nature of multiple individual interactions. For example, benefits to human health and well-being may be a key outcome (Clark et al. 2014). Yet the benefits may arise from single rare interactions or multiple common interactions and depend heavily on the characteristics of the encounter (Curtin 2010; McIntosh & Wright 2017). These include whom you share the encounter with; sharing an encounter with children, can for example act as an emotional amplifier (Hicks 2016).

From an animal perspective, broad impacts may stem from changes in human behaviours. These might include changes in persecution level or habitat destruction (Dickman 2010) or increased coexistence (e.g. Irrawaddy dolphins (*Orcaella brevirostris*), D'Lima et al. 2014). Thus, impacts are critical in shaping potential management, making them an important factor that feeds back into earlier parts of frameworks. Doing this allows better management of existing or emerging forms of human–wildlife conflict as well as human–wildlife coexistence (White & Ward 2010).

6.4 SHORTCOMINGS OF EXISTING FRAMEWORKS

Though most studies deal with encounters between humans and wild-life, a few address co-occurrence (Table 6.2). Not to be confused with coexistence (Harihar et al. 2013) or connectedness (Schultz 2001), co-occurrence is when humans and animals share some form of spatial proximity though they may be temporally separated in such a way as to avoid any form of interaction (e.g. tigers–humans, Carter et al. 2012). The lack of inclusion of co-occurrence in most frameworks is puzzling, because for any form of encounter to take place, there must be some form of co-occurrence. Existing frameworks possibly assume that some co-occurrence is implicit as they are addressing some form of inter-action; however, this does not address the fact that changing spatial proximity (either the human or the animal) seems an obvious choice for management either to reduce human–wildlife conflict or to promote human–wildlife benefits. Broad-scale management techniques such as fencing (Treves et al. 2009; DeVault et al. 2011), removal of attractant resources (Leblond et al. 2007) or habitat management (Belant 1997; White et al. 2006; Treves et al. 2009) can be used to reduce co-occurrence between humans and wildlife.

A second key limitation of the frameworks stems from the under-standing of the outcomes and impacts (Table 6.2). Most do not consider the time frame of impact possibly because the duration of studies and/or funding is too short. This possibly can be addressed with more careful study design that targets these longer-term impacts. This is not to say that impact itself is not considered, but rather that there is a difference between the immediacy of certain interactions versus the longer-term changes, which are harder to measure (Barua et al. 2013). For example, Ogra (2008) separates direct impacts of elephant damage such as prop-erty damage, from those that are *hidden* (i.e. delayed or indirect). These *hidden* impacts include increased workloads and psychological trauma. Similarly, the interactions between children and nature lead to a greater emotional affinity to biodiversity and in turn an increased willingness to conserve it (Soga et al. 2016b). Few frameworks consider how the impacts can take time to manifest and have consequences stretching into the future; some wildlife diseases for example may take decades to manifest (e.g. *Echinococcosis*: Davidson et al. 2012) or, in some cases, positive interactions have long-term positive benefits. Childhood frequencies of contact with nature were positively related not only to students' emotional connectedness to nature but also to their perceptions of neighbourhood nature (Soga et al. 2016a). One reason that there were few studies considering a longer-term time frame of some impacts may stem simply from the difficulty in studying it and the lack of funding for long-term research. The further the interactions from the ultimate impact, then the greater the influence of multiple confounding variables. Ultimately, until researchers of human–wildlife interactions consider longer-term impacts and therefore plan studies that can examine these appropriately, our understanding of the importance and relevance of these will be limited.

Similarly, the ultimate direction of impact is rarely considered and those that are considered are typically negative (Table 6.2). The emphasis on negative aspects probably stems from the most frameworks being applied within the specific area of human–wildlife conflict. In particular, human–wildlife conflict typically has easily quantifiable costs and clearer direct impacts. However, considering human–wildlife coexistence has numerous potentially beneficial outcomes for both humans and animals. These can include motivation to participate in conservation (Zeppel & Muloin 2008). Urban conservation projects for example have been shown to motivate participants' involvement with nature and engage in activities such as biological recording, wildlife gardening and local volunteering (Hobbs & White 2016). Similarly, people participating in bird feeding or ecotourism show a positive willingness to pay for conservation (Powell & Hamm 2008; Clucas et al. 2015). Ultimately, the impact of interactions can lead to an increased willingness and likelihood of coexistence, either by reducing negative outcomes or increasing positive ones.

6.5 A NEW COMBINED FRAMEWORK

Based on the review of existing frameworks, we present a framework that contains four dimensions: Socio-Environment, Intervention, Outcomes and Effects (Figure 6.1). In addition, the framework can potentially be used for human–wildlife interactions from the perspective of humans as well as wildlife (Bath & Enck 2003).

This framework combines elements of current frameworks with areas currently not addressed. The first dimension is *Socio-Environment*, within which we incorporate co-occurrence of humans and wildlife, encounter probability, attitudes and both real and perceived risk. All are important targets for *Intervention* as education about the benefit and likelihood of risks is an important modifier of the effects of interactions.

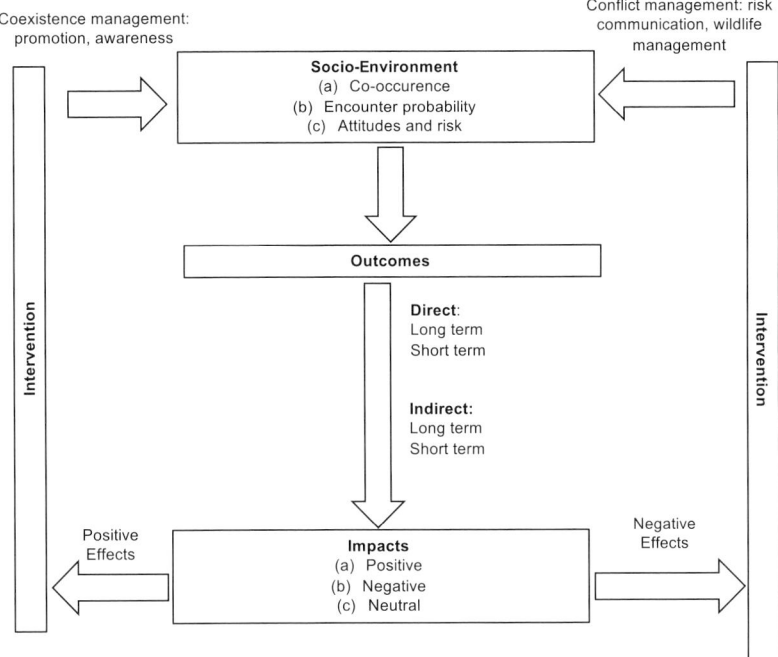

Figure 6.1 New framework containing Socio-Environment, Intervention, Outcomes and Effects of interactions. Arrows indicate the link between elements within an interaction.

The second dimension is *Outcomes*, which is subdivided into both short- and long-term effects and direct and indirect effects. Outcomes ultimately modify the third dimension, *Impacts*, which can be positive, negative or neutral (see 'benign awareness' in Manfredo et al. 2009). The type of *Impact* ultimately determines the need for *Intervention*.

Our framework is not designed to provide a specific recommendation for a certain situation. Instead, its purpose is to provide a structured approach to the problem, ensuring that all aspects, positive, negative and neutral, can be considered together. It is designed in a way that can be scaled from the local level through to broader-level interactions and on a continuum of types and scale of interactions, including the conflict-to-coexistence continuum. In the section below, we apply the framework to a well-studied human–wildlife interaction: human–red fox interactions in urban areas. We illustrate the use of the framework as a conceptual instrument of analysis to understand interactions and outline its potential for future modelling exercises (Text Box 6.1).

Text Box 6.1 Case Study: Urban Red Foxes (*Vulpes vulpes*)

Red foxes are a globally widespread urban species (Soulsbury et al. 2010). They undergo a number of interactions with humans. These include human–wildlife conflict such as rare, but severe encounters, e.g. attacks on people (Cassidy & Mills 2012), attacks on pets (Plumer et al. 2014) and potential source of disease for humans and pets (Soulsbury et al. 2007; Davidson et al. 2012). Conflict also includes more frequent, but relatively minor forms such as noise, fouling and other minor damage during denning (Harris 1985). At the same time, people derive pleasure and enjoyment from seeing foxes in urban areas, including putting food out for them in much the same way as they feed birds (Baker et al. 2004; Baker & Harris 2007).

In Figure 6.2, we populate the framework with interactions based on human–urban fox conflict. In our framework, the likelihood of an encounter is high. This is because urban foxes typically occur at high densities in urban areas, across a range of habitats (Soulsbury et al. 2010). Interactions themselves are not spaced evenly in time. Sightings often peak in autumn/winter (Plumer et al. 2014), which coincides with the dispersal and breeding seasons (Soulsbury et al. 2011), whereas conflict is generally highest in spring/summer, which coincides with denning and cub rearing activities (Harris 1985; Soulsbury et al. 2010). Most problems are direct and short term (Figure 6.2), but have serious outcomes. Recent years have seen a growing perception of urban foxes as pests, fuelled by the media (Cassidy & Mills 2012; Stewart & Cole 2015). Consequently, intervention can take three forms. The first is to reduce the likelihood of encounter probability, e.g. by making gardens less attractive for foxes by reducing potential denning locations, removing food sources and using repellents. The second option is to manage the conflict by providing information to address perceptions of risk and to promote understanding of key periods of conflict (e.g. denning). The third option is that some householders may want to resort to lethal control, e.g. by bringing in a private contractor to remove the foxes.

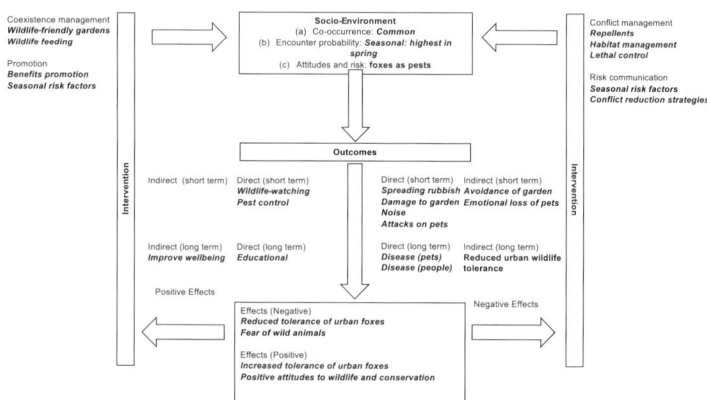

Figure 6.2 Framework populated with human–urban fox interactions.

In Figure 6.2, we also populate the framework for human–urban fox coexistence. Like many interactions, the benefits of human–wildlife interaction are much less well understood and are harder to measure (Soulsbury & White 2015). For example, the pleasure and educational benefits of wildlife watching are hard to quantify (Curtin 2009). Wildlife watching can have significant direct benefits, such as improving physical and mental health (Cox & Gaston 2016; Shanahan et al. 2016). Management of positive interactions is also challenging. Promoting wildlife-friendly gardens is one potential way to enhance interactions (Baker & Harris 2007; Goddard et al. 2013; Hand et al. 2017), but any promotion must come with education about potential risks. Interactions can easily and quickly move from being positive to negative in an urban area. For example, attracting wildlife into one garden may increase conflict in neighbouring households. Appropriate risk communication is therefore key both to positive and negative interactions.

The positive and negative interactions should not be analysed separately, but instead combined to present an overview of interaction as a *whole*. The type and nature of outcomes occur on a continuum, and bringing them together allows a greater understanding of how they influence each other and how any future management or intervention must balance positive and negative outcomes of interactions.

Our framework identifies key elements of the interaction that allows quantification and costing to occur (Table 6.3). Whilst data on cost will inevitably be incomplete, the consideration of financial costs alongside non-monetary costs and benefits will allow for more informed decisions to be made. For example, the costs of control of urban foxes is substantial and is balanced by the considerable benefits derived from wildlife watching (Table 6.3). Where possible, economic valuation can contribute to understanding the relative benefits and costs, but it is recognized that not all the impacts (positive or negative) can be valued robustly in an economic sense, and hence subjective judgements inevitably come into play. The strength of the framework is that it allows these objective and subjective elements to be considered alongside one another in a transparent way to inform decision-making.

6.6 CONCLUSIONS

Human–wildlife interactions are multifaceted and stretch from conflict to the benefits of interactions and coexistence. Assessing and quantifying these interactions can be challenging. After reviewing existing frameworks, we propose a new one that addresses deficiencies in the assessment of

Table 6.3 *Estimates of values related to benefits and costs of urban foxes and associated management*

Element	Values	Comments
Positive effects		
Direct: wildlife watching	$11–$70 per household (willingness to pay)	No economic estimates for urban foxes, but other studies show a considerable willingness to pay for watching garden birds (Brock et al. 2015; Clucas et al. 2015).
Direct: educational	–	Difficult to quantify. Potentially important, but robust methods need to be generated.
Indirect: improve well-being	$40–$86 per person/per hour; $7,300–$15,695/year (based on 30 minutes per day)	No economic estimates for urban foxes or wildlife, but other studies have used proxies to value mental health and stress of access to green space (Munoz & Nimegeer 2012).
Negative effects		
Direct: spreading rubbish	–	Localized problem so cost likely to be minimal.
Direct: damage to gardens	Cost of $0–$600	No estimates, but likely to be low, ranging from inconvenience to significant damage, costed according to replacement costs of plants and structures damaged or undermined (White author data).
Direct: noise	–	No estimates, but likely to be minimal; main noise issues associated with fox breeding season, so temporary impact only.
Direct: attacks on pets	Cost of $4–$12 per household for deaths, and $5 per household for injuries	Fox attacks recorded on cats, guinea pigs, tortoises, rabbits and dogs. Survey in north-west Bristol reported 28 deaths and 69 attacks across 5,500 households (Harris 1981). Overall costs across these households estimated as $20,500–$63,500 for deaths and $25,000 for injuries (White author data)
Direct: disease (pets)	Cost of treatment of dog for sarcoptic mange up to $370	There is a range of parasites carried by foxes that are of potential harm to humans, many of which are increasing in distribution and abundance. Sarcoptic mange can be transmitted from foxes to dogs; infection in cats is exceptionally rare. Risk of *Echinococcus* from foxes to pets is perceived to have increased across Europe with growth in fox populations but overall it is still considered low. Most pets are not adversely affected by infection but it can cause serious illness in humans. Some targeted management through anti-helminthic baiting has been successful in reducing *Echinococcus* in urban foxes on the European mainland (Hegglin et al. 2015).

Direct: disease (people)	Cost of treatment per infected person per year $3,700	See above. Potential pathway of infection from foxes through pets to humans, and some evidence from European studies for an association between human infections and prevalence of these parasites in the fox population at a general level. $0–$12 benefit in saved medical costs per person resulting from baiting with praziquantel (Frank et al. author data).
Indirect: avoidance of garden	–	Infrequent impact, so likely to be minimal.
Indirect: emotional loss of pets	–	Not possible to quantify robustly.
Interventions (management)		
Lethal control	Cost of $600–$12,200 per fox killed	Highly variable cost depending on method. Cost per fox cage-trapped in Bristol in early 1990s estimated at average of $500–$1,300 in spring and summer, and $1,200–$12,200 in autumn and winter, depending on fox density, based on cost per trap night of $16 (White author data). Cost to household to hire pest controller to kill fox in London is $500 for a cull session plus $90 per fox killed.
Repellent	Cost of up to $610	Repellents vary in their effectiveness, and may be physical, electrical, chemical or ultrasonic, but there is no completely effective deterrent available. Price varies according to method used (White author data).
Habitat modification	–	Not usually carried out specifically for fox control, but some habitat modification, e.g. increased provision of green infrastructure, can result in more favourable fox habitat and hence higher risk of contact with people.

interactions. The proposed framework does not provide a specific recommendation for a certain situation, but instead gives a structured approach to the problem, ensuring that all aspects, positive, neutral and negative, can be considered together. Including economic valuation where possible provides a subjective assessment of the value of the interaction and allows relevant interventions to be considered if necessary. Ultimately, this allows both conflict and coexistence to be combined and the ability to mitigate and reduce conflict whilst at the same time promoting coexistence.

6.7 RECOMMENDATIONS AND FUTURE DIRECTIONS

- Human–wildlife interactions should be considered beyond short-term direct impacts to a longer-term perspective, with greater attempts to assess, evaluate and plan for indirect and positive outcomes of interactions that may occur beyond the immediate time frame. Stop, think and educate is a key motto to think about when addressing any type of human–wildlife interactions.
- The importance of attitudes and perceptions in shaping the outcomes of human–wildlife interactions should be an important consideration in decision-making. Education to promote more informed attitudes surrounding interventions is likely to have important benefits for wildlife management and wildlife coexistence.
- Finding ways in which human–wildlife co-occurrence can occur with minimal conflict should be a key part of management, both to reduce conflict overall but also to promote positive benefits. Greater research effort is needed to explore the positive benefits of human–wildlife interactions in terms of their specific benefits and the mechanisms behind these.
- Urban wildlife is a critical component of urban ecosystems. Promoting coexistence based on improved understanding is therefore essential to ensure wildlife persists in these environments and that the positive role of wildlife can continue.

6.8 References

Baker, P. J., Dowding, C. V., Molony, S. E., White, P. C. L. & Harris, S. (2007). Activity patterns of urban red foxes (*Vulpes vulpes*) reduce the risk of traffic-induced mortality. *Behavioral Ecology*, 18(4), 716–24.

Baker, P., Funk, S., Harris, S., Newman, T., Saunders, G. & White, P. C. L. (2004). The impact of human attitudes on the social and spatial organization of urban foxes (*Vulpes vulpes*) before and after an outbreak of sarcoptic mange. In W. W. Shaw, L. K. Harris & L. VanDruff, eds., *Proceedings of the 4th International Symposium on Urban Wildlife Conservation*, School of Natural Resources, College of Agriculture and Life Sciences. Tucson: University of Arizona, pp. 153–63.

Baker, P. J. & Harris, S. (2007). Urban mammals: What does the future hold? An analysis of the factors affecting patterns of use of residential gardens in Great Britain. *Mammal Review*, 37, 297–315.

Barua, M., Bhagwat, S. A. & Jadhav, S. (2013). The hidden dimensions of human–wildlife conflict: Health impacts, opportunity and transaction costs. *Biological Conservation*, 157, 309–16.

Bath, A. J. & Enck, J. W. (2003). Wildlife–human interactions in National Parks in Canada and the USA. National Park Service; and USA Department of the Interior *All U.S. Government Documents (Utah Regional Depository)*. Paper 424. Available from http://digitalcommons.usu.edu/govdocs/424 (accessed 24 October 2018).

Belant, J. L. (1997). Gulls in urban environments: Landscape-level management to reduce conflict. *Landscape & Urban Planning*, 38(3–4), 245–58.

Brock, M., Perino, G. & Sugden, R. (2015). The warden attitude: An investigation of the value of interaction with everyday wildlife. *Environmental & Resource Economics*, 67(1), 127–55.

Carter, N. H., Shrestha, B. K., Karki, J. B., Pradhan, N. M. B. & Liu, J. (2012). Coexistence between wildlife and humans at fine spatial scales. *Proceedings of the National Academy of Sciences*, 109(38), 15360–5.

Cassidy, A. & Mills, B. (2012). Fox tots attack shock: Urban foxes, mass media and boundary-breaching. *Environmental Communication*, 6(4), 494–511.

Chen, X. & Wu, S. (2014). Examining patterns of animal–vehicle collisions in Alabama, USA. *Human–Wildlife Interactions*, 8(2), 235–44.

Chowdhury, A. N., Mondal, R., Brahma, A. & Biswas, M. K. (2016). Ecopsychosocial aspects of human–tiger conflict: An ethnographic study of tiger widows of Sundarban Delta, India. *Environmental Health Insights*, 10, 1–29.

Clark, N. E., Lovell, R., Wheeler, B. W., Higgins, S. L., Depledge, M. H. & Norris, K. (2014). Biodiversity, cultural pathways, and human health: A framework. *Trends in Ecology & Evolution*, 29(4), 198–204.

Clucas, B., Rabotyagov, S. & Marzluff, J. M. (2015). How much is that birdie in my backyard? A cross-continental economic valuation of native urban songbirds. *Urban Ecosystems*, 18(1), 251–66.

Cox, D. T. C. & Gaston, K. J. (2016). Urban bird feeding: Connecting people with nature. *PLoS ONE*, 11(7), e0158717.

Curtin, S. (2009). Wildlife tourism: The intangible, psychological benefits of human–wildlife encounters. *Current Issues in Tourism*, 12(5), 451–74.

Curtin, S. (2010). What makes for memorable wildlife encounters? Revelations from 'serious' wildlife tourists. *Journal of Ecotourism*, 9(2), 149–68.

Damerell, P., Howe, C. & Milner-Gulland, E. J. (2013). Child-orientated environmental education influences adult knowledge and household behaviour. *Environmental Research Letters*, 8(1), 015016.

Davidson, R. K., Romig, T., Jenkins, E., Tryland, M. & Robertson, L. J. (2012). The impact of globalisation on the distribution of *Echinococcus multilocularis*. *Trends in Parasitology*, 28(6), 239–47.

Davies, Z. G., Fuller, R. A., Dallimer, M., Loram, A. & Gaston, K. J. (2012). Household factors influencing participation in bird feeding activity: A national scale analysis. *PLoS ONE*, 7(6), e39692.

Decker, D. J., Evensen, D. T., Siemer, W. F., Leong, K. M., Riley, S. J., Wild, M. A., Castle, K. T. & Higgins, C. L. (2010). Understanding risk perceptions to enhance communication about human–wildlife interactions and the impacts of zoonotic disease. *ILAR Journal*, 51(3), 255–61.

DeVault, T. L., Belant, J. L., Blackwell, B. F. & Seamans, T. W. (2011). Interspecific variation in wildlife hazards to aircraft: Implications for airport wildlife management. *Wildlife Society Bulletin*, 35(4), 394–402.

Dickman, A. J. (2010). Complexities of conflict: The importance of considering social factors for effectively resolving human–wildlife conflict. *Animal Conservation*, 13(5), 458–66.

D'Lima, C., Marsh, H., Hamann, M., Sinha, A. & Arthur, R. (2014). Positive interactions between Irrawaddy dolphins and artisanal fishers in the Chilika Lagoon of Eastern India are driven by ecology, socioeconomics, and culture. *Ambio*, 43(5), 614–24.

Dudo, A. D., Dahlstrom, M. F. & Brossard, D. (2007). A risk-related assessment of avian influenza coverage in U.S. newspapers. *Science Communication*, 28 (4), 429–54.

Frank, B. (2016). Human–wildlife conflicts and the need to include tolerance and coexistence: An introductory comment. *Society & Natural Resources*, 29, 738–43.

Fuller, R. A., Irvine, K. N., Davies, Z. G., Armsworth, P. R. & Gaston, K. J. (2012). Interactions between people and birds in urban landscapes. *Studies in Avian Biology*, 45, 249–66.

Goddard, M. A., Dougill, A. J. & Benton, T. G. (2013). Why garden for wildlife? Social and ecological drivers, motivations and barriers for biodiversity management in residential landscapes. *Ecological Economics*, 86, 258–73.

Gore, M. L. & Knuth, B. A. (2009). Mass media effect on the operating environment of a wildlife-related risk-communication campaign. *Journal of Wildlife Management*, 73(8), 1407–13.

Gore, M. L., Siemer, W. F., Shanahan, J. E., Schuefele, D. & Decker, D. J. (2005). Effects on risk perception of media coverage of a black bear-related human fatality. *Wildlife Society Bulletin*, 33(2), 507–16.

Hand, K. L., Freeman, C., Seddon, P. J., Recio, M. R., Stein, A. & van Heezik, Y. (2017). The importance of urban gardens in supporting children's biophilia. *Proceedings of the National Academy of Sciences*, 114(2), 274–9.

Harihar, A., Chanchani, P., Sharma, R. K., Vattakaven, J., Gubbi, S., Pandav, B. & Noon, B. (2013). Conflating 'co-occurrence' with 'coexistence'. *Proceedings of the National Academy of Sciences*, 110(11), E109.

Harris, S. (1981). The food of suburban foxes (*Vulpes vulpes*), with special reference to London. *Mammal Review*, 11, 151–68.

Harris, S. (1985). Humane control of foxes. In D. Britt, ed., *Humane Control of Land Mammals and Birds*. Potters Bar: UFAW, pp. 63–74.

Hedblom, M., Heyman, E., Antonsson, H. & Gunnarsson, B. (2014). Bird song diversity influences young people's appreciation of urban landscapes. *Urban Forestry & Urban Greening*, 13, 469–74.

Hegglin, D., Bontadina, F. & Deplazes, P. (2015).Human–wildlife interactions and zoonotic transmission of *Echinococcus multilocularis. Trends in Parasitology*, 31(5), 167–73.

Hicks, J. R. (2016). Children as an emotional amplifier in Northern Illinois white-tailed deer interactions. *The Journal of Environmental Education*, 47(4), 287–95.

Hill, N. J., Carbery, K. A. & Deane, E. M. (2007). Human–possum conflict in urban Sydney, Australia: Public perceptions and implications for species management. *Human Dimensions of Wildlife*, 12(2), 101–13.

Hobbs, S. J. & White, P. C. L. (2016). Achieving positive social outcomes through participatory urban wildlife conservation projects. *Wildlife Research*, 42(7), 607–17.

Hosaka, T. & Numata, S. (2016). Spatiotemporal dynamics of urban green spaces and human–wildlife conflicts in Tokyo. *Scientific Reports*, 6, art. 30911.

Hosaka, T., Sugimoto, K. & Numata, S. (2017). Effects of childhood experience with nature on tolerance of urban residents toward hornets and wild boars in Japan. *PLoS ONE*, 12(4), e0175243.

Jones, D. N. & Reynolds, J. S. (2008). Feeding birds in our towns and cities: A global research opportunity. *Journal of Avian Biology*, 39, 265–71.

Kaplan, B. S., O'Riain, M. J., van Eeden, R. & King, A. J. (2011). A low-cost manipulation of food resources reduces spatial overlap between baboons (*Papioursinus*) and humans in conflict. *International Journal of Primatology*, 32(6), 1397–1412.

Kansky, R., Kidd, M. & Knight, A. T. (2014). Meta-analysis of attitudes toward damage-causing mammalian wildlife. *Conservation Biology*, 28(4), 924–38.

Kansky, R., Kidd, M. & Knight, A. T. (2016). A wildlife tolerance model and case study for understanding human wildlife conflicts. *Biological Conservation*, 201, 137–45.

Kretser, H. E., Sullivan, P. J. & Knuth, B. A. (2008). Housing density as an indicator of spatial patterns of reported human–wildlife interactions in Northern New York. *Landscape & Urban Planning*, 84, 282–92.

Kurosawa, R., Kanai, Y., Matsuda, M. & Okuyama, M. (2003). Conflict between humans and crows in greater Tokyo: Garbage management as a possible solution. *Global Environmental Research*, 7(2), 139–48.

Jones, D. N. & Reynolds, J. S.(2008). Feeding birds in our towns and cities: A global research opportunity. *Journal of Avian Biology*, 39, 265–71.

Leblond, M., Dussault, C., Ouellet, J. P., Poulin, M., Courtois, R. & Fortin, J. (2007). Management of roadside salt pools to reduce moose–vehicle collisions. *Journal of Wildlife Management*, 71(7), 2304–10.

Leong, K. M. (2009). The tragedy of becoming common: Landscape change and perceptions of wildlife. *Society & Natural Resources*, 23, 111–27.

Loram, A., Tratalos, J., Warren, P. H. & Gaston, K. J. (2007). Urban domestic gardens (X): The extent & structure of the resource in five major cities. *Landscape Ecology*, 22, 601–15.

Magle, S. B., Lehrer, E. W. & Fidino, M. (2016). Urban mesopredator distribution: Examining the relative effects of landscape and socioeconomic factors. *Animal Conservation*, 19, 163–75.

Manfredo, M. J. & Dayer, A. A. (2004). Concepts for exploring the social aspects of human–wildlife conflict in a global context. *Human Dimensions of Wildlife*, 9(4), 1–20.

Manfredo, M. J., Teel, T. L. & Henry, K. L. (2009). Linking society and environment: A multilevel model of shifting wildlife value orientations in the Western United States. *Social Science Quarterly*, 90(2), 407–27.

McCleery, R. A. (2009). Improving attitudinal frameworks to predict behaviors in human–wildlife conflicts. *Society & Natural Resources*, 22(4), 353–68.

McKinney, M. L. (2008). Effects of urbanization on species richness: A review of plants and animals. *Urban Ecosystems*, 11(2), 161–76.

McIntosh, D. & Wright, P. A. (2017). Emotional processing as an important part of the wildlife viewing experience. *Journal of Outdoor Recreation & Tourism*, 18, 1–9.

Morandi, N. & Williams, J. (1997). Snakebite injuries: Contributing factors and intentionality of exposure. *Wilderness Environment Medicine*, 8(3), 152–5.

Morzillo, A., de Beurs, K. & Martin-Mikle, C. (2014). A conceptual framework to evaluate human–wildlife interactions within coupled human and natural systems. *Ecology & Society*, 19, art. 44.

Munoz, S. & Nimegeer, A. (2012). *Evaluating and Measuring the Impact of Health Promotion Activities in Hospital Grounds*. Inverness: Centre for Rural Health, University of the Highlands and Islands.

Nyhus, P. J. (2016). Human–wildlife conflict and coexistence. *Annual Review of Environment and Resources*, 41, 143–71.

Ogra, M. V. (2008). Human–wildlife conflict and gender in protected area borderlands: A case study of costs, perceptions, and vulnerabilities from Uttarakhand (Uttaranchal), India. *Geoforum*, 39(3), 1408–22.

Plumer, L, Davison, J. & Saarma, U. (2014). Rapid urbanization of red foxes in Estonia: Distribution, behaviour, attacks on domestic animals, and health-risks related to zoonotic diseases. *PLoS ONE*, 9, e115124.

Poessel, S. A., Gese, E. M. & Young, J. K. (2017). Environmental factors influencing the occurrence of coyotes and conflicts in urban areas. *Landscape & Urban Planning*, 157, 259–69.

Powell, R. B. & Ham, S. H. (2008). Can ecotourism interpretation really lead to pro-conservation knowledge, attitudes and behaviour? Evidence from the Galapagos Islands. *Journal of Sustainable Tourism*, 16(4), 467–89.

Ramalho, C. E. & Hobbs, R. J. (2012). Time for a change: Dynamic urban ecology. *Trends in Ecology & Evolution*, 27(3), 179–88.

Reynolds, P. C. & Braithwaite, D. (2001). Towards a conceptual framework for wildlife tourism. *Tourism Management*, 22(1), 31–42.

Sandifer, P. A., Sutton-Grier, A. E. & Ward, B. P. (2015). Exploring connections among nature, biodiversity, ecosystem services, and human health and well-being: Opportunities to enhance health and biodiversity conservation. *Ecosystem Services*, 12, 1–15.

Schultz, P. W. (2001). Assessing the structure of environmental concern: Concern for self, other people, and the biosphere. *Journal of Environmental Psychology*, 21(4), 1–13.

Seto, K. C., Sánchez-Rodríguez, R. & Fragkias, M. (2010). The new geography of contemporary urbanization and the environment. *Annual Review of Environment & Resources*, 35, 167–94.

Sha, J. C. M., Gumert, M. D., Lee, B. P.Y.-H., Jones-Engel, L., Chan, S. & Fuentes, A. (2009). Macaque–human interactions and the societal perceptions of macaques in Singapore. *American Journal of Primatology*, 71, 825–39.

Shanahan, D. F., Bush, R., Gaston, K. J., Lin, B. B., Dean, J., Barber, E. & Fuller, R. A. (2016). Health benefits from nature experiences depend on dose. *Scientific Reports*, 6, art. 28551.

Shumway, N., Seabrook, L., McAlpine, C. & Ward, P. (2014). A mismatch of community attitudes and actions: A study of koalas. *Landscape & Urban Planning*, 126, 42–52.

Siemer, W. F., Decker, D. J., Shanahan, J. E. & Wieczorek-Hudenko, H. A. (2014). How do suburban coyote attacks affect residents' perceptions? Insights from a New York case study. *Cities and the Environment (CATE)*, 7, 7.

Smith, R. M., Gaston, K. J., Warren, P. H. & Thompson, K. (2005). Urban domestic gardens (V): Relationships between land cover composition, housing and landscape. *Landscape Ecology*, 20(2), 235–53.

Soga, M., Gaston, K. J., Koyanagi, T. F., Kurisu, K. & Hanaki, K. (2016a). Urban residents' perceptions of neighbourhood nature: Does the extinction of experience matter? *Biological Conservation*, 203, 143–50.

Soga, M., Gaston, K. J., Yamaura, Y., Kurisu, K. & Hanaki, K. (2016b). Both direct and vicarious experiences of nature affect children's willingness to conserve biodiversity. *International Journal of Environmental Research & Public Health*, 13(6), 529.

Soulsbury, C. D., Baker, P. J., Iossa, G. & Harris, S. (2010). Red foxes (*Vulpes vulpes*). In S. D. Gehrt, S. P. D. Riley and B. L. Cypher, eds., *Urban Carnivores: Ecology, Conflict, and Conservation*. Baltimore, MD: John Hopkins University Press, pp. 63–75.

Soulsbury, C. D., Iossa, G., Baker, P. J., Cole, N. C., Funk, S. M. & Harris, S. (2007). The impact of sarcoptic mange *Sarcoptes scabiei* on the British fox *Vulpes vulpes* population. *Mammal Review*, 37, 278–96.

Soulsbury, C. D., Iossa, G., Baker, P. J., White, P. C.L. & Harris, S. (2011). Behavioral and spatial analysis of extraterritorial movements in red foxes (*Vulpes vulpes*). *Journal of Mammalogy*, 92(1), 190–9.

Soulsbury, C. D. & White, P. C. L. (2015). Human–wildlife interactions in urban areas: A review of conflicts, benefits and opportunities. *Wildlife Research*, 42 (7), 541–53.

Sparkes, J., Ballard, G., Fleming, P. & Brown, W. (2017). Social, conservation and economic implications of rabies in Australia. *Australian Zoologist*, 38(3), 457–63.

Sponarski, C. C., Vaske, J. J., Bath, A. J. & Loeffler, T. A. (2016). Changing attitudes and emotions toward coyotes with experiential education. *Journal of Environmental Education*, 47(4), 296–306.

Stewart, K. & Cole, M. (2015). The creation of a killer species. *Critical Animal and Media Studies: Communication for Nonhuman Animal Advocacy*, 77, 124.

Sun, C., Mingie, J. C., Petrolia, D. R. & Jones, W. D. (2015). Economic impacts of nonresidential wildlife watching in the United States. *Forest Science*, 61, 46–54.

Teixeira, B., Hirsch, A., Goulart, V. D., Passos, L., Teixeira, C. P., James, P. & Young, R. (2015a). Good neighbours: Distribution of black-tufted marmoset (*Callithrix penicillata*) in an urban environment. *Wildlife Research*, 42(7), 579–89.

Teixeira, C. P., Passos, L., Goulart, V. D., Hirsch, A., Rodrigues, M. & Young, R. J. (2015b). Evaluating patterns of human–reptile conflicts in an urban environment. *Wildlife Research*, 42(7), 570–8.

Thomas, K. L. & Jones, N. D. (1997). Control versus chaos: A unique suburban wildlife conflict in Australia. *Human Dimensions of Wildlife*, 2, 50–1.

Treves, A., Naughton-Treves, L. & Shelley, V. (2013). Longitudinal analysis of attitudes toward wolves. *Conservation Biology*, 27(2), 315–23.

Treves, A., Wallace, R. B. & White, S. (2009). Participatory planning of interventions to mitigate human–wildlife conflicts. *Conservation Biology*, 23, 1577–87.

United Nations. (2014). *World Urbanization Prospects: The 2014 Revision, Highlights*. New York: United Nations (ST/ESA/SER.A/352).

Whitaker, P. B. & Shine, R. (1999). When, where and why do people encounter Australian brownsnakes (*Pseudonaja textilis: Elapidae*)? *Wildlife Research*, 26 (5), 675–88.

White, J. G., Gubiani, R., Smallman, N., Snell, K. & Morton, A. (2006). Home range, habitat selection and diet of foxes (*Vulpes vulpes*) in a semi-urban riparian environment. *Wildlife Research*, 33(3), 75–180.

White, P. C. L. & Ward, A. I. (2010). Interdisciplinary approaches for the management of existing and emerging human–wildlife conflicts. *Wildlife Research*, 37(8), 623–9.

White, R. M., Fischer, A., Marshall, K., Travis, J. M., Webb, T. J., Di Falco, S., Redpath, S. M. & Van der Wal, R. (2009). Developing an integrated conceptual framework to understand biodiversity conflicts. *Land Use Policy*, 26(2), 242–53.

Wine, S., Gagné, S. A. & Meentemeyer, R. K. (2015). Understanding human–coyote encounters in urban ecosystems using citizen science data: What do socioeconomics tell us? *Environmental Management*, 55(1), 159–70.

Zelenski, J. M., Dopko, R. L. & Capaldi, C. A. (2015). Cooperation is in our nature: Nature exposure may promote cooperative and environmentally sustainable behavior. *Journal of Environmental Psychology*, 42, 24–31.

Zeppel, H. & Muloin, S. (2008). Conservation benefits of interpretation on marine wildlife tours. *Human Dimensions of Wildlife*, 13(4), 280–94.

7

Predators in Human Landscapes

KETIL SKOGEN, SUNETRO GHOSAL, SILJE SKULAND
AND SIDDHARTHA KRISHNAN

Much of the literature on human–wildlife interactions focuses on biology and the term *conflict* is often used as a synonym for the material impact on agriculture and livestock (see e.g. Treves & Karanth 2003; Inskip & Zimmermann 2009). Social science studies often concentrate on attitudes towards specific species (e.g. Bruskotter et al. 2007; Ericsson et al. 2008; Gusset et al. 2008), without necessarily accounting for the social and cultural contexts that shape them. People's views are measured as *negative* or *positive, for* or *against*, etc. (e.g. Saberwal et al. 1994; Bjerke et al. 1998; Ericsson & Heberlein 2003; Frank 2016). However, people's views on large carnivores – in general, their presence in a specific place or their conservation – are characterized by ambivalence, internal dilemmas and ambiguity rather than stable valuations (e.g. Skogen et al. 2017, pp 76–114).

Studies have shown that conflicts over large carnivores are embedded in deeper societal tensions – tensions that may have little to do with the predators per se (e.g. Skogen 2001; Sjölander-Lindqvist 2007; Skogen et al. 2008; Skogen et al. 2017). Some people in rural areas view protection of predators as part of a series of assaults on their

This chapter is partly a synthesis of the following articles:

Ghosal, S., Skogen, K. and Krishnan, S. (2015). Locating human–wildlife interactions: Landscape constructions and responses to large carnivore conservation in India and Norway. *Conservation and Society*, 13(3), 265–74.

Skuland, S. E. and Skogen, K. (2014). Rovdyr i menneskenes landskap. *Tidsskriftet Utmark*. http://utmark.nina.no/Portals/utmark/utmark_old/utgivelser/pub/2014–1% 262%26S/ordin/Skuland_Skogen_UTMARK_1%262.html.

communities and ways of life, whereas others see the return of predators as welcome rewilding of a degraded nature. In this chapter, we
address one particular aspect and discuss how human–carnivore relations are shaped by the social construction of landscapes (how the land
is perceived and attributed meaning) through which the same piece of
land may be infused with different values by different groups of people.
We will show how these social constructions reflect and respond to
change – physical and cultural – and how this impinges on the perceived
place for large carnivores. An important message is that views on the
place of large carnivores in a particular area may depend as much on
the understanding of that area as a landscape as on attitudes towards the
carnivore species in question.

Although some publications have introduced interpretations of the
landscape as a significant factor influencing views on large carnivore
presence (e.g. Sjölander-Lindqvist 2007; Figari & Skogen 2011), we
argue that a specific landscape perspective has mostly been lacking. To
understand the potential for coexistence between humans and large
carnivores, we will engage with an array of literature from social sciences (e.g. human geography, social anthropology) on the social construction of landscapes (e.g. Bender 1993; Cronon 1995; Olwig 1996;
Ingold 2000) to explore implications for large carnivore presence, and
thus for large carnivore conservation. We will do this based on studies
carried out in places as different as India and Norway. (For comprehensive descriptions of study sites and methods, see Ghosal et al. 2015;
Skogen et al. 2017.) However, while we explore the impact of social
constructions of landscapes on responses to large carnivores, we do not
claim that social constructions alone account for these responses, and
we recognize the significance of the material impact of changes in land
use and conservation policies. This obviously includes economic and
other practical problems that large carnivores may cause, for example to
livestock herders and – as in the case of Norway – hunters. Our purpose
here is to highlight one dimension of people's valuation of large carnivores that is crucial to understanding human–carnivore interactions,
and yet has received limited research attention.

The study sites are different in many ways. The socio-economic
contrasts between India and Norway are extreme, and cultural diversity
in India is very different from rural Norway. However, one rationale for
comparative research is to identify 'social mechanisms' by juxtaposing
different social contexts, and look for similar social processes that lead to
similar outcomes. A social mechanism in this sense may be described

broadly as a constellation of factors that regularly – but not necessarily – produce specific outcomes (Hedström 2005). With this background, the main question we ask is this: How do social constructions of landscapes influence the responses to the presence – and conservation – of large carnivores? To answer, we draw on studies that dealt with the social construction of landscapes in four study sites (two in each country), and how these constructions were tied – in various ways – to tasks performed on the land. The studies were conducted against a background of change (physical, social and cultural) in all study locations, and we observed how the presence of large carnivores was perceived in different social groups from conflict to coexistence.

One Indian case study dealt with the Toda, an indigenous tribal community in the Nilgiri Mountains of Tamil Nadu. Here, buffalo (*Bubalus bubalis*) herding on the grasslands of the Nilgiri plateau used to be the economic backbone and a cultural fundament. However, wattle, eucalyptus and pine plantations have created dense forests where the landscape used to be open, and buffalo herding is now difficult. Tiger (*Panthera tigris*) and leopard (*Panthera pardus*) attacks on buffalo have become more common, although not necessarily intense, and the Todas associate the carnivore problem with the afforestation. Once nonchalant towards tigers, they are now more hostile to these carnivores, which they often claim have been secretly introduced by the government (Ghosal et al. 2015). The other Indian case is Akole in Maharashtra, where the expansion of irrigation has enabled development of sugarcane plantations and a sugar processing industry from the 1980s. The socio-economic and ethnic composition of the region is complex, but many social groups have seen improved living conditions following the industrialization of the valley. Leopards have found the cane fields a suitable habitat, where they prey exclusively on dogs and small livestock. Despite this, the leopards are generally tolerated in Akole, and there have been no demands for their removal.

Trysil is one of the Norwegian cases. Situated in the south-east against the Swedish border, Trysil is traditionally a logging community with some wood processing industry and limited livestock production. Recent decades have seen a formidable winter tourism development, but despite this, the population has declined sharply. Wolves returned to this part of Norway around 1990, and in Trysil we see that social groups with roots in traditional land use do not welcome them. On the one hand, rumours about secret reintroduction flourish, strikingly similar to the stories told by the Todas (Ghosal et al. 2015). One the other hand, leisure home owners (many who have owned their leisure homes/cabins in Trysil for generations) and people with weaker ties to the resource economy (e.g. many youngsters) have a much more positive view.

Finally, Halden, on the extreme southern end of Norway's border with Sweden, is a fair-sized industrial town surrounded by a rural landscape with forests and agriculture. Wolves returned to this area in the 1980s. As the forested parts of Halden are close to the town and the whole region is much more urbanized than the region where Trysil is found, a considerable part of the current rural population is not rooted in the old resource economy, and enjoys the forest as *wilderness*. The residents see the wolves as an organic part of the wild nature they cherish, and *wildness* is part of the reason why they have chosen to live where they do.

The four sites host large carnivores in multi-use landscapes, with a complex mix of distinct and shared features. All sites have seen relatively limited material damage from large carnivores but different levels of conflict. There are strong anti-predator sentiments with more or less open conflicts in Trysil and the Nilgiris. The other two sites present a mixed picture – Akole appears to have the lowest overall conflict level, but strong pro-carnivore sentiments prevail in Halden too. One might say that they represent different positions on the conflict-to-coexistence continuum, and we set out to explain how landscape interpretations contribute to this.

In everyday speech, the word *landscape* often refers to qualities of the physical environment such as mountains or forests, and distinguishes between different areas based on physical differences (*mountain landscape, forest landscape*, etc.). The term thus helps us draw up spatial boundaries (Wylie 2007). Several authors explain how the relation between humans and animals can also unfold as a spatial relationship (see e.g. Philo & Wilbert 2000). Establishing zones is typical of the way in which we organize the place of animals – wild or domestic – in relation to humans: some spaces, buildings, etc. are suitable for animals, while others are not. People usually have opinions on which spaces are appropriate for different animal groups (Philo & Wilbert 2000). Consequently, we must explore how conflicts ensuing from the return of large carnivores may be connected to different understandings of landscapes: landscapes as spaces that are – or are not – appropriate for certain types of animals.

7.1 LANDSCAPE IN THE SOCIAL SCIENCES

To explore the links between landscape interpretations and views on large carnivores, we lean on a conceptual framework that treats landscape as embodied practice, i.e. landscapes being constructed through valuation of tasks and activities performed on the land (Bender 1993; Olwig 1996; Ingold 2000). Thus, while the physicality of the land and its wildlife certainly influence different interpretations, there are also important social, cultural and historical factors that shape the landscape as a social construction. British social anthropologist Tim Ingold claims that to understand how people see their landscape, we must first understand how they relate to it through their practices. When people 'read' *their* landscape, they see something that outsiders do not see. To dwell in a landscape means possessing knowledge about it. Ingold writes: 'If ... every object is to be regarded as a *collapsed act*, then the landscape as a whole must likewise be understood as ... a pattern of activities *collapsed* into an array of features' (Ingold 2000, p. 198). Even if the landscape is primarily perceived as a completed form, we should understand it as a process. A landscape is continuously a work in progress.

Ingold (2000) introduces the concept of *taskscape* to emphasize the dynamic nature of the landscape. Dwelling in the landscape and using it means that life histories blend with the texture of the landscape and with the life cycles of plants and animals. Because the landscape is primarily experienced as form (one observes and interprets the physical land-scape), it is possible to read enduring signs long after the activity that created them has ceased. According to Ingold (2000), the landscape remains a material expression of the lives and work of generations that dwelled in it, and who offset something of themselves in it. The land-scape tells a story of life and work.

In our analysis, the concept of *taskscape* serves as a gateway to under-standing people's relationship with the landscape. The concept is useful because it embeds practices and meanings in the landscape, and shows that a landscape is not simply a piece of land. Using the landscape and taking it into possession (in a symbolic if not legal meaning) are two sides to the same coin. The concept of taskscape transcends the antagonism between concrete practices, or work, and the mental representation of the landscape – between practice and meaning. Yet the concept of taskscape as we understand it here is not restricted to practices in the landscape tied to work and everyday tasks. Recreational activities, some of which are not easily distinguished from everyday practices, in the landscape also con-tribute to the constitution of taskscapes (Ingold 2000).

Although the concept of taskscape has a significant analytical poten-tial, not least in relation to this chapter's topic, there are also some limitations if we want to explain how interpretations of landscapes may develop in *late modernity* (e.g. Giddens 1991). In an epoch where a growing portion of the population is not engaged in any form of every-day practice in the landscapes that we might understand as *natural*, and where ties to the practices of earlier generations are weakened, it is not always reasonable to link interpretations of landscapes to tasks. Outdoor recreation activities may be understood in this way; however, landscapes are also experienced and interpreted on a different basis, as media and extreme mobility enable people to relate to landscapes where they have no tangible or self-experienced relation. For example, people may feel very strongly about rainforests in parts of the world they will never visit – or if they do visit, it is only to briefly observe.

Simultaneously, a view of nature where resource utilization is seen as fundamentally harmful has emerged. Here, *tasks* leading to human impact on a vulnerable nature may be understood as detrimental (Mac-naghten & Urry 1998; Figari & Skogen 2011). Human practices in landscapes are not irrelevant seen from this perspective, but it may be that nature is most highly valued if it is not affected by such practices. Thus, human transformation of the land is still important as a compon-ent in the landscape interpretation, but ultimately understood as the *negation* of what constitutes a good or benevolent landscape. Perhaps we can think of it as an *anti-taskscape* or a *non-taskscape*, that is, a landscape not perverted by humans.

In this chapter, we will identify different landscape interpretations (different taskscapes), that is, different ways to perceive and think about the landscape. We must think about these as *ideal types*. An ideal type is a description of a phenomenon that is in a sense distilled, so that signifi-cant features are emphasized (see e.g. Calhoun et al. 2007). Reality tends to be messy, and the phenomena that are represented by our ideal types rarely exist in a *pure* form. However, ideal types are useful tools in order to systematize observations – as long as we remember that they are just that: analytical tools and not exact reflections of reality.

We understand the social construction of landscapes to have a distinct normative dimension. Barbara Bender (1998) points out that people who are concerned about landscape change contract normative landscapes. They have strong ideas about the landscape as they think it *should* be. This means that perception and interpretation of a landscape will draw on images of what the landscape ought to be, and not only what it is. James D. Proctor (1998) writes that all landscapes are *moral* because we cannot describe landscapes in their essence. By introducing the term *moral landscapes*, Proctor emphasizes that ideas about *the benign* are embedded in all socially constructed landscapes – either as dimensions of the landscape such as it is, or as its negation. As we shall see, images of benign landscapes figure prominently in the landscape interpretations we came across in our studies.

7.2 CONSTRUCTING LANDSCAPES

The demographic composition of the upper Nilgiri plateau is complex. The state government of Tamil Nadu repatriated Tamil refugees from Sri Lanka in the 1970s. Supported by the local tea board, the Tamils spearheaded a shift from small-scale vegetable cultivation to growing tea as a staple economic activity. However, few Toda communities engage primarily in tea cultivation. Rather, they converted to diverse forms of agriculture (but including tea) both for subsistence and for the market. While Toda youth are generally reconciled to being farmers, elders often express nostalgia about their pastoral past, but this does not translate to neat generational depositories of tradition and modernity. To clarify, there is a tendency for younger people to be more adapted to the current situation, yet the longing for the lost pastoralist lifestyle and corresponding landscape spans generations. Hence, the physical alteration of the plateau has fused the sense across generations of a 'lost buffalo-herding landscape' from an idealized past.

The Toda taskscape emerges not only from the agro-pastoral tasks they perform today, but also just as much from pastoral tasks they are unable to perform in a drastically altered physical landscape. Afforestation of grasslands on the upper Nilgiris has been a cause for concern. Toda lands were included in afforestation efforts and these were not resisted, as forest personnel told Todas they could benefit from felling mature trees; a decision many regret given the bureaucratic delays in obtaining felling permits. Todas are nostalgic about the openness and visibility that characterized their landscape. Open landscapes helped protect buffaloes, as carnivores were conspicuous. Shrinkage of grasslands and agricultural adaptation has contributed to a decline of buffalo herds, while also providing tigers and leopards with cover to hunt buffaloes. The dense plantations have made it nearly impossible to supervise herds. While there are intermittent reports of tigers and leopards preying on Toda buffaloes, there are no official records as Todas rarely report losses or claim compensation, due to time constraints and bureaucratic hurdles. Even as Todas adapt to the afforested land and its predatory risks, they complain that plantations have shrunk grasslands, desiccated swamps and drastically reduced visibility. The Todas' pastoral landscape of the past is clearly a lost taskscape, and indeed a moral landscape with a strong normative dimension.

The Todas acknowledge that tigers have historically been present in the Nilgiris, and some Toda elders recall the British hunting tigers in Wenlock Downs. However, they claim that depredation was occasional. They say the Forest Department released tigers and leopards in the area during the 1990s and 2000s. Some youngsters are said to have witnessed these clandestine acts. In these narratives, zoos and the Mudumalai Tiger Reserve emerge as source areas. Authorities could no longer feed the zoo animals, so they released them in Toda lands. The *zoo hypothesis* along with the *closing of pastures helps predators hunt* conjecture are important factors that the Todas identify to claim an increase in depredation by tigers. The Todas also point to an apparent behavioural difference between old forest tigers and the allegedly introduced ones. The forest tigers were shy, while the *new* tigers are extroverted, easily observed and do not fear people. Furthermore, they are claimed to be clumsy when attacking buffaloes. There are references to states of *naturalness* and *wildness* in such invocations of tigers, as the *new* tigers are not seen as *natural* or *wild* and thus do not belong.

Historically, people in Akole were materially impoverished and engaged in subsistence and seasonal agro-pastoral activities. The landscape was interpreted through moral and religious practices, which helped negotiate socio-economic and ecological challenges. Leopards featured prominently in this moral landscape. While intensive agricultural practices have improved material conditions, significant elements of the social construction of the land remain intact. In a complex social system, political, socio-economic and historical processes as well as narratives divide and bind people, connecting identities to the land and livelihoods in multiple ways. Many people have decreased their pastoral practices to focus on agriculture. Tribal groups, living in peripheral areas with no irrigation, continue to herd livestock. However, there is a widespread understanding that Akole is a production landscape (though 'production' itself has variable interpretations). The changes since the 1980s have only reinforced this view. Yet leopards maintain a place in social constructions of the landscape. Several tribal communities have traditionally revered big cats (e.g. tigers and leopards) as deities, and continue to do so. Throughout a history of coexistence, the majority population has adopted elements of tribal interpretations of big cats, including ceremonial aspects of tribal reverence for these animals (Ghosal & Kjosavik 2015).

In Akole, change is widely interpreted as intensification of historic resource use and is predominantly seen as representing continuity. Though benefits are not evenly spread, the change is valued as desirable for having lifted people from *abject poverty* and *backwardness*. This perception is located within larger narratives of progress, so the sugarcane transformation helps constitute a benign moral landscape. Significantly, the very landscape changes welcomed by people also improved its ecological potential for leopards.

There are accounts of leopard releases in Akole, along with an acknowledgement of the link between the intensification of agricultural practices and increase in leopard numbers. Conservation managers admit that leopards have been trapped and relocated locally under political pressure. Popular accounts attribute these releases to various factors, ranging from leopards being released to prevent collection of firewood from forestry plantations to lack of infrastructure to accommodate trapped animals. These narratives are employed to explain a perceived increase in leopard numbers and their *tameness*. Interestingly, the tameness is interpreted very differently in Akole as compared to the Nilgiris, where lack of fear in tigers was deemed as dangerous. In Akole, leopards, like humans and other animals, are recognized as being social actors, i.e. humans and leopards can share reciprocal social relations. The leopards' legitimate presence and the fact that they need to eat are widely recognized. Most people are primarily engaged in agricultural activities (and work directly with the land) in an area where leopards feed on small livestock and dogs (Athreya et al. 2014). So as long as leopards do not harm humans, they are tolerated.

Rumours of leopard introduction provide important insights into the dynamic relationship between residents and the state, especially conservation managers, in the context of changes being negotiated. For instance, people and leopards continue to be intertwined in complex social relations in Akole. On the one hand, leopards are integrated in the relationship between people and the state, which protects and manages leopard populations. There is widespread awareness that leopards are legally protected. Managers face different degrees of pressure, the most intense being in the wake of an attack on a human. While people do fear leopards, they also recognize relations of reciprocity – that leopards do not harm humans unless provoked. On the other hand, indigenous groups living in peripheral areas of the valley, with marginal political influence, have institutions that socially integrate leopards as village deities. Here leopards play a sacred-moral role in the form of *Waghoba*, the benign deity 'who never harms the righteous'. These groups are aware of leopards in the area and diligently protect their livestock on which they depend for their livelihood. In both cases, leopards are regarded as an integral part of the landscape. In tribal communities, rare depredation losses are regarded as moral acts, and this allows them to exert a degree of control over the situation. Though these beliefs are dominant amongst indigenous communities, they are also important to others.

However, there are exceptions to this in Akole town, especially amongst a small group of social elite involved in large-scale sugarcane farming who interpret the situation differently. They agree that leopards must be conserved, but argue that this must be done in protected areas and not in Akole. They regularly petition the department to trap leopards and demand compensation for depredation losses. These individuals no longer work the land themselves and form part of a socio-political elite. They subscribe to certain aspects of local belief systems, but acknowledge a disconnection from tasks they (and their ancestors) once performed. Thus, leopards present a socio-political challenge, which the local elite addresses by exercising their greater access to political influence. While they interpret change in Akole positively, having derived relatively greater benefits from it, and agree that leopards must be conserved, they insist leopards should be kept away from humans.

In Trysil and in some parts of rural Halden, the dominant narrative among people with cultural ties to the resource-based economy is one of economic decline, leading to depopulation and dismantling of private and public services. The forest industry employs only a handful of people, and agriculture is on a downhill slope at least in terms of employment. In Trysil, farm abandonment leads to spontaneous reforestation. Importantly, this happens in an age when conservation ethos has achieved a hegemonic position in public discourse, and increasingly manifests itself in practical land management. Some social groups interpret these changes in the cultural valuation of nature (of which wolf protection is one expression) as driving forces behind the decline in resource industries, and as threats to a traditional rural lifestyle that rests on harvesting resources (Skogen et al. 2017).

The sense of 'losing the landscape' that we saw among the Todas is observed here too. Farmers and many local hunters claim that conservation measures – in concert with a negative economic development –

are ruining the beauty of their managed landscape. They fear that if the land is not managed, it will soon be overgrown. While Trysil is a naturally forested area, the open spaces created by agriculture and grazing are even more cherished. They are seen as aesthetically pleasing, and as strong symbols of the relationship between people and nature, and the toil of the ancestors. Agriculture in Halden is more robust than in Trysil, being situated in a fertile area with larger farms. Yet small holdings are abandoned here too, and forest owners (often farmers) are subject to the same environmental restrictions as elsewhere. They strongly sympathize with their colleagues in marginal areas, and subscribe to the general notion of farming and resource industries as being under serious threat.

Changes are interpreted not as the return of true wilderness, but the onslaught of chaos. Yet as much as people worry about the physical landscape changes, they are even more concerned by taskscape changes. While the physical changes to the landscape in south-eastern Norway are limited compared to the Indian cases, many people feel that the land management rationale has shifted dramatically, from production to protection. They see the wilderness discourse as having achieved hegemony, so that traditional ways to use and manage their land gradually become impossible. Taking possession of the landscape and using the land for productive purposes is a core element in this ideal type understanding. The work that is laid down in the fields, forests and mountains confers meaning upon the landscape. The landscape is crucially *used* for agriculture, grazing, logging and other forms of resource utilization, including hunting and fishing. People who see the landscape in this way – as a used landscape – understand it through the tasks performed in it. The tasks materialize as a cultural landscape – the physical expression of the tasks. The landscape contains many traces of the practices that have taken place in it, and the concept of taskscape is clearly useful in order to capture this. Ingold (2000) points out that the landscape, with its physical features including all kinds of traces of human activity, is the congealed form of the taskscape. The used landscape, or the cultural landscape, is clearly a moral landscape. The social construction of the used landscape contains ideas about what the landscape should be and what a good landscape is. Not using the landscape means breaking off links to the toil of earlier generations, and ruining the landscape they created.

Hunting may also be understood as a form of use through harvesting, and not least as management. Hunters in the studies in Trysil and

Halden emphasized their responsibility as stewards (see also Fischer et al. 2013), because nature has been tampered with for so long that it can no longer be left to itself.

It may seem that understanding the landscape as a used landscape implies a utilitarian view of nature (that nature is there solely for the benefit of humans). This is not untrue, but it seems that the cultural landscape is imagined as a product of reciprocal exchange. For example, grazing animals are thought to produce high biodiversity and a healthy forest, and so are beneficial for the forest itself. Understanding nature as something outside society – as wild, pristine and *separate* – is alien to this ideal type landscape construction.

An alternative taskscape – indeed a non-taskscape – is on the offensive in both Norwegian study areas. We will now discuss this taskscape change in more depth. It is – we think – particularly important to understand the context for carnivore conservation in developed countries such as Norway, where the local economic impact (one way or the other) of physical landscape change is often limited. While this perspective is important to understand conservation conflicts also in India, it is even more present in our Norwegian case studies than in the Indian ones. That is not to say that it is not crucial in the larger context in India and elsewhere.

Pro-wolf attitudes are certainly present in our study areas. Earlier work has shown that for example younger people and particularly those with what we may term an *outward orientation* tend to have a positive view of wolves even in Trysil (see Skogen 2001). For the present analysis, we have chosen to focus on two groups of interviewees that help us throw as clear a light as possible on the ideal type landscape construction where wolves fit in nicely. It is indeed present also in other groups, although in less clear-cut versions.

In Trysil, we rely on interviews with cabin (second home) owners who have deep roots in the area, either through family ties or owning the cabin through several generations or both. Their cabins are simple, mostly without electricity from the grid and without running water. These cabins are far removed from the modern ski resort with its fancy leisure homes and apartments – both in a cultural sense and because they are located many kilometres apart. Like many of the permanent residents included in our Halden case, they nurture a construction of the landscape that embraces wilderness (and that is why we have selected them to represent this landscape interpretation). From this perspective, resource extraction as performed today is harmful. Typically, this view prevails among people who are generally not culturally rooted in traditional land use, which is the case for the Trysil cabin owners as well as the Halden residents we rely on for the following account.

A desire to escape from a stressful life and everyday materialism was a predominant theme among the cabin owners. They sought an environment that represented a sharp contrast to their regular life, with qualities that contemporary society has largely lost. A key element in their cabin life, indeed a *task*, is *the hike*.[1] Going on a hike, preferably a long one, is the primary mode of landscape appropriation. The hike is certainly a practice in the landscape. The hike is tied to the ideal type interpretation of the landscape as *authentic nature*. As we have indicated, the authentic nature can perhaps best be labelled a non-taskscape. The uncontaminated nature is a negation of tasks. Seeking the authentic landscape is to pay nature a *visit*, and nature itself should be as unspoiled by humans as possible. This landscape interpretation is tied to a landscape that one is normally not surrounded by. The hike is a mode of appreciating the landscape and infusing it with meaning, but this is a landscape seen as separate from the activities the hiker is normally involved in.

The wolf-friendly residents of rural Halden have taken the pursuit of authentic (uncontaminated) nature a step further: they live close by it. The authentic landscape is clearly a moral landscape, because it should not be contaminated by human materialism and destructive exploitation. As an ideal type, it includes notions of endemic fauna and flora, but also ideas about historic human activities. Traditional land use is perceived as benevolent. Old buildings, summer farms and grazing animals, as well as other traces of historic impact, are included in so far as they can be seen as *authentic*. In particular, large carnivores contribute to establishing the landscape as authentic, that is, with limited human impact. Eileen O'Rourke (2000) claims that some charismatic animals, such as big predators, will label the landscape as wild. Carnivores strengthen the experience of an intact, unspoiled landscape, approaching a non-taskscape. If the landscape is not entirely authentic today, they would like to see it more so in the future. They revere the hike as an ultimate form of landscape appropriation, and have actively chosen a life that is as sheltered as possible from the commotion of

---

[1] The Norwegian word *tur* could mean both a hike and a shorter walk, indeed also a big expedition. It is a core element in the typical Norwegian conceptualization of outdoor recreation, but lacks an English equivalent. We think the word 'hike' comes closest.

modern society. Nevertheless, their work-related tasks are not performed in their cherished landscape. It is for them a non-taskscape, just as it is for the cabin owners.

7.3 PREDATORS IN THE LANDSCAPES

The four case studies present different forms of change, and we observe contestations and continuities in landscape constructions. In the Nilgiris and Norway, we find contested interpretations of changes and corresponding threats to taskscapes, as well as new taskscapes emerging. Changes in Akole do not challenge continuity and could be seen as intensification of earlier practices, bringing progress, and so changes are widely regarded as desirable. In these emotionally charged landscapes, we locate the large carnivores.

In Norway, wolf supporters and sceptics speak about the wolf in ways that are similar to each other (Figari & Skogen 2011). Very few see themselves as wolf haters. Wolves in their natural environment are thought of as impressive and fascinating; they are intelligent, social – and above all – wild. The disagreement boils down to whether wolves currently belong in Norwegian landscapes (which are attributed very different meanings by different groups), and whether the wolves that are present now are real, wild wolves. Those who adhere to a traditional construction of the used landscape see a symbolic mismatch between (wild) wolves and the (humanized) local landscape. Consequently, the wolves living in the forests of eastern Norway cannot be understood – or treated – as natural. Many interviewees were even convinced they were hybrids, or *bastards*. These wolves, when observed in the neighbourhood and approaching buildings and people, come far too close and are not shy enough to be real wolves. Instead, they are perceived as unnatural animals with unnatural behaviour, showing all the signs of being polluted by humans. Because of this, they are perceived as dangerous, like the tigers in the Nilgiris. There are rumours about how captive-bred wolves have been secretly introduced by the government (Skogen et al. 2008). These rumours are strikingly similar to the tiger introduction stories circulated among the Todas and serve the same purpose – to underscore that current large carnivore presence is unnatural, and to place the responsibility firmly on actors of flesh and blood, rather than on diffuse bureaucratic systems (Skogen et al. 2008).

The case studies have revealed diverse interpretations of large carnivores, but the conflicts are not centred on the carnivores alone. Instead, as the Nilgiris and Norwegian examples illustrate, responses seem to be rooted in negative interpretations of changes in the physical landscapes, and power structures that are seen as drivers behind that change. Furthermore, these changes are seen as grave threats to taskscapes, and in the Norwegian cases, this threat of a taskscape change indeed appears to be a more forceful driver of conflict than the physical landscape change. Taskscapes are key features to consider in the conflict-to-coexistence continuum as the position held by an individual or a group may not only depend on individual negative-to-positive mental dispositions towards wildlife, but also on the perceived taskscape changes the presence of a species is associated with.

We are not denying that predators cause material damage, or that the physical landscape change (particularly in the Toda case) can have a substantial negative economic impact. Instead, we argue that also the responses to these tangible effects may be more fully understood in the context of social constructions of landscapes that define people's relations to their environment. The large carnivores thus find themselves in an environment fraught with competing interpretations as well as socio-economic and cultural conflicts. Those dwelling in the case study sites must engage with change and the forces behind it. The perception of conflict – by certain groups but not others – observed in the Norwegian sites, particularly Trysil, and in the Nilgiris can be traced to negative interpretations of change – whether mostly material or mostly at the taskscape level. These are conflicts arising from historical discontinuities in the tasks performed in the landscape, but also from shifts in the hegemonic discourse on land management and conservation.

Supporters of wolf presence in rural Norway engage with the land in new ways. Their landscape is also connected to tasks, namely their own low-impact, non-commercial practices, supported by narratives of a more sustainable, small-scale resource use in the past. They use the forest for recreational outdoor activities. For them, the landscape is a wilderness in which the wolves belong. Interestingly, people with cultural ties to the resource economy, and who oppose wolf protection, also use outdoor activities as a bridge to the past. The typical case is hunting, which symbolically links contemporary rural culture to the managed production landscape that formed the basis for settlement. Hunting as a leisure activity for larger segments of the population is only a few decades old in Norway, as in many other advanced societies. While subsistence hunting has always been of some importance in marginal areas, and sport hunting has been a pastime for the upper class, the increase in wealth and available leisure time following the Second World War has been a crucial driver of hunting as a modern form of outdoor recreation (Brottveit & Aagedal 1999). Yet it is socially constructed as an ancient tradition in rural areas. One explanation is that there are currently few other culturally significant harvesting activities for people to engage in, and fewer people are economically dependent on the forest today. Hunting becomes an *invented tradition* (Hobsbawm & Ranger 1992) of great significance. Wolves threaten typical Scandinavian hunting with free-ranging dogs, and so become an even stronger symbol of threat to traditional rural culture.

For Todas, pastures hold similar recreational value, given their past range-herding practices. Men still go to look at remaining pastures and even to watch sacred buffaloes, which have now turned feral. Todas value the recreational aspects of annual activities like collecting grass from swamps to thatch their temples. The changed land curbs these activities while seeming to facilitate the presence of large carnivores. While this change is positive for large carnivore conservation, Todas contrast it with their lost taskscape. Thus, disagreements are not only about large carnivores or their conservation, but also about where they belong. On the other hand, a positive interpretation of change in Akole encourages a more benign interpretation of large carnivores, especially since leopards were already integrated into social constructions of the landscape. The interpretation of the affluent minority, who claim that leopards do not belong to Akole, can partly be traced to their changed relations with the land and their political engagements with the state. According to them, Akole is meant for humans alone and leopards should be kept in protected areas. This echoes conservation practice in India, which seeks to create boundaries between natural and social spaces, with leopards belonging to the former and humans to the latter (Ghosal et al. 2013).

This is surely a simplified interpretation of a complex reality, but it provides a useful contribution to our understanding of the differing responses to the presence of large carnivores. Communities and social groups have inherent power dynamics, which favour specific ideas of nature – linked to general world-views – that shape or even drive conflicts over the interpretations and use of the land (Peterson et al. 2010a, 2010b). The Nilgiris and Norwegian cases illustrate how unresolved conflicts of interpretation, along with the material basis for these interpretations, have deep implications. Akole provides a contrast, where change has benefited both people and leopards. Changes do affect the social construction of the landscape, but the interpretation is benevolent, accommodating leopards and institutions built around them. The way landscapes and their uses are interpreted can influence human–wildlife interactions, and thus determine whether a species will generate conflict or be tolerated to the point of coexistence.

7.4 CONCLUSION

We will argue that the experience of physical landscape change, as well as perceptions of changing management regimes and shifting power relations will influence social constructions of the landscapes in different ways, and determine whether changes are seen as desirable or undesirable – leading to coexistence or conflict with wildlife. These interpretations of change are related to production and recreational tasks performed or hindered, but also to how broader processes of economic and cultural change are experienced by different social groups. Constructions of the landscape resemble 'act[s] of remembrance, of engaging perceptually with an environment that is itself pregnant with the past' (Ingold 2000, p. 189).

How change relates to landscape constructions influences responses to large carnivore presence. Notions of belonging, polluted identity and behavioural anomaly in animals also suffuse contestations of carnivore presence. Negative interpretations of physical and cultural change thus bear upon similar interpretations of carnivore presence. Nostalgia prevails for a more aesthetic past that was also a controlled past, where hunting, grazing and monitoring of livestock were possible in a benevolent landscape. Yet the notion of an authentic landscape held by friends of large carnivores also includes images of the past, understood as unspoiled or pristine; a landscape where predators fit perfectly. Positive interpretations of the physical landscape change are accompanied by benign responses towards carnivore presence.

While material damage inflicted by carnivores generally has substantial bearing upon conflicts, we argue that responses towards carnivores need not be driven by material loss alone. It has been documented in previous research that strong anti-carnivore sentiments may develop independently of material damage (Skogen & Krange 2003; Figari & Skogen 2011; Skogen et al. 2017). This can be discussed in the conceptual context of social mechanisms. Here we may perhaps claim to be on the track of a very simple social mechanism:

If the changes that bring predators are seen as threatening and imposed by malevolent outside forces, then predators will not be welcome, and easily become symbols of the wider processes of change, even if they cause limited material damage. If the changes that bring predators are seen as benevolent, regardless of their origin, then predators may be tolerated or even revered, as long as material damage is still limited. If social constructions of the landscape are already contested, the presence of large carnivores will become embedded in these conflicting ideas and may drive people towards the negative side of the conflict-to-coexistence continuum.

The opposition to large carnivore conservation in specific areas is rooted in historical, socio-economic and physical engagements with the land, and hence cannot be separated from the social constructions of the landscape that emerge from these engagements. Conservation laws may impose a 'new set of meanings on the land, a landscape of nature consumption, devoid of human history, that clash with locally constructed meanings' (Neumann 1998, p. 202). While applying perspectives such as ours is no guarantee of success, we are convinced that treating so-called human–wildlife conflicts exclusively as conflicts between people and animals is a certain road to failure. Considering social constructions of landscapes, and how they influence human–wildlife relationships, represents a crucial step towards understanding a broader array of mental dispositions and behaviours, which can span beyond conflict to include tolerance and coexistence.

7.5 RECOMMENDATIONS AND FUTURE DIRECTIONS

While extensive literature exists on the social construction of landscapes, little is known about how interpretation of landscapes, particularly their interpretation as taskscapes, influence human–wildlife conflict-to-coexistence interactions. The following research efforts are therefore needed:

- There is a need for more studies of the social context in which conflicts over wildlife play out. We already know that some conflicts that *involve* large carnivores are driven as much by external issues as by the actual interaction between humans and animals. We need more research on conflicts over other species and particularly in the Global South, in order to identify the issues at stake besides the tangible impact of wildlife on human activities.

- In particular, we need studies that seek to unravel people's understandings of the relationship between wildlife and the landscapes in which wildlife exist. If conflicts are as much about landscapes – or even about taskscapes – as they are about wildlife per se, then this may render some conservation and communication strategies useless and call for other approaches.

Wildlife managers and policy makers who want to address human-wildlife interactions should:

- Be aware that landscapes are not just physical areas but also socially constructed spaces where people draw lines between the places for humans and wildlife. It is therefore fundamental to acknowledge that socially constructed landscapes – for example understood as taskscapes – can influence people's mental disposition towards conservation and wildlife issues. Understanding the difference between what the landscape should be and what a good landscape is from the perspective of a particular social group can make the difference between failure and success of a conservation project.

7.6 References

Athreya, V., Odden, M., Linnell, J. D. C., Krishnaswamy, J. & Karanth, K. U. (2014). A cat among the dogs: Leopard *Panthera pardus* diet in a human-dominated landscape in western Maharashtra, India. *Oryx*, 50(1), 156–62.

Bender, B. (1993). *Landscape: Politics and Perspectives*. Oxford: Berg Publishers.

Bender, B. (1998). *Stonehenge: Making Space*. New York: Oxford International Publishers.

Bjerke, T., Reitan, O. & Kellert, S. R. (1998). Attitudes toward wolves in south eastern Norway. *Society & Natural Resources*, 11(2), 169–78.

Brottveit, Å. & Aagedal, O. (1999). *Jaktapåelgjaktkulturen* (Hunting for the moose-hunting culture). Oslo: Abstraktforlag.

Bruskotter, J. T., Schmidt, R. H. & Teel, T. L. (2007). Are attitudes toward wolves changing? A case study in Utah. *Biological Conservation*, 139, 211–18.

Calhoun, C. J., Gerteis, J., Moody, J., Pfaff, S. & Virk, I. (2007). *Contemporary Sociological Theory*. Malden, MA: Blackwell.

Cronon, W. (1995). *Uncommon Ground: Toward Reinventing Nature*. New York: W.W. Norton and Co.

Ericsson, G., Bosdet, G. & Kindberg, J. (2008). Wolves as a symbol of people's willingness to pay for large carnivore conservation. *Society & Natural Resources*, 21, 294–309.

Ericsson, G. & Heberlein, T. A. (2003). Attitudes of hunters, locals and the general public in Sweden now that the wolves are back. *Biological Conservation*, 111(2), 149–59.

Figari, H. & Skogen, K. (2011). Social representations of the wolf: A core of shared understanding. *Acta Sociologica*, 54(4), 317–32.

Fischer, A., Kereži, V., Arroyo, B., Mateos-Delibes, M., Tadie, D., Lowassa, A., Krange, O. & Skogen, K. (2013). (De-)legitimising hunting: Discourses over the morality of hunting in Europe and eastern Africa. *Land Use Policy*, 32, 261–70.

Frank, B. (2016). Human–wildlife conflicts and the need to include tolerance and coexistence: An introductory comment. *Society & Natural Resources*, 29 (6), 738–43.

Ghosal, S., Athreya, V. R., Linnell, J. D. C. & Vedeld, P. O. (2013). An ontological crisis? A review of large felid conservation in India. *Biodiversity & Conservation*, 22(11), 2665–81.

Ghosal, S. & Kjosavik, D. J. (2015). Living with leopards: Negotiating morality and modernity in Western India. *Society & Natural Resources*, 28(10), 1092–1107.

Ghosal, S., Skogen, K. & Krishnan, S. (2015). Locating human–wildlife interactions: Landscape constructions and responses to large carnivore conservation in India and Norway. *Conservation & Society*, 13(3), 265–74.

Giddens, A. (1991). *Modernity and Self-Identity: Self and Society in the Late Modern Age*. Cambridge: Polity Press.

Gusset, M., Maddock, A. H., Gunther, G. J., Szykman, M., Slotow, R., Walters, M. & Somers, M. J. (2008). Conflicting human interests over the reintroduction of endangered wild dogs in South Africa. *Biodiversity Conservation*, 17, 83–101.

Hedström, P. (2005). *Dissecting the Social: On Principles of Analytical Sociology*. Cambridge: Cambridge University Press.

Hobsbawm, E. J. & Ranger, T. (1992). *The Invention of Tradition*. Cambridge: Cambridge University Press.

Ingold, T. (2000). *The Perception of the Environment: Essays in Livelihood, Dwelling and Skill*. London: Routledge.

Inskip, C. & Zimmermann, A. (2009). Human–felid conflict: A review of patterns and priorities worldwide. *Oryx*, 41, 18–34.

Macnaghten, P. & Urry, J. (1998). *Contested Natures*. London: Sage Publications Ltd.

Neumann, R. P. (1998). *Imposing Wilderness: Struggles over Livelihood and Nature Preservation in Africa*. Berkeley: University of California Press.

Olwig, K. R. (1996). Recovering the substantive nature of landscape. *Annals of the Association of American Geographers*, 86, 630–53.

O'Rourke, E. (2000). The reintroduction and reinterpretation of the wild. *Journal of Agricultural & Environmental Ethics*, 13, 145–65.

Peterson, M. N., Birckhead, J. L., Leong, K., Peterson, M. L. & Peterson, T. R. (2010a). Rearticulating the myth of human–wildlife conflict. *Conservation Letters*, 3, 74–82.

Peterson, R. B., Russel, D., West, P. & Brosius, P. J. (2010b). Seeing (and doing) conservation through cultural lenses. *Environmental Management*, 45, 5–18.

Philo, C. & Wilbert, C. (2000). Animal spaces, beastly places: New geographies of human–animal relations. An introduction. In C. Philo & C. Wilbert, eds., *Animal Spaces, Beastly Places: New Geographies of Human–Animal Relations*. London: Routledge.

Proctor, J. (1998). The spotted owl and the contested moral landscape of the Pacific Northwest. In J. Wolch & J. Emel, eds., *Animal Geographies*. London: New Left Books.

Saberwal, V. K., Gibbs, J. P., Chellam, R. & Johnsingh, A. J. T. (1994). Lion–human conflict in the Gir forest, India. *Conservation Biology*, 8, 501–7.

Sjölander-Lindqvist, A. (2007). Local identity, science and politics indivisible: The Swedish wolf controversy deconstructed. *Journal of Environmental Policy & Planning*, 10, 71–94.

Skogen, K. (2001). Who's afraid of the big, bad wolf? Young people's responses to the conflicts over large carnivores in Eastern Norway. *Rural Sociology*, 66 (2), 203–26.

Skogen, K. & Krange, O. (2003). A wolf at the gate: The anti-carnivore alliance and the symbolic construction of community. *Sociologia Ruralis*, 43(3), 309–25.

Skogen, K., Krange, O. & Figari, H. (2017). *Wolf Conflicts: A Sociological Study*. New York: Berghahn Books.

Skogen, K., Mauz, I. & Krange, O. (2008). Cry wolf! Narratives of wolf recovery in France and Norway. *Rural Sociology*, 73(1), 105–33.

Treves, A. & Karanth, K. U. (2003). Human–carnivore conflict and perspectives on carnivore management worldwide. *Conservation Biology*, 17, 1491–9.

Wylie, J. (2007). *Landscape*. New York: Routledge.

8

Corridor of Conflict

Learning to Coexist with Long-Distance Mule Deer Migrations, Wyoming, United States

JOSHUA MORSE AND SUSAN G. CLARK

A major effort is under way to conserve the longest large mammal migration in the lower forty-eight United States. In south-western Wyoming, a portion of the Sublette County mule deer herd (*Odocoileus hemionus*) was discovered to migrate over 150 miles (240 km) between the winter habitat in the Red Desert and summering grounds in the Hoback, just south of Jackson Hole and Grand Teton National Park (Sawyer et al. 2014). The Red Desert to Hoback migration crosses many barriers, including a mix of private, state and federal lands, livestock and elk (*Cervus canadensis*) fences, three major highways and land open to energy development. A host of cultural and political factors originating in long-standing regional tensions over how to use and manage natural resources also threaten the migration.

The interest groups involved in this migration are committed to coexisting with migratory mule deer in principle, and many actively promote a coexistence outcome (see Frank 2016). However, conflict among stakeholders remains a potent risk that could stymie efforts to maintain this migration. A clarified, secured and sustained common-interest management policy has yet to be achieved (see McDougal et al. 1981; Clark 2002; Steelman & DuMond 2009 for discussion of shared, special, common interests). Common interests are characterized by agreement upon the desired goal, appropriate strategies to achieve it and the successful implementation of those strategies for a sustained period (Brunner & Steelman 2005; Castree 2013; Clark & Wallace 2015).

In this chapter, we focus on organization, management and policy issues to conserve the mule deer herd and migration route. First, we describe the ongoing conservation effort based on our observations, interviews and site visits. Second, we analyse those efforts using an integrated approach that focuses on the regional context, ecological setting and social and decision-making processes (see Clark & Wallace 2015). Finally, we offer recommendations to enhance prospects for human–wildlife coexistence in the common interest. Identifying and safeguarding a common-interest outcome in this case requires that efforts be creative, timely, constructive and ameliorative (Lasswell 1971, p. 96; Clark 2002, p. 60). If these standards are met, this case – and the stakeholder community, and the public invested in it – has the potential to achieve coexistence and serve as a model effort.

8.1 CONTEXT AND METHODS

The Red Desert to Hoback migration falls within the south-eastern region of the Greater Yellowstone Ecosystem (GYE). At 22 million acres, the GYE is the largest and most intact ecosystem in the lower forty-eight United States (Primm & Clark 1996), providing core habitat for grizzly bears (*Ursus arctos*), elk, grey wolves (*Canis lupus*) and mule deer. However, the needs of these populations cannot be fully met within the boundaries of Yellowstone National Park or the adjacent Grand Teton National Park (Hohl et al. 2015). Seasonal ungulate migrations and associated predator movements cross multiple jurisdictions, each with different rules and regulations dictating what modes of human interaction with wildlife are appropriate. In this context, the needs of wildlife to access an expanded, intact landscape are unambiguous. However, securing such access sparks tension among diverse human interest groups. Industrial concerns fear increased federal regulation of land use in migratory corridors, agricultural interests perceive potential threats to their herds in the form of disease transmission and increased predator activity outside of the national parks, and numerous non-profits and advocacy groups hold highly specific and sometimes conflicting conservation goals related to sustaining large mammal migrations. Thus, perceived human–wildlife conflict emerges principally from the competing human interests apparent in the diversity of jurisdictions and management policies throughout the GYE. Many of these are incompatible with traditional conservation, and with one another (Clark & Rutherford 2014; Farrell 2015; Jaicks 2016).

8.1.1 Biophysical Context

The migration runs 150 miles (240 km) from the dry Red Desert low-lands along interstate 80 in south-western Wyoming to the high-

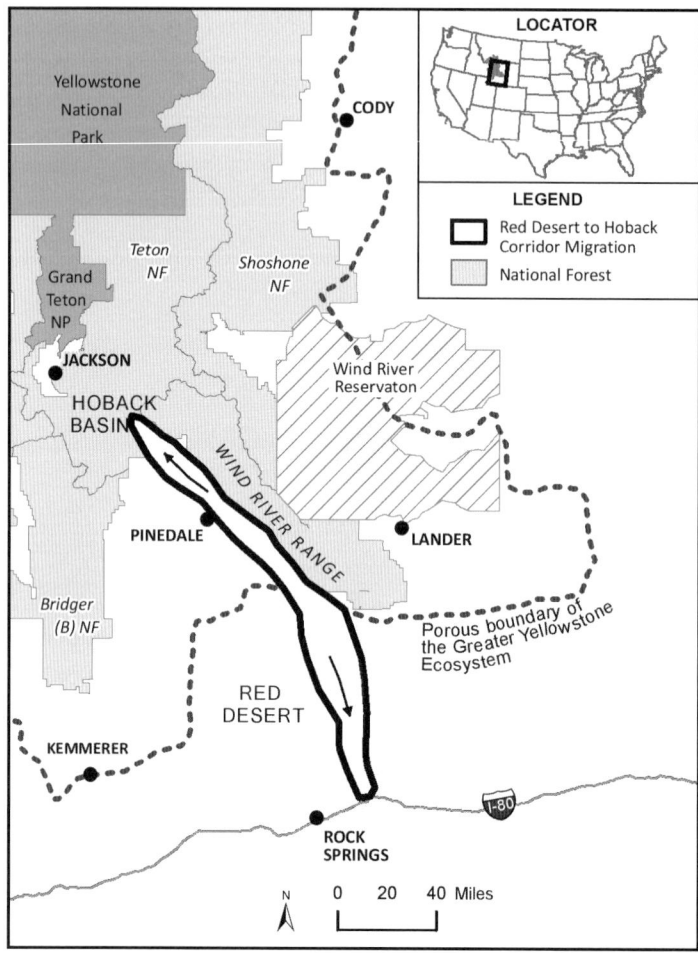

Figure 8.1 The Red Desert to Hoback migration corridor, and surrounding Wyoming sections of the Greater Yellowstone Ecosystem.

elevation forests and meadows of the Hoback Basin just south of Jackson Hole (Figure 8.1). Along this route, migrating deer exploit a 3,000–4,000 ft. (900–1200 m) gradient maximizing access to nutritious forage during the spring green-up (Sawyer et al. 2014). Within the broad corridor, the migration is not uniform in width or use patterns. Global Positioning System (GPS) studies using collared deer show that individual animals join and leave the corridor at different points, and that certain areas of the corridor receive much higher use than other areas (Sawyer 2014).

The migration faces many physical obstacles. State routes 28 and 352 and US highway 189/191 bisect the route. Eight foot (2.4 m)-high fencing erected to separate wild elk herds from domestic cattle criss-crosses the corridor. Although elk fences have some permeable points like gates, navigating long impassable stretches of fencing taxes migrating animals (Sawyer et al. 2016). Shorter livestock fences on private and public rangeland also pose obstacles for migrating deer. Furthermore, the migration crosses private lands where residential development is profitable, and state and federal parcels that can be leased for large-scale energy development. Another nearby long-distance migration, Wyoming's Path of the Pronghorn (*Antilocapra americana*, Cherney & Clark 2008) recently came under threat of such development, highlighting this risk (Thuermer 2017). Although well-planned development can be permeable to wildlife, deer moving quickly through densely developed areas suffer a heightened energetic cost (Sawyer et al. 2016). Additionally, energy development has been found to alter habitat use over long periods with potentially harmful consequences at the population level (Sawyer et al. 2017).

Minimizing physical barriers to migration has been a major focus for researchers, conservationists, developers, managers and policymakers, with some success to date. First, environmental non-government organizations (NGOs) interested in the migration corridor continue their long-term efforts to replace high-risk livestock fencing with newer, *wildlife-friendly* fencing. Wildlife-friendly fencing typically removes the lowest wire, reduces fence height and offers sites (e.g. jumps) to facilitate movement. Second, a privately owned parcel encompassing a narrow river crossing on the corridor was placed under easement to prevent development and protect the migration. Third, new regulatory definitions for wildlife migrations have been drafted and adopted by the Wyoming Department of Game and Fish. This has implications for management of public lands in the migration corridor. Efforts to reduce present and forestall future obstacles to migration continue, largely championed by a loose network of NGOs – the Red Desert to Hoback Partnership.

8.1.2 Social Context

The area of the Red Desert to Hoback is embroiled in cultural conflict. Increasingly, newer transplants to the region who hold very different values (Taylor & Clark 2013) are challenging the conceptions of wildlife and the value of wildlife conservation established by early American pioneers and their descendants. *Old West* interests, stereotypically rallying around the ranching and energy industries, and other utilitarian uses, often see wildlife conservation as a challenge to consumptive uses, despite their high regard for certain iconic species, especially those that can be hunted. *New West* communities and interests, stereotypically populated by coastal transplants, second homeowners, academics and passive-recreationists often champion active wildlife conservation, and may have a more all-encompassing regard for the region's wildlife (Clark 2008; Farrell 2015; Jaicks 2016). The conflicting practical agendas and moral claims held by these diverse groups complicate the challenges of clarifying, securing and sustaining the common interest in this case. These competing interests – although evident to local laypeople, professionals and researchers – are rarely formally addressed explicitly or analytically in the management policy process. Yet they exert a complex, tacit influence on the Red Desert to Hoback migration. Finding a common interest is key to allow individuals and groups in different positions along the conflict-to-coexistence continuum to overcome dissimilarities and conflicts over resource uses, thus allow mule deer migration to continue and coexistence to move from a principle to a successful outcome.

8.1.3 Methods

We used the integrative, analytic approach of the *policy sciences* (Text Box 8.1) as a research framework for fieldwork, analysis and as a basis for recommendations (Lasswell 1971). The policy sciences have been successfully used in a wide range of natural resource management challenges over the decades (e.g. elk feedgrounds, Clark & Vernon 2015, 2017; Vernon et al. 2015; Vernon & Clark 2016). This approach, well over half a century in use in diverse cases worldwide, is described briefly below and detailed more thoroughly in Clark (2002).

The approach guides users to be problem-oriented in both content and process terms, identifying problems based on difference between goals and the likely future given conditions (see Weiss 1989), while attending to both matters of social process (especially value dynamics) and decision-making process. It asks if the social and decision processes are in the common interest and it offers three partial tests analysts can use to make a determination (see cites above on common interest). Policy sciences analyses take into account all methods and data that

Text Box 8.1 The Policy Sciences

The policy sciences are an analytic meta-framework that provides a systematic guide for observing, describing and interpreting complex human and environmental interactions. The policy sciences, also called the *configurative* approach, have a long history of use and consist of three related functions:

(1) Social process mapping provides a structure for analysing the value demands of participants in any case, with attention to basic beliefs (the doctrines, formulae and symbols) around which people structure their identities, expectations and demands. On this basis people pursue strategies to achieve their desired outcomes.
(2) Decision process mapping provides a system and standards to appraise the adequacy of the decision-making process. It draws attention to how policies arise, are debated, responded to and appraised and adjusted over time.
(3) Problem orientation provides a system for diagnosing and proposing solutions to problems, including mapping social and decision processes, with attention to discrepancies between the current situation and desired conditions that fuel those discrepancies as the basis for analysis. After defining a problem, possible solutions are invented, appraised and selected for implementation.

Multiple-method analysis complements the policy science framework. For a review of the application of this meta-framework to natural resource management, see Susan Clark's *The Policy Process* (Clark 2002).

elucidate these dimensions in a case. Finally, the approach requires that the person using the policy sciences be self-aware of their own standpoint in their research, analytic and recommending work.

In this study, we attended to all elements of the policy sciences but focused on two major elements in particular – social process mapping and problem definition. For the former, we looked at the perspectives, values and strategies driving the engagements of interviewees and interest groups with the Red Desert to Hoback migration. For the latter, we identified the goals sought and the forces and factors preventing or limiting goal attainment for participants in the case. Our assessment took into account the interactive analysis of goals (values), trends (history), conditions (explanations) and projection (futuring).

We used document analysis (Bowen 2009), semi-structured interviews and participant observation (Weiss 1995), to identify and map human perspectives about the Red Desert to Hoback migration and to assess the adequacy of current conservation efforts. Document analysis helped to identify interest groups and map social process, drawing from organization web pages, popular press articles, conference proceedings and grey-literature reports as primary data. Interviews and participant observation yielded our core data on social process and problem orientation, and were facilitated by *snowball* networking through a small group of key informants (Weiss 1995) that we surveyed to compile a list of principal interest groups. Key informants were also solicited for additional contacts. Interviews ranged from 30 to 90 minutes, and were recorded with permission, or documented in shorthand. Of the fifty-five interviews we conducted, forty-five proved sufficiently detailed for analysis.

Interviews were structured around three themes: (1) what interests (values) and agendas (strategies) were being pursued by participants, (2) how did participants identify and respond to the challenges that they saw facing the migration and (3) what steps they believed must be taken to maintain the migration, and also meet the demands of the individuals and organizations whose interests were at stake. We used questions based on these themes to prompt interviewees to provide data that could be coded for the analytic dimensions of social process and problem definition as introduced by Lasswell (1971) and Clark (2002).

8.2 RESULTS

We organized our description of contextual matters following social process mapping analytic categories from Lasswell (1971). In doing so, we found diverse perspectives, values, situational issues and four different and conflicting problem definitions at play in the migration case. To clarify, secure and sustain the common interest, these definitions need to be recognized, integrated and jointly resolved (Terway 2017) to maximize the likelihood of a successful cooperative effort that makes possible human–wildlife coexistence.

8.2.1 Social Context

Mapping social context requires investigating participants, their perspectives and interests (values), areas of interactions, strategies, outcomes sought and long-term effects of participant interactions. We summarize our findings with respects to these categories.

8.2.1.1 *Participants*

The idea of maintaining functioning large mammal migrations has seized the public's attention. National and international media, including *National Geographic* (Wilkinson 2016), *The New York Times* (Gorman 2015) and *The Atlantic* (Phippen 2016) have covered Red Desert to Hoback and other migrations, focusing on the awe that these spectacles inspire and the threats that they face. This national media coverage, and the fact that Red Desert to Hoback occurs partly on public lands, complicated the task of defining our participant community. Interest from individuals and groups operating well outside of the migration corridor, and the overarching philosophy of public trust as pertaining to natural resources on federal lands, created a context in which the entire American public could be considered a part of the stakeholder community. To be most useful to local and regional policy-making and citizens, we restricted our analysis to interest groups that are consistently and actively engaged in the policy process in the region. This focus yielded eight groups (Table 8.1).

8.2.1.2 *Perspectives*

Across all interviewees and interest groups, individuals invested in the Red Desert to Hoback case were in agreement over a very broad, shared goal: ensuring that the migration continues for future generations. However, despite this shared interest, polarization over specific issues was evident. In August 2016, two headlines in the *Casper Star Tribune* capture this tension. 'Let's Stand Together on Migration Corridors', reads one op-ed by a local NGO director, going on to call for a unified approach to protecting the awe-inspiring phenomenon. In contrast, another headline reads 'Industry Group Expresses Concerns about Migration Corridor

Table 8.1 *Interest groups involved in the Red Desert to Hoback case.*

Note that some member entities can fall within multiple interest groups.

Interest group (interviewees)
State agencies (5)
Federal agencies (3)
Researchers (6)
Environmental NGOs (12)
Ranchers (5)
Private landowners (3)
Energy industry (3)
Hunters (outfitters – 4; sportsmen – 1)

Effort', following public statements by the Petroleum Association of Wyoming that question the need for sweeping conservation measures for migrations and the validity of the processes by which conservation policy is being reached. These articles reflect two overarching, competing perspectives that occurred throughout our study: (1) that migrations are imperilled and require immediate, *collaborative* efforts to ensure their continuation, and (2) that migrations, while important, are not at immediate risk, and instead pose a policy challenge primarily because of the utility of the deer to humans (in both economic and intangible terms) and the consequences of conservation for various human interests.

8.2.1.3 Areas of Interaction

During our study, a number of initiatives, conferences and policy meetings took place, mostly focusing on making sense of Sawyer et al.'s (2014) released ecological data on the migration. Meetings ranged from private and relatively secret, to well-publicized public meetings (Table 8.2). Outside of these forums, most people involved in the migration for reasons other than research or advocacy (e.g. hunters, non-environmental professionals, ranchers) showed only sporadic engagement with the case. Individuals affiliated with organizations with a mandate or mission that aligned with the general goal maintained stronger, more sustained involvement such as NGOs focusing on promoting wildlife-friendly fencing.

8.2.1.4 Value Claims

Participants in any social process pursue desired values, including friendship, alliances and events that favour them (see Mattson et al. 2012). We grouped value claims into eight categories based on Lasswell (1971; Table 8.3). Values can be understood both in terms of interests that are already abundantly held, and in terms of those that are sought by participants (Clark 2002). In social interactions and decision-making processes, participants leverage the values available to them to gain those values they want more of. For example, the ranching community's attachment to the land it manages is an expression of many values that its members already hold in abundance, primarily affection for the natural resources they steward. They leverage this value as a basis for resisting what they perceive as state and federal policies that undermine their interests and to readdress a deficiency in power. Similar examples can be drawn for each of the principal individual and organized actors in this case as noted below (Table 8.3).

Table 8.2 *Timeline of public and selective forums for engagement with Red Desert to Hoback, compiled based on interviews and document analysis*

Forum (date)	Accessibility	Topic
NGO & Funders Meeting (June 2014)	Private; NGOs and funders	Strategic planning for land conservation along the corridor
RD2H Coalition Meetings (June 2014, ongoing)	Private; participating NGOs and invited guests	Ongoing coordinating for conservation and advocacy along the corridor
Wyoming Department of Game & Fish (September 2015, ongoing)	Public	Discussions of available information and its management implications for big game migrations
Wyoming Department of Game & Fish Commission (November 2015; January 2016)	Public	Deliberations on new regulatory measures and strategic goals for big game migrations
Ruckelshaus Emerging Issues Form: Migration (November 2015)	Public; widely publicized	Promotion of migrations as a policy issue, dissemination of scientific knowledge
Concerned Stakeholders Meetings (December 2015)	Private; organized by selected participants	Energy, ranching and agencies discuss concerns with blanket conservation along corridor
Agency Meeting (January 2016)	Private: state and federal agencies, NGOs	Mile-by-mile assessment of conservation priorities along the corridor
Rock Springs Bureau of Land Management Resource Management Plan Revision (February 2011, ongoing)	Public; formal public comment process	Ongoing public comment process to revise federal management plans for public lands encompassing RD2H corridor, with a comprehensive focus that exceeds RD2H

We assessed the interests (values) of the groups in our study in terms of both values held and values sought based on their interview statements and written materials. The core value claims being made by interviewees in each interest group are summarized in Table 8.4.

Trends were clear at both the level of individual interest groups and within the broader population. Affection, encompassing a wide range of positive experiences of and feelings for the resource (i.e. the deer,

Table 8.3 *The eight functional values used to understand social processes in a policy-sciences analysis*

Value	Definition	Sample value claim
Affection	Emotional warmth and openness towards a person, resource, place, etc. Example: a strong family attachment to the landscape of the Hoback region	'Culturally, mule deer are one of those iconic, big reasons why I live here'
Enlightenment	Knowledge, technical or theoretical. Example: an academic understanding of migration ecology	'Nobody had known these animals were doing that, so that sense of discovery was neat'
Power	The ability to influence the outcome of decision-making processes. Example: agency to dictate land use on a portion of the migration route	'…but [landowners] don't want folks telling them what to do'
Rectitude	Ethical merit or justification. Example: belief in the inherent right to use and enjoy public lands shared by all US citizens	'I think the prize is the accountability to know that we did right in our effort as not only those that revere wildlife but just as human beings'
Respect	Acknowledgement of agency, freedom and merit in others or one's self. Example: the ability to gain the trust of a guarded community	'I guess it's just because I'm known in both communities so that's probably why I get involved'
Skill	The ability to exercise specific behaviours to achieve a desired end. Example: the ability to stalk and harvest a mule deer with primitive firearms	'Their ability to offer a challenge to a hunter, you're truly matching your wits against an incredible beast'
Wealth	Access to material richness and goods. Example: the capacity and financial resources to implement a management policy	'We were very involved … just given that we had a lot to lose, as far as value for our shareholders, if that species were to get listed'
Well-being	Material, mental and spiritual comfort and stability. Example: reliable access to food, shelter and fulfilling work and recreation	'Always wanted to come back [to the Wind Rivers]. Just had to figure out how to do it and provide for my family'

Table 8.4 *Value claims by interest group.*

Claims that occurred in more than half of the interviews within a given interest group are reported for each category. A dash (-) indicates that no value claims were made by more than half of interviewees within a given participant group.

Interest group	Values held	Values sought
NGO	Enlightenment, Affection, Respect	Respect, Enlightenment, Well-being, Wealth
Research	Affection, Enlightenment	Enlightenment, Rectitude
State	Skill, Enlightenment	Enlightenment, Wealth
Federal	Respect, Affection, Skill	Rectitude, Wealth, Affection, Well-being
Ranching	Respect, Affection, Power	Power, Respect
Landowners	Affection, Skill	Wealth, Affection, Well-being
Hunting	Affection, Skill	–
Energy	–	Power, Wealth, Rectitude

migrations) that Frank (2016) identifies as a precursor for true coexistence, was held by six of the eight stakeholder groups. As one Pinedale rancher observed, "We all have a common ground [share interest] that we care about the resource, and we want to be able to utilize the resource" (interview, 2016). Recognition of a shared *affection* value for big game migrations (i.e. *the resource*) is frequently cited as a point of commonality underlying attempts at collaboration around the migration, like the 2015 Ruckelshaus Emerging Issues Forum (Table 8.2). Also noteworthy, four interest groups cited 'skill' as a core value for them, reflecting confidence in their methods and ability to carry them out.

In addition to analysis at the level of interest groups, the frequency of value claims made across the entire population was also informative (Table 8.5). First, affection and wealth were the most frequently cited values held and pursued, respectively. Second, a relatively even number of participants (19–24) cited each value, with the exception of skill. Third, three values were identified by less than a quarter of participants as values held (i.e. power, wealth and well-being), reflecting a perception of disenfranchisement and impoverishment among participants in the case.

8.2.1.5 Strategies

People leverage their value holdings to pursue their interests through strategies. Lasswell (1971) lists four strategies that participants apply in

Table 8.5 *Four competing problem definitions exist in the Greater Yellowstone Ecosystem, each supported by different base values and advancing different scope values claims. Values in each category claimed by half of the population orienting towards a given definition are included below.*

Definition	Key challenges	Example data	Values held	Values sought
Biophysical	Identifying and removing material barriers to the migration	'As a transportation engineer, my goal is to reduce collisions. As a hunter, my goal is to get the animals through with a minimum amount of fatalities' (anonymous Wyoming Department of Transportation engineer, 2016)	Enlightenment, Affection	Enlightenment, Affection, Wealth
Social	Engaging the diverse range of interest groups involved in this case in a productive policy process	'I think one of the obstacles is going to be, as I mentioned early in this, polarization of the issue. The fact that you don't have a lot of people in the middle' (anonymous energy industry employee, 2016)	Respect, Enlightenment, Affection, Skill	Respect, Rectitude, Enlightenment, Skill
Governance	Disparities of power within the broad stakeholder community complicate the task of establishing and enforcing a common-interest policy	'Wyoming has problems with people outside of Wyoming trying to dictate what we should or shouldn't do… The wildlife belong to the state of Wyoming… Why should someone from NYC have a say in this' (anonymous Pinedale area outfitter, 2016)	Affection, Skill	Power
Non-Issue	None	'I think you're at the point where you've addressed a lot of the main threats and it will continue unimpeded, right?' (anonymous Jackson NGO staff, 2016)	Affection, Skill	Wealth

social processes: (1) ideological strategies that leverage the spread of ideas to shift decision-making, (2) diplomatic strategies that leverage one-on-one or small group interactions with decision-makers to directly communicate value claims, (3) economic strategies that leverage control of material resources to encourage a desired outcome and (4) coercive strategies that use force, such as litigation. In this deer migration case, all interest groups appear to be relying primarily on ideological, diplomatic and economic strategies at this stage. The state agencies, the NGO community, the energy industry and hunting outfitters are using economic strategies overtly.

8.2.1.6 Outcomes and Effects of Engagement in the Social Process

Changing dynamics of value claims and strategies produce short-term outcomes and longer-term effects in a social process. Outcomes represent changes in the distribution of values, whereas effects represent the creation of new institutions and practices responding to value redistribution. Because the Red Desert to Hoback migration case is relatively young, our observations in this category were limited to three outcomes.

First, increased knowledge and awareness of this migration and others is broadly cited as an important ideological outcome. Both the NGOs and the academic research community publicize their work on migrations and its conservation benefits (e.g. the Wyoming Migration Initiative's (WMI) Migration Assessment, Sawyer et al. 2014). Awareness of the migration has also been raised through artistic media, most notably the photography of Joe Riis, a WMI fellow, which has been published widely. Generally, heightened awareness of the migration has increased the affection that participants in the case hold for the resource, moving them towards a preference for coexistence, rather than conflict (Frank 2016). Second, however, many participants also described increasing polarization around migrations as a policy problem. Whereas no participants voiced opposition to conserving mammal migrations, many did express concern that the issue may become hotly politically contentious, indicating the importance of human–human conflict as a barrier to securing the broadly desired condition of human–wildlife coexistence. Finally, many interviewees saw the establishment of new migration-centric management policy as an important outcome. For example, in February 2016, the Wyoming Game and Fish Department proposed a new Ungulate Migration Corridor Strategy, with management policy implications for both state and federal lands. This formal acknowledgement of migration as a management priority is seen by many as an important precedent, comparable to the prioritization of winter range issues decades earlier.

8.3 PROBLEM DEFINITIONS

Problem definition is the diagnosis of discrepancies between sought-after goals and present (and historic) conditions (Clark 2002). Often, problem definitions are assumed to be widely visible and agreed upon by their adherents, but in fact reality is seldom so simple (Cherney & Clark 2008; Vernon et al. 2015). Conflict often results from competition among tacit, competing problem definitions held by different participants within a case.

Our results showed that nearly all participants oriented to the high-level goal of ensuring that mule deer are able to continue their migration from the Red Desert to the Hoback. However, reaching that goal (and setting smaller, incremental goals towards that ultimate purpose) evoked a wide range of proposed strategies and actions among interviewees. We identified four implicit competing problem definitions, three of which acknowledged a disparity between current conditions and desired future conditions. However, each definition identified different categories of obstacles to be overcome. A fourth problem definition showed a lack of distinction between the current condition and the desired future conditions. Although there was no evident relationship between an interviewee's interest group identification and problem definition, we did find that the four overlapping problem definitions rely on specific value claims (Table 8.5). The four competing problem definitions are described below.

8.3.1 Biophysical Definition

The biophysical definition addresses the broad goal of keeping the Red Desert to Hoback corridor materially permeable to migrating deer. It was the most frequently cited problem definition (Table 8.5). As one University of Wyoming researcher framed it, 'Well, I'd like for it [the corridor] to remain intact, that's the primary goal' (interview, 2016). Because of its focus on material obstacles to migration, this definition invokes strategies such as installing wildlife-friendly fencing, improving highway crossing conditions and securing private lands along the route with conservation easements.

In many cases, the strategies advocated by adherents to this definition were seen as *win–win* solutions across the participant community. In cases such as highway crossings and fence modifications, approaching the migration from a biophysical definition addresses a wide range of interests in a simple tangible manner with minimal conflict. At the time of our fieldwork, a range of participant groups, including state agencies, NGOs and private landowners were actively organizing to collaborate on these sorts of projects along the migration corridor.

The biophysical definition, however, also encompasses challenges that cannot be resolved without asymmetrical impact to the various interest groups involved. A major goal of many interviewees in the research and NGOs interest groups is achieving permanent conservation status for the entire corridor. One University of Wyoming researcher described an ideal outcome to the case as an overlapping series of conservation policies tailored to the many jurisdictions and existing land uses of different corridor segments. Ideally, this would include easements on private land, conservation designation on state-owned parcels and conservation stipulations in the management guidelines for federal lands (interview, 2016). Other interviewees advocated for an even stricter umbrella approach to protecting the corridor: 'We want the corridor off limits to oil and gas. We want the corridor off limits to any other development threat that could exist' (NGO staff member, 2016). Many in the ranching and energy industries reject this proposal, which would have consequences for their ability to do business, despite sharing a biophysical definition of the problem.

8.3.2 Social Definition

The social problem definition, acknowledging the legacy of human–human conflict surrounding many wildlife management cases in the GYE, was invoked by eleven of our forty-five interviewees. Participants focusing on the social definition recognized the range of conflicting interests raised by the biophysical realities of the Red Desert to Hoback migration. They saw the main challenges facing a desirable outcome as social – getting the diverse interest groups to the table to find a common interest path forward. Specifically, participants concerned with the social problem definition see tension building between different interest groups despite widely shared affection for the resource. Tensions between engaged participants can arise even in cases when the species in question are generally well regarded, like elk or mule deer (Vernon &

Clark 2016), in addition to more contentious species like wolves (Farrell 2015). Our research highlighted several key areas of tension.

First, our research confirmed concerns that the policy-making process surrounding the migration might exclude certain interest groups. Many participants in the case are already aware of this risk, and actively seeking to address it. However, their efforts have only been partially successful. The University of Wyoming Ruckelshaus Institute held a large, multi-stakeholder conference in the autumn of 2015 with the express intent of bringing the full range of interest groups to the table to discuss migration as a policy issue. One University of Wyoming researcher described the situation thusly: 'If we were going to have a conference around conservation and migration corridors, I wanted to make sure that it was relevant to all parties, right? That we were not just focusing on what was interesting and relevant to those people that had a mission to protect these corridors' (interview, 2016). At same time, several other interviewees observed that despite this good intention, the forum structure presented barriers to inclusivity, including its considerable distance from the migration route itself, and its incompatibility with the busy late autumn schedule of ranchers living along the corridor. Additionally, many other gatherings of participants mentioned in our interviews were exclusive (Table 8.2).

Second, many interviewees noted a tendency towards polarization in the participant community (Table 8.2). Similar to participant awareness of the risks of exclusivity, many interviewees recognized that the risk of polarization should be considered too. Awareness alone is insufficient to prevent polarization. Trends towards polarization were compounded by lack of expertise in identifying shared interests and clarifying and securing common-interest policy. One NGO staff member described chronic gridlock among members of relatively value-aligned NGOs as evidence of this point, noting 'The partnership coalition [of NGOs working to protect the migration] was always ... a little over-structured, and a little under-action' (interview 2016). This element of the social problem definition is interesting given that many participants and interest groups viewed themselves as fully capable collaborators, and viewed Wyoming as a state in which the ability to work together was possible due to its small population and homogenous interest groups.

8.3.3 Governance Definition

The governance definition is concerned with the imbalance of power (authority and control) among participants. This view questions the issue of who gets to decide how decisions regarding the migration should be made, and was cited by eleven of our forty-five interviewees. Participants focusing on this problem definition orient towards ensuring that the decision-making arenas constitute a relatively level playing field for all parties involved.

One of the most common expressions of this view was awareness that not all interest groups involved need to collaborate in order to achieve their goals. Both state and federal agencies were often perceived as being privileged in this way: 'they [Wyoming Game and Fish Department] have public meetings and being a typical government agency . . . You go in there and when you go out of the meeting they do what they're going to do anyway' (Pinedale area landowner, interview, 2016). The energy industry was also seen as possessing disproportionate influence. Despite what conference organizers describe as many genuine overtures, the energy industry largely refused to participate in the University of Wyoming's 2015 stakeholder forum on wildlife migrations. One University of Wyoming researcher observed that 'they have a better [alternative] than engaging in a collaborative process', specifically, a direct line to the governor's office (interview 2016).

Power imbalance was a major challenge facing a common-interest policy outcome for the migration. Many participants – often in the energy, landowner and ranching communities – voiced a concern that out-of-state interests were attempting to influence the outcome of migration policy in the case, and natural resource policy across the American West more generally. As one hunting outfitter put it, 'Wyoming has a problem with people from outside of Wyoming trying to dictate what we should do' (interview, 2016).

8.3.4 Non-Issue Definition

In addition, a fourth problem definition was at play in this case. Seven of our forty-five interviewees believed that participants have largely achieved the goal of ensuring the migration. This perspective rests on the ideas that (1) challenges facing the migration have been adequately addressed, and that (2) other challenges cited by certain interest groups were irrelevant to the goal of sustaining the migration and did not need further attention (Table 8.5). This second perspective rests on the perception that Wyoming remains flush with abundant wildlife. For example, 'we have great resources here so it's kind of hard to mess up a good thing' reported one Wyoming Game and Fish Department employee (interview, 2016).

8.4 DISCUSSION

Our findings capture complex social dynamics in the Red Desert to Hoback mule deer migration case. Although all participant groups are individually committed to coexistence with mule deer, competing value claims and differing problem definitions threaten to undermine collaborative efforts to clarify and secure a common-interest management policy. These points of tension among people and groups underscore the point that human–wildlife conflict is a symptom of human–human conflict (Rocheleau 2017; Wallace 2017), a perspective of increasing academic and policy significance that will be crucial to advancing the theory and practice of human–wildlife coexistence. In this section, we offer a problem-oriented perspective that takes the above into explicit account. We also offer recommendations for improving the policy dynamic in this case.

8.4.1 Competing Perspectives: Goals

"The idea of corridors in general, and maintaining wildlife corridors, has percolated in all participants. Nobody is against them conserving corridors. There is no opposition to [the] idea of corridors" (interview 2016).

In this case, understanding competing perspectives, values and strategies is challenging, because to a degree all participants truly do share an interest in deer conservation. However, a shared interest is not necessarily the common interest (Lasswell 1971); while a shared interest reflects broad alignment around an overarching normative claim (e.g. the migration should be allowed to persist), it does not meet the criteria of alignment of specific goals, methods and execution that characterize a common interest. Surrounding the shared interest of maintaining the Red Desert to Hoback migration, there is a diversity of perspectives, value claims and competing problem definitions. This gives rise to a range of potentially competing strategies (e.g. achieving a no-development conservation status on the entire corridor versus keeping the corridor open to energy development). Recognizing and addressing these foundational differences – in this case with the aid of a comprehensive analytical framework like the policy sciences – is an essential step in clarifying, securing and sustaining a common-interest outcome, and thereby creating the conditions of harmony between human stakeholders needed to ensure human–wildlife coexistence.

8.4.2 Competing Value Outlooks

Overt conflicts of interest were relatively rare. Even interest groups with largely oppositional agendas like the NGO community and the energy industry acknowledged some validity to each other's claims. More prevalent, and more problematic from a policy perspective, were cases in which interest groups seeking the same value – power in the decision-making processes surrounding the migration, for example – viewed the claims of others as harmful to their own interests.

Tension among the ranching community, NGOs and researchers over representation (a power concern) exemplifies this point. More than half of the ranchers that we interviewed expressed a desire for greater influence in decision-making (Table 8.2). NGOs and researchers did not orient overtly to power concerns (Table 8.2). However, they readily participated in the Red Desert to Hoback Partnership, the WMI and the University of Wyoming Ruckelshaus Institute's 2015 Emerging Issues Forum (Gorman 2015; Phippen 2016; Wilkinson 2016) – all arenas that amplified their perspectives and gained the attention of policy-makers. Many ranchers that we interviewed saw the rise of NGOs and researchers to public prominence as a zero-sum power game. They expressed a view that they would be diminished in the eyes of policy-makers and the public to the extent that the NGOs and research communities gained prominence.

This value conflict is difficult to perceive and manage constructively. Ranchers, researchers and NGOs all voiced strong affection for the resource, despite their conflicts with each other. We suggest that careful, systematic attention to social context mapping at the outset of policy and media campaigns is crucial to formulating policy that addresses the needs of all stakeholders and avoids creating conditions for human–human conflict. In this case, such an approach could yield a more open and inclusive decision-making process, better communication between stakeholder groups and greater clarity on shared goals, making it the best strategy to promote human–wildlife coexistence.

8.4.3 Competing Problem Definitions

Competing problem definitions are another largely unacknowledged challenge facing the Red Desert to Hoback migration conservation effort. Significantly, few interviewees acknowledged the existence of multiple competing problems at play in the case. Such perspectives are problematic because interviewees even within the same interest group or organizations often unknowingly oriented towards different problem definitions, limiting their ability to identify and advocate for the common interest. The NGO community illustrates this point most clearly. The NGO staff across all organizations differentially promoted one or more of the four problem definitions. While interviewees in this community acknowledged the diversity of challenges facing the migration, few recognized that their peers were orienting to fundamentally different aspects of this suit of challenges. However, many observed that the NGOs partnership organized to address the migration was plagued by disorganization and lack of momentum – symptoms of the tacit competition between diverse problem definitions.

Competing problem definitions in natural resource management cases in the GYE are not a new phenomenon (Clark 2008), and resolving the conflicts they create remains an elusive policy challenge. For example, Cherney and Clark (2008) document three problem definitions in efforts to protect a pronghorn antelope (*Antilocapra americana*) migration originating in Grand Teton National Park and running south towards the Red Desert to Hoback corridor. A decade before that, competing problem definitions in the endangered black-footed ferret (*Mustela nigripes*) case seriously hampered species recovery (Clark 1997). More currently, Vernon and Clark (2016) detailed the competing problem definitions in elk management in western Wyoming. Likewise, Farrell (2015) and Jaicks (2016), respectively, documented controversy surrounding wolf (*Canis lupus*) reintroduction and other wildlife issues in the GYE. In all cases, domination of the biophysical problem definition precludes recognition and effective management of co-occurring social and governance problem definitions, which are central to conservation. Conflicting perspectives, value claims and problem definitions limit the potential of participants in the Red Desert to Hoback to achieve a lasting human–wildlife coexistence outcome.

8.4.4 Underlying Conditions

How can we understand the competition among multiple, unacknowledged perspectives, value claims and problem definitions, evident in this case, and in wildlife management in the GYE more generally? We suggest that two central, overarching conditions must be taken into account. First, there is a tendency to reduce issues that are fundamentally and profoundly human to a limited set of physical elements, externalized in language and thought as being *out there* (e.g. fences as barriers to migration). This view puts ecology and wildlife biology in the driver's seat, and forces social and cultural considerations into the rear. It also maintains a positivist view of knowledge, knowing and problem-solving as dominant (Castree 2013). Second, and closely related to this first element, there is a tendency to reduce human values discourses to the quantifiable.

The first condition – scientism – has a strong appeal in natural resource management cases. By reducing a complex human and policy challenge to a limited set of biophysical elements, the challenge can be most easily navigated within established thinking and by a single, dominant world-view: predictive science (Brunner 2013). This approach dominates the Red Desert to Hoback migration case, and was voiced by a plurality of interviewees (sixteen out of forty-five), who oriented towards the biophysical problem definition. As long as the biophysical challenges remain the dominant focus of attention for policy-makers and others, the social and governance challenges underlying this case, and similar cases, will remain unacknowledged and inadequately addressed (Schon 1983).

The second condition – the tendency to reduce value discourses to the quantifiable (Chan et al. 2012) – has particular consequences for social process in a case like Red Desert to Hoback. In this case, values animating participant engagement range from concrete (e.g. financial interest in lands along the corridor) to intangible concerns (e.g. identity with being a steward of the land). However, the policy process of this case is not formally equipped to address the latter class of concern. Consequently, stakeholders with values that are easily quantifiable are well represented, while those pursuing less tangible values are chronically disadvantaged. This outcome limits the range of people, perspectives, knowledge and skills available to address the conservation challenges, at all levels and concerns. It may even jeopardize finding a common interest that can be secured.

8.4.5 Moving Forward

The fragmentation, competition and reactivity in the deer migration case evident from our study complicate efforts to ensure human–wildlife coexistence. We suggest that a more comprehensive problem definition is needed to address the full complexity of this case. For example, McDougal et al. (1981) describe arenas that can address differing value claims, and other issues we identified at play in the deer case. Such arenas provide sufficient opportunity for all participants to exercise influence over the process, and hold all participants accountable to the decisions they make. We argue that the trends towards competition among largely unacknowledged value claims and problem definitions in the Red Desert to Hoback case are symptomatic of a social and decision-making process that is neither fully factual, comprehensive or ameliorative nor sufficiently accountable and pragmatic (see Clark 2002).

We offer two suggestions for improving the quality of the decision-making arenas in this case. First, we recommend that an overarching problem definition, well grounded in social concerns (e.g. recognizing the social process and governance definitions) as well as biophysical concerns, be established. This should allow social and governance issues to receive the same level of attention as the biophysical definition, which has dominated and limited the process to date. Such a recommendation proved successful in similarly complex cases of wildlife conservation (Oppenheimer & Richie 2014). Second, our findings highlight the need for greater accountability in this case, by ensuring that all interests that contribute to decisions are held to the terms and consequences of those decisions. Such a structure would encourage decision-makers to attend more realistically and equitably to competing value claims and problem definitions. These steps should facilitate a process that identifies and secures an enduring common interest that achieves human–human and human–wildlife coexistence in practice (see Richie et al. 2012).

8.5 CONCLUSION

Securing a common interest management policy for human–wildlife coexistence requires a comprehensive understanding of the social and value context and content as in this case. Achieving this requires moving well beyond the ecology of the migration and the geography of the corridor. In this case, lack of systematic attention to implicit value claims, divergent problem definitions and untested assumptions of shared interests threaten conserving the herd and its migration. Whereas participants must continue to address the biophysical challenges, so too should they attend to the social and governance challenge.

We suggest that a systematic, data-grounded understanding of the social context and decision process offers a path forward. An integrative approach can help in understanding where along the conflict-to-coexistence continuum individuals and groups stand and harmonize people's differences, hopefully securing a common interest.

8.6 RECOMMENDATIONS AND FUTURE DIRECTIONS

- This case demonstrates that the linkage between lack of attention to social context in wildlife management and human–human conflict can complicate efforts towards human–wildlife coexistence. Further research should look for and seek to test linkages between high degrees of attention to/amelioration of human–human conflict and desirable human–wildlife coexistence outcomes.
- This study focuses on establishing the social context baseline needed for participants in the Red Desert to Hoback case to create a more effective decision-making process towards the goal of common-interest policy. To this end, we mapped the social process and problem definition functions of Lasswell's policy sciences (1971). However, a complete policy science study will also include an analysis of the decision process function as pertaining to current policy initiatives in this case. Such analysis will be critical to offering concrete suggestions for improved practice beyond the broad strokes outlined in this chapter.
- Practitioners working towards human–wildlife coexistence in this and related cases should incorporate the idea of a conflict-to-coexistence spectrum as outlined by Frank (2016) into their analysis of both human–wildlife and human–human interactions. Recognition that the willingness of human participants in such cases to work together towards coexistence goals can have a direct bearing on the success of human–wildlife coexistence efforts is a critical next step in the development and application of conflict–coexistence theory.
- Practitioners in the Red Desert to Hoback migration exhibited varying degrees of awareness of the social context underlying this management challenge, and varying levels of expertise in incorporating that awareness into their decision-making processes. We suggest that practitioners in this case and related cases seek resources and training in integrative, interdisciplinary decision-making to better incorporate social context and ecological context into their management decisions.

8.7 ACKNOWLEDGEMENTS

We thank the Knobloch Family Foundation, Berkley Conservation Scholarship and UCross High Plains Stewardship Initiative for funding, and the Northern Rockies Conservation Cooperative for assistance in the field. Justin Farrell, Amity Doolittle and Timothy Terway offered critical advice, and Katherine Panek assisted with cartography. We are indebted to Jessica Johnson, Albert Sommers and Julia Stuble for support on the ground in Wyoming.

8.8 References

Bowen, G. A. (2009). Document analysis as a qualitative research method. *Qualitative Research Journal*, 9(2), 27–40.

Brunner, R. D. (2013). Introduction to the Transaction edition. In H. D. Lasswell & A. Kaplan, eds., *Power and Society: A Framework for Political Inquiry*. New Brunswick, NJ: Transaction Publishers.

Brunner, R. D. & Steelman, T. A. (2005). Beyond scientific management. In R. D. Brunner, T. A. Steelman, L. Coe-Juell, C. M. Cromley, C. M. Edwards & D. W. Tucker, eds., *Adaptive Governance: Integrating Science, Policy, and Decision Making*. New York: Columbia University Press, pp. 1–47.

Castree, N. (2013). *Making Sense of Nature*. Abingdon: Routledge.

Chan, K. M. A., Guerry, A. D., Balvanera, P., Klain, S., Satterfield, T., Basurto, X., Bostrom, A., Chuenpagdee, R., Gould, R. & Halpern, B. S. (2012). Where are cultural and social in ecosystem services? A framework for constructive engagement. *BioScience*, 62(8), 744–56.

Cherney, D. N. & Clark, S. G. (2008). The American West's longest large mammal migration: Clarifying and securing the common interest. *Policy Sciences*, 42(2), 95–111.

Clark, S. G. (2002). *The Policy Process: A Practical Guide for Natural Resources Professionals*. New Haven, CT: Yale University Press.

Clark, S. G. (2008). *Ensuring Greater Yellowstone's Future*. New Haven, CT: Yale University Press.

Clark. S. G. & Rutherford, M. B. (2014). *Large Carnivore Conservation: Integrating Science and Policy in North American West*. Chicago: Chicago University Press.

Clark, S. G. & Vernon, M. (2015). Governance challenges in joint inter-jurisdictional management: Grand Teton National Park, Wyoming, elk case. *Environmental Management*, 56, 286–99.

Clark, S. G. & Vernon, M. (2017). Elk management and policy in Greater Yellowstone: Assessing the constitutive process. *Policy Sciences*, 50(2), 295–316, 1–22.

Clark, S. G. & Wallace, R. L. (2015). Integration and interdisciplinarity: Concepts, frameworks, and education. *Policy Sciences*, 48(2), 233–55.

Clark, T. W. (1997). *Averting Extinction: Reconstructing Endangered Species Recovery*. New Haven, CT: Yale University Press.

Farrell, J. (2015). *The Battle for Yellowstone: Morality and the Sacred Roots of Environmental Conflict*. Princeton, NJ: Princeton University Press.

Frank, B. (2016). Human–wildlife conflict and the need to include tolerance and coexistence: An introductory comment. *Society & Natural Resources*, 29(6), 738–43.

Gorman, J. (2015). For mule deer, an incredible 150-mile migration. *The New York Times*, 2 February. Available from www.nytimes.com/2015/02/03/science/deer-on-the-move.html (accessed May 2016).

Hohl, A. M., Picard, C. H., Clark, S. G. & Middleton, A. (2015). Approaches to large-scale conservation: A survey. In S. G. Clark, A. M. Hohl, C. H. Picard & E. Thomas, eds., *Large-Scale Conservation in the Common Interest*. Springer Series on Environmental Management. New York: Springer International Publishing, pp. 29–51.

Jaicks, H. F. (2016). *The Conflicts of Coexistence: Rethinking Humans' Placement and Connections with Predators in the Greater Yellowstone Ecosystem*. PhD Thesis. New York: New York University, chapter 4.

Lasswell, H. D. (1971). *A Preview of the Policy Sciences*. New York: American Elsevier.

Mattson, D., Karl, H. A. & Clark, S. G. (2012). Values in natural resource management and policy. In H. A. Karl, L. Scarlett, J. C. Vargas-Moreno & M. Flaxman, eds., *Restoring Lands: Coordinating Science, Politics and Action*. Dordrecht: Springer Netherlands, pp. 239–59.

McDougal, M. S., Lasswell, H. D. & Reisman, M. W. (1981). The world constitutive process of authorative decision. In M. S. McDougal & W. M. Reisman, eds., *International Law Essays: A Supplement to International Law in Contemporary Practice*. New York: Foundation Press.

Oppenheimer, D. & Richie, L. (2014). Collaborative grizzly bear management in Banff National Park: Learning from a prototype. In S. G Clark & M. B. Rutherford, eds., *Large Carnivore Conservation: Integrating Science and Policy in the North American West*. Chicago: Chicago University Press, pp. 215–44.

Phippen, W. J. (2016). America's wildlife corridors are in danger. *The Atlantic*, 14 December. Available from www.theatlantic.com/science/archive/2016/12/deer-migration/509033/?utm_source=atltwThe (accessed May 2016).

Primm, S. A. & Clark, T. W. (1996). The Greater Yellowstone policy debate: What is the policy problem? *Policy Sciences*, 29(2), 137–66.

Richie, L. J., Oppenheimer, D. & Clark. S. G. (2012). Social process in grizzly bear management: Lessons for collaborative governance and natural resource policy. *Policy Sciences*, 45, 265–91.

Rocheleau, B. (2017). *Wildlife Politics*. Cambridge: Cambridge University Press.

Sawyer, H. (2014). *Seasonal Distribution Patterns and Migration Routes of Mule Deer in the Red Desert and Jack Morrow Hills Planning Area*. Laramie, WY: Western Ecosystems Technology, Inc.

Sawyer, H., Hayes, M., Rudd, B. & Kauffman, M. (2014). *The Red Desert to Hoback Mule Deer Migration Assessment*. Laramie: Wyoming Migration Initiative, University of Wyoming.

Sawyer, H., Korfanta, N. M., Nielson, R. M., Monteith, K. L. & Strickland, D. (2017). Mule deer and energy development: Long-term trends of habituation and abundance. *Global Change Biology*, 23(11), 4521–29.

Sawyer, H., Middleton, A. D., Hayes, M. M., Kauffman, M. & Monteith, K. L. (2016). The extra mile: Ungulate migration distance alters the use of seasonal range and exposure to anthropogenic risk. *Ecosphere*, 7(10), e01534.

Schon, D. A. (1983). *The Reflective Practitioner: How Professionals Think in Action*. New York: Basic Books.

Steelman, T. A. & DuMond, M. E. (2009). Serving the common interest in U.S. forest policy: A case study of the Healthy Forests Restoration Act. *Environmental Management*, 43, 396–410.

Taylor, D. & Clark, S. G. (2013). Management context: People, animals, and institutions. In T. Clark, M. Rutherford & D. Casey, eds., *Coexisting with Large Carnivores: Lessons from Greater Yellowstone*. Washington, DC: Island Press.

Terway, T. (2017). *Sustained in Significance with(out) Context and Ourselves: Expert Environmental Knowledge and 'Social-Ecological-Systems'*. PhD Thesis, New Haven, CT: Yale University.

Thuermer, A. M. (2017). Ranch owner builds in path of pronghorn. *WyoFile*, 3 January. Available from www.wyofile.com/ranch-owner-builds-path-pronghorn/ (accessed January 2017).

Vernon, M. E., Bischoff-Mattson, Z. & Clark, S. G. (2015). Discourses of elk hunting and grizzly bear incidents in Grand Teton National Park, Wyoming. *Human Dimensions of Wildlife*, 21(1), 65–85.

Vernon, M. E. & Clark, S. G. (2016). Addressing a persistent policy problem: The elk hunt in Grand Teton National Park, Wyoming. *Society & Natural Resources*, 29(7), 836–51.

Wallace, R. (2017). *Welcome and Opening Remarks*. Paper presented at the 6th Annual Jackson Hole Wildlife Symposium, Jackson, WY, 10 March.

Weiss, J. (1989). The powers of problem definition: The case of government paperwork. *Policy Sciences*, 22, 92–121.

Weiss, R. S. (1995). *Learning from Strangers: The Art and Method of Qualitative Interview Studies*. New York: Simon and Schuster.

Wilkinson, T. (2016). Yellowstone National Park: Great migrations – keeping Yellowstone's lifeblood flowing. *National Geographic Magazine*, 18 April. Available from www.nationalgeographic.com/magazine/2016/05/yellowstone-national-parks-animal-migration/ (accessed April 2016).

9

A Collaborative Approach for Coexistence with Wildlife in Rural Regions of Japan

RYO SAKURAI

Human–wildlife conflicts, including agricultural and property damage as well as occasional human casualties, occur all over the world where wildlife and people live in close proximity (Conover 2002; Manfredo et al. 2009). Yet the background of how and why conflicts occur differs among countries with varied cultural and social settings (Woodroffe et al. 2005; Manfredo 2008). In Japan, human–wildlife conflicts are one of the biggest issues that people face, especially those living in rural areas. The average annual cost of agricultural and forestry damages caused by wildlife was around 20 billion yen ($200 million) in 2008–14 (Ministry of Agriculture, Forestry and Fisheries 2014a). In 2014 alone, 6.5 billion yen in damages were caused by deer (*Cervus nippon*), 5.5 billion yen by wild boars (*Sus scrofa*) and 1.3 billion yen by macaques (*Macaca fuscata*) (Ministry of Agriculture, Forestry and Fisheries 2014a). Casualties caused by wildlife, such as the Asiatic black bear (*Ursus thibetanu*; hereafter bear), also create human–wildlife conflicts. This species, with its estimated population of 15,000–20,000 individuals, is listed as Vulnerable (Ministry of the Environment 2013; International Union for Conservation of Nature (IUCN) 2016). In the drought of 2010 (fiscal year from April 2010 to the end of March 2011), when large numbers of bears appeared around human settlements, 3,538 bears were captured and 3,074 of them were killed to prevent conflicts with local residents and impacts on agriculture and forestry (Ministry of the Environment 2016a). In the same year, more than 142 people were injured and 2 people were killed by bears (Ministry of the Environment 2016b).

There is a diversity of reasons that may have led to such a high amount of damages and casualties. The depopulation and ageing of

rural communities have caused changes in land use where the abandonment of agricultural fields has led to the increase of wildlife habitats near
human settlements (Ministry of Agriculture, Forestry and Fisheries
2008). Since the 1970s, deer habitat has increased 2.5 times, while that
of wild boars has risen 1.7 times (Ministry of the Environment 2015).
Additionally, the total population of Japan is declining, and is estimated
to be less than 100 million in 2050 compared to 127 million in 2013
(Cabinet Office Japan 2014). A decrease in hunters further favours the
spread of certain overabundant species across the country (Tsunoda &
Enari 2012). In the 1970s, there were more than 500,000 hunters in
Japan, whereas by 2009 there were fewer than 150,000 because of the
strict gun control policy enforced by the government (Ueda et al. 2010;
Kaji et al. 2013). All these factors contribute to human–wildlife conflicts
in Japan.

Within Japan, human–wildlife conflicts are especially concentrated
in mountainous regions. Diverse interventions have been undertaken in
such areas to reduce damages, such as installing fences to prevent
damage by wildlife, removing underbrush or cutting grass to form
buffer zones between wildlife and humans and removing unnecessary
fruit trees that attract wildlife (Agricultural Production Bureau of Ministry of Agriculture, Forestry and Fisheries 2009). Because of extended
agricultural damage and occasional casualties, the majority of local
residents across rural areas of Japan hold negative dispositions towards
wildlife (Knight 2006; Sakurai et al. 2014b). Previous studies revealed
that local residents felt that there were too many deer, wild boars and/or
bears around their regions, and the majority of them wanted to get rid
of/destroy wild species living around their villages (Ministry of the
Environment 2007; Sakurai et al. 2014a, 2014b). A study conducted in
the remote area of Hyogo Prefecture (western part of Japan) revealed
that, out of 868 residents surveyed, nearly 40 per cent believed that deer
and wild boars should be culled. Another 30 per cent preferred that
macaques and bears be culled, showing little tolerance towards species
living near human settlements (Sakurai et al. 2014b). In rural areas of
Japan where human–wildlife conflicts are high, residents may have
extremely negative attitudes towards wildlife, which positions them on
the *conflict* side of the conflict-to-coexistence continuum (Frank 2016). If
people hold negative perceptions towards wildlife, it is challenging to
foster coexistence. Only a change in one's mindset towards wildlife will
allow for movement of mental dispositions from the negative to the
neutral/positive side of the continuum.

9.1 MOVING TOWARDS HUMAN–WILDLIFE COEXISTENCE

Previous studies revealed the importance of people's tolerance for co-existing with wildlife. Some studies examined the degree of local people's tolerance towards certain species such as cougars (e.g. Riley & Decker 2000), wolves (e.g. Slagle et al. 2012; Vaske et al. 2013) and bears (e.g. Zajac et al. 2012; Sakurai et al. 2013b) by asking stakeholders if there were too many of these species, while others looked at factors influencing tolerance such as risk perceptions and social trust (e.g. Gore et al. 2006; Bruskotter & Wilson 2014). Other researchers focused on wildlife-based tourism and revenues generated by attracting visitors from all over the world to communities living with wildlife as an approach to change local people's mindset from conflict to coexistence (e.g. Child 1996; Gandiwa et al. 2013). Effective communication was another theme explored by researchers in relation to increasing people's tolerance towards wildlife. Bruskotter and Wilson (2014), for example, discussed how communication should not only focus on the risks associated with species, but also on the benefits of living with wildlife. Sakurai et al. (2013b) further revealed that in order to increase residents' trust towards agencies regarding wildlife management, a two-way dia-logue is needed, where agencies disseminate messages and listen to people's concerns and necessities.

The success of the approaches discussed in the previous paragraph in fostering coexistence depends upon the region and country, cultural and historical background and society in which they are carried out, and might not necessary be a feasible approach everywhere (Waylen et al. 2010). In Japan, for example, local residents might not want to attract foreigners to their community to avoid disrupting their traditional life-style. In comparison to Western countries, Japanese are known to traditionally have strong subjective norms and bonds within community members, and attachment to their group of belonging – also known as *ittaikan*, which can be translated as oneness with fellow members of the community (Abrams et al. 1998; Gudykunst & Nishida 1999/2000). Hints and tips for enhancing human–wildlife coexistence, especially in unique cultural contexts such as Japan, can be found by looking at specific situations in which people are trying to share the landscape with wildlife. Such an understanding can help further develop the conflict-to-coexistence continuum by offering an example about how culture and unique social contexts influence people's mental dispos-itions towards wildlife.

In order to take the cultural aspect of the study sites into consideration, research methods need to be chosen carefully. To better understand how social contexts influence mental dispositions towards wildlife, qualitative data were collected through two case studies on collaborative approaches for coexisting with wildlife in rural areas of Japan. The methods used for these studies were, basically, interview (as well as survey) and observation of activities. In Hyogo Prefecture, students living in urban areas visited rural villages to revitalize the region and help local residents prevent wildlife damages (Ichikawa & Sakurai 2012). In Tochigi Prefecture, local residents and the local government hosted a 'Parent and Child Meeting on Interventions towards Wild Boars', which consisted of cooking and eating wild boars hunted around the region to increase awareness about the species (Sakurai et al. 2013a). These collaborative activities led by various stakeholders, including local residents, urban students, government and researchers, are discussed further in the following section. These examples show how the presence of wildlife can benefit community development by providing opportunities for residents to work together towards human–wildlife coexistence.

9.2 CASE STUDY OF HYOGO PREFECTURE: COLLABORATION BETWEEN STUDENTS AND LOCAL RESIDENTS

In Toyooka City, in the northern part of Hyogo Prefecture (Figure 9.1), damage to forestry and agriculture industries by wildlife has become a severe problem. A breakdown of the agricultural damage in the prefecture shows that 34 per cent is caused by wild boars and 29 per cent by deer, indicating that harm by these two species alone accounts for more than 60 per cent of the entire impact (Hyogo Prefecture 2010). Within this prefecture, macaques also damage agriculture and threaten daily lifestyle by intruding into houses to search for food (Hyogo Prefecture 2009). Human–bear conflict is also increasing due to the recent rise in the bear population (Yokohama et al. 2008). In Japan, the smallest administrative division is called a *district* and each district typically has a district leader. In small districts where depopulation and ageing of society have advanced, it is difficult for local residents to take measures to prevent damage by wildlife (Kawai & Hayashi 2009).

To overcome a labour shortage, joint university and government programmes were developed across Japan. These programmes involved city students interested in human–wildlife conflicts and local residents implementing damage-prevention countermeasures. Through such an

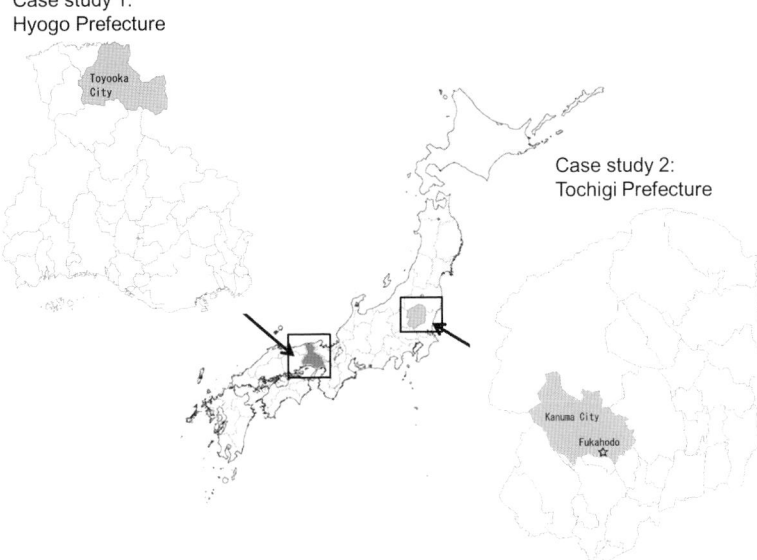

Case study 1:
Hyogo Prefecture

Toyooka City

Case study 2:
Tochigi Prefecture

Kanuma City
Fukahodo
☆

Figure 9.1 Location of Toyooka City in Hyogo Prefecture (case study 1) and Fukahodo within Kanuma City in Tochigi Prefecture (case study 2).

approach, opportunities for personal contact between city-dwellers and residents of agricultural communities were created as students travelled to regions that actually suffered from human–wildlife conflicts to study real wildlife issues and gain valuable experiences that were impossible to obtain in the city. Such programmes are conducted nationwide and include: the Monkey-Persimmon Environmental Education Programme in Nagano Prefecture (Japan Bear Network 2010), the Underbrush Cutting Project in Karuizawa Town in Nagano Prefecture (Karuizawa Web 2011) and the Jyugai Ranger Programme in Hyogo Prefecture. The latter programme will be discussed in depth in the following paragraphs.

The Jyugai (meaning 'wildlife issues') Ranger Programme of Hyogo Prefecture focuses on implementing measures to prevent agricultural and forestry damage. Another goal of this programme is to teach students about the attractive features of rural regions by participating in traditional regional craftmaking and farming. The Tajima District Administration Office of Hyogo Prefecture and the faculty of a technical college in Osaka City launched the Jyugai Ranger Programme in 2009. The Tajima District Administration Office of Hyogo Prefecture is in charge of finding new districts and handling the preliminary discussions with the local residents, while the students taking part in this programme are recruited and managed by the faculty of the technical college. By January 2013, seventeen two-day / one-night programmes in ten districts were conducted – mainly in the northern parts of Hyogo Prefecture. Since it was assumed that the Jyugai Ranger Programme could revitalize rural areas through active interaction between local residents and urban students, research was conducted in four of those districts to understand the effects of the programme by measuring awareness of local residents and students who participated in the programme.

9.2.1 Study Sites and Methods

Interviews were conducted with local residents and questionnaires were administered to students in four districts where damage by wildlife was particularly severe in comparison with other parts of Toyooka City (personal communication, Division of Oriental White Stork and Human Coexistence of Toyooka City Government) (Ichikawa & Sakurai 2012). The numbers of households in each district were: twenty-five in the Ogouchi district, thirty in the Mihara district, fifty-seven in the Hirata district and sixty-four in the Akabana district. A total of twenty-six students participated in the Jyugai Ranger Programme in these four districts. Consistent with other rural areas across Japan, Toyooka City is facing ageing and depopulation. In this area, there was a decrease of 3,727 individuals from 2006 to 2011 (Toyooka City Government 2012). Those who were more than 75 years old increased by 748 individuals, and 1 out of 3.7 persons were more than 65 years old in 2011 (Toyooka City Government 2012). The situation of ageing and depopulation is even more severe in mountainous villages. Among the four districts where this study was conducted, more than 30 per cent of residents were older than 65 years of age; in one district, about half of the residents were older than 65 years.

Activities implemented in the Jyugai Ranger Programme included:

- cutting grass, underbrush and bamboo that provide hiding places for wildlife around farmlands and residential areas to make a buffer zone (Figure 9.3);
- repairing and erecting fences around farmlands to prevent wildlife damage;
- holding regional get-togethers (e.g. barbecues, dinners in the community hall);

Figure 9.2 Landscape of one of the districts where the Jyugai Ranger Programme was implemented. *A black and white version of this figure will appear in some formats. For the colour version, please refer to the plate section.*

Figure 9.3 Activities of cutting brush. *A black and white version of this figure will appear in some formats. For the colour version, please refer to the plate section.*

Figure 9.4 Activities of making *mochi*; traditional rice cakes.

- making traditional crafts and *mochi* (traditional rice cakes) (Figure 9.4); and
- cutting persimmon and chestnut trees to prevent wildlife from coming into the village.

After the completion of activities on the second day, local residents were asked open-ended questions about (1) their impression about undertaking damage interventions with students, (2) benefits and disadvantages of the activities, (3) impacts of the activities on the district and (4) whether or not they wanted to perform such activities again in the future. Each interview took approximately 5–10 minutes to complete and a total of thirty-eight local residents participated in the survey. Since the survey was conducted on those who participated in the activities, these thirty-eight respondents do not necessarily represent the whole of the residents in the area. Accordingly, on the second day, all participating students (n = 26) were asked to answer a questionnaire after damage countermeasure activities. The questionnaire included items focused on understanding students' impressions of participating in the activities, changes in their awareness on rural areas, etc., throughout the activity and their desire to participate in future activities. For most items, students were asked to choose among the following response categories in the questionnaire: *I think so, Neither option* and *I do not think so.*

9.2.2 Local Residents' Perceptions of the Programme

Regarding the activities of the Jyugai Ranger Programme, most of the residents interviewed were positively impressed by the students and stated 'they helped us' (eighteen comments) or 'the young people worked hard' (thirteen comments). One resident mentioned, 'I am grateful that this programme brought students here. They were a big help.' Another resident answered, 'by coming, students stimulated [the district through their involvement in activities] and helped out the region. They exchanged information with the people of the region.' Someone also mentioned, 'young people coming here help the development of the district. I thought that most young people were not interested in such things, but I am delighted they came. We want to create a homey atmosphere and get to know them. I want this programme to continue forever.' Several of the residents believed that being able to get to know young people was an attractive feature of the programme and all of them wanted the Jyugai Rangers Programme to occur again in their district.

In response to the question about benefits and disadvantages of the project, most respondents cited positive aspects of interacting with students. When asked about the biggest problem currently facing the region, most of respondents who answered this question (n = 15) cited depopulation and ageing, and a fear that the next generation will not farm those areas. One resident mentioned, 'the fact that we are all old people is a dilemma. I am worried about the future. We do not even have a doctor. I want young people to come here.'

When asked if the activities of the Jyugai Rangers had revitalized their district, almost all (n = 32) answered that they had. One resident mentioned, 'students revitalized the village. Talking to young people cheers me up.' A minority (n = 6) felt that 'it did not revitalize our district', or stated that they did not know. A resident explained 'because they only came once, it was not effective. The next time they come will be important and will stimulate the village.'

9.2.3 Students' Perceptions about the Programme

Among students, fourteen of them were male and twelve were female; their average age was 19.5 years old. In response to the three items: *was it worthwhile?*, *do you want to participate in the Jyugai Ranger Programme again?* and *did you understand the hardship of damage countermeasures?* almost 80 per cent of the students answered positively, revealing that the participants hold a good impression of the activities undertaken. In response to, *did you understand the actual state of damage caused by wildlife in the region?*, 39 per cent answered 'I think so'. This result is likely due to the lack of explanation of the actual state of damage in the region prior to carrying out the activities. This implies that solely doing damage prevention activities would not necessarily increase students' under-standing of the damage status of sites. Opportunities such as lectures by local residents, government officials or faculties regarding the current situation of human–wildlife conflicts in the region might be necessary to increase the educational impacts on students. In response to the ques-tion, *do you want to participate in any activities undertaken in a rural area (not only the Jyugai Ranger Programme) in the future?*, approximately 60 per cent answered, 'I think so'.

As for reasons for participating in the Jyugai Ranger Programme, 27 per cent answered that they joined in because they were interested and/or in response to a request of a college faculty member. Twenty-three per cent of students specified 'I wanted to volunteer' and 'I wanted to learn about human–wildlife conflicts'. Four participants answered, 'I participated previously' and 'I wanted to meet people'. Their impres-sions of participating in the activities were often positive (e.g. 'it was a good experience' or 'I enjoyed it'). As for changes in their awareness, some students reported that it boosted their knowledge about human–wildlife conflicts or strengthened their desire to take part in other types of volunteering. Examples of these students' answers included, 'I learned how serious human–wildlife conflicts are, and it encouraged me to consider what I could do to address the problem. I hope I can contribute in the future', and 'I became interested by learning how difficult it is for people and wildlife to coexist.' Ichikawa and Sakurai (2012) offer a more in-depth analysis of the case study presented.

9.2.4 Discussion

The Jyugai Ranger Programme aimed at connecting urban students and local residents on wildlife issues. The positive answers given by local residents during the interviews suggest that accepting and interacting with city students through the programme could represent a way to increase tolerance towards wildlife by attracting young people to remote areas. Before the Jyugai Ranger Programme was carried out in the district, local residents had been implementing damage prevention activities by themselves. For local residents, wildlife such as deer, wild boars, macaques and bears were nothing more than pests, and, therefore, they had been asking the government to eliminate those species around the districts. In that situation, local residents had no tolerance towards wildlife and their mindset was, likely, consistent with the statement *it is impossible to coexist with wildlife*. By having the Jyugai Ranger Programme in the district, the damage prevention activities became a collaborative work with the young generation (with whom local residents rarely interact in daily life). During the programme, local residents felt they were stimulated and the district was revitalized through active interactions and collaborative work with students. Students came to the district because of wildlife issues, and local residents realized that wildlife and/or having wildlife issues could attract people into the district.

Knowing the benefits of having wildlife around the district could potentially increase local residents' tolerance towards wildlife. If all wildlife were to be eliminated from the district, there would be no Jyugai Ranger Programme, and no interaction with the young generation, under the theme of preventing wildlife damage. Local residents' mindsets could potentially be changed to a belief in coexistence with wildlife if young people such as Jyugai Rangers come to help them and revitalize the district.

9.3 CASE STUDY OF TOCHIGI PREFECTURE: THE PARENT AND CHILD MEETING ON INTERVENTIONS TOWARDS WILD BOARS

Tochigi Prefecture (Figure 9.1), located in central Japan, had the third-highest amount of agricultural damage caused by wildlife in the country in 2007 (Ministry of Agriculture, Forestry and Fisheries 2007), and it remains high as of 2014 (Ministry of Agriculture, Forestry and Fisheries 2014b). Species that mainly cause agricultural damage in Tochigi are wild boars (56%), deer (21%), masked palm civets (*Paguma larvata*) (13%) and macaques (7%) (Division of Rural Development of Tochigi Prefecture 2011). The Nature Preservation Division of the prefectural government is responsible for managing wildlife and preventing damage. The division started a wildlife damage prevention project, the *Model District Programme*, in 2010 to reduce wildlife damage by increasing residents' awareness towards wildlife issues and encourage damage prevention activities. In selected districts, this programme offers a variety of activities every few months, including seminars focused on regional wildlife issues, field trips to check wildlife damage sites (Figure 9.5) and implementation of interventions such as cutting abandoned fruit trees and brushes that can attract wildlife. By implementing these activities, the division aims at increasing residents' knowledge of wildlife issues, and their confidence and willingness to engage in damage prevention interventions. Specifically, the goal of the division is to provide the skills needed by participants to prompt them to implement wildlife damage prevention activities on their own.

Figure 9.5 Field trip to damage site.

9.3.1 Collaborative Activity of Fukahodo District

Fukahodo District is located in the south-west part of the prefecture, and has 189 households. It was designated as a Model District in 2010 because the degree of wildlife damage was high compared to other districts in the city. A Satoyama (*sato* means 'villages' and *yama* means 'mountains' in Japanese) Improvement Project, undertaken by the Department of Agriculture of the prefectural government, was launched in the district in the same year. Through this project, city, town and village administrations implemented regional environmental diagnoses and countermeasures, and verified their effectiveness while the prefecture subsidized half the cost for these activities. Residents responded to this initiative by establishing the Fukahodo Satoyama Protection Association, which focused on preventing agricultural damage and casualties caused by wildlife. Fukahodo is bordered by mountains on its west side, and in 2010 funds from the Satoyama Improvement Project were used to contract a company to cut the undergrowth vegetation on the slopes of the mountains. This was done to prevent children from encountering wild boars and other kinds of wildlife as they commute to school on roads along the base of the mountains. In neighbouring districts, for example, junior high school pupils had suddenly encountered wild boars on their way to school (Sakurai et al. 2013a). During the same year, damage to paddy rice by wild boars became a conspicuous problem; therefore, in 2011, approximately ten farmland owners in the district cooperatively installed electric fences with half of the cost subsidized by the government. Nevertheless, these preventive measures were not enough to reduce human–wildlife conflicts across the district, as solving the problem in these few areas did not result in fewer human–wild boar encounters or less agricultural damage (Agricultural Production Bureau of Ministry of Agriculture, Forestry and Fisheries 2009; Yamabata 2011).

Because of agricultural damage and threat of encountering wild boars, many local residents held negative attitudes towards them. The resident survey conducted in Fukahodo in 2011 revealed that the majority of respondents felt there were too many wild boars around the district and worried about the damage caused by this species (Sakurai et al. 2015). Furthermore, there was a difference in people's risk perceptions towards wildlife issues as well as their motivation to prevent damage. For example, those residents who were involved in cultivating crops had high awareness and willingness to prevent damage, while those who were not involved in agriculture had less awareness about wildlife issues

(Sakurai et al. 2013a). Similarly, those residents who had children who walked to school worried about them encountering wild boars, while those without children did not.

To address human–wildlife conflicts in the whole district, there was a need to implement interventions collaboratively and to encourage participation in damage prevention activities among local residents. The challenge was to find ways to increase awareness towards wildlife throughout the district, particularly for those not associated with agriculture. Specifically, the general public needed to become more aware of wildlife issues in the Fukahodo district and participate in preventative interventions to address human–wildlife conflicts to garner empathy for the situation at hand and realize the severity of the problem (personal communication, Chairman of the Fukahodo Satoyama Protection Association). To address this issue, the 'Parent and Child Meeting on Interventions towards Wild Boars' was held in July 2011 (Figure 9.6). The aim of this initiative was to hold an event for both children and their parents to help residents of the district – including those not associated with agricultural activities – to understand the problems caused by wildlife and the measures to counter these problems. Throughout Japan,

Figure 9.6 Parent and Child Meeting on Interventions towards Wild Boars.

and especially in rural areas suffering from human–wildlife conflicts, it is challenging to foster awareness about wildlife issues and to encourage residents to take part in damage prevention. The Parent and Child Meeting was the first step in making residents personally aware of the wildlife problem across the entire district by:

1. focusing on children, which allowed for the participation of many parents (as in many cases, children come with their parents) and the use of lectures and activities comprehensible to everyone (including the children);
2. emphasizing that the wildlife problem is a regional one by offering a lecture on the ecology of wild boars and the methods to prevent damage by these animals. Additionally, a wildlife video recorded in the district was shown;
3. offering a wild boar hot pot to allow people to taste the wildlife living around the district.

The Chairman of the Fukahodo Satoyama Protection Association worried; he stated, 'because of people's low concern with wildlife interventions, people might not show up to the activity' (Sakurai et al. 2013a). However, as it turned out, it was a well-attended gathering that attracted approximately eighty local residents, and twenty officials from the local government, for a total of a hundred participants.

The Parent and Child Meeting played a significant role in increasing the awareness of local residents, showing that human–wildlife interactions concern the whole district and not just a limited stakeholder group (i.e. farmers). Commenting on the meeting, the Chairman of the Fukahodo Residents Association mentioned, 'originally, only people who had actually suffered from human–wildlife conflicts were aware of the wildlife problems, so I am sure that the Parent and Child Meeting increased residents' concern to some extent. In addition, the hot pot of wild boar meat was provided, making it an effective resident awareness-raising event' (personal communication, Chairman of the Fukahodo Residents Association). He explained further, 'the Parent and Child Meeting was a big success. Many women came to prepare the hot pot, so it went well. Food is the best way to bring people together.' Furthermore, the meeting offered a valuable opportunity to educate the next generation by involving young parents and children. When older, such participants can potentially take initiative to prevent wildlife damage in the district.

To further increase public engagement in human–wildlife issues in Fukahodo, the Fukahodo Residents Association decided to contract half of the vegetation undergrowth cutting to a contractor in 2011, while the other half was undertaken by volunteer residents (personal communication, Chairman of the Fukahodo Residents Association). In districts with a fairly large number of households, such as Fukahodo (with about 200 households), it is challenging to promote awareness of wildlife among residents. Another challenge is to sustain and continue collaborative interventions throughout the district. If the financial support provided by the government is used to pay the company to cut underbrush around the district (as done in Fukahodo in 2010), preventative activities may not continue once the government stops providing financial aid. It is important to acknowledge that the meeting was hosted voluntarily and did not receive financial support. Yet the event succeeded in gathering residents and various stakeholders, and potentially represents an important step in creating a sustainable participatory community by allowing residents across groups (e.g. farmers and non-farmers) and age to discuss the future of their region.

9.4 CONCLUSION

While the importance of residents' tolerance towards and/or acceptance of wildlife for fostering coexistence has been studied previously all over the world (Riley & Decker 2000; Decker et al. 2001; Sakurai et al. 2013b; Bruskotter & Wilson 2014; Frank 2016), processes and measures regarding how to increase people's tolerance are still largely unexplored. Several studies emphasized the importance of communicating benefits associated with wildlife in order to increase tolerance; however, those benefits are in most cases explained as *ecological benefits* (Bruskotter & Wilson 2014). In the case of the two examples reported in this chapter, the areas where preventative interventions were implemented were limited in space, and activities were performed for no more than two days. Hence, ecological benefits and a decrease in wildlife damage were not significant. Other studies used economic benefits such as revenues generated through wildlife-based tourism as means to increase coexistence (Child 1996; Gandiwa et al. 2013). Yet neither of the programmes discussed generated significant economic revenue for the district, as these were implemented voluntarily and no fee was requested of participants that attended the programme.

In addition, recent studies identified or proposed essential principles for effective community-based programmes in environmental management

(Reed et al. 2014; Mishra et al. 2017). While these studies analysed principles based on a standpoint of how researchers could build partnership with communities to effect biodiversity conservation, the case studies shown in this chapter do not necessarily include researchers as an indispensable part of the project. In fact, the case in Hyogo was about interaction between local residents and urban students. Yet both case studies represented opportunities to attract people to learn about human–wildlife conflicts and unite the region in community development; they were platforms to foster collaborative approaches for regional revival and future endeavours. In that sense, although the study approach was different, the outcomes that were revealed in the case studies introduced in this chapter overlapped with key factors of the principles developed by Mishra et al. (2017); understanding that problems are opportunities to improve the situation, enhancing social capital and providing transparency in interaction between stakeholders.

Wildlife conflicts do not only create negative human–species interactions. These events can also provide opportunities by allowing for the creation of collaborations between local residents and urban students and/or by connecting different generations and various stakeholders. In other words, having wildlife around human settlements does not only create conflicts; it also generates cultural and social benefits by fostering positive human–human interactions and good social relations. These collaborative activities can represent the first step towards enhancing coexistence between human and wild species by potentially changing people's mindset from *conflict* to *passive tolerance* to *coexistence*.

It is important to acknowledge that some cases and programmes can be effective only in certain cultural and societal contexts. For example, some Asian countries (such as Japan) are known to have a collectivistic culture. In such collectivistic countries, meeting public expectation is an important part of people's daily lives and people are more likely to care about and follow what others do (Abrams et al. 1998; Gudykunst & Nishida 1999/2000; Hashimoto et al. 2008; Sakurai et al. 2015). Activities like the Parent and Child Meeting where various residents met are important since they foster a collaborative approach to implementing damage prevention actions. Through collaborative approaches, coexistence with wildlife may became a collectively accepted norm in the district, and a larger section of society may move towards the more positive side of the conflict-to-coexistence continuum. Hence, studies on the influence of cultural and historical context on tolerance and coexistence are needed. This study was able to explore only certain parts of the cultural aspect relating to wildlife issues in Japan. To deepen further the understanding of cultural contexts, different research methods and approaches such as anthropological studies (e.g. ethnographic studies) are necessary. In addition, it should be noted that through the Jyugai Ranger Programme, local residents taught students how to make traditional crafts or traditional rice cakes and, therefore, it became an important opportunity where local and traditional knowledge was transmitted to the younger generation.

Through the two programmes in the Hyogo and Tochigi Prefectures, wildlife can be seen as useful and not just as a negative issue to be addressed. Knowing the benefits of having wildlife in one's region could be the first step towards achieving coexistence with wildlife, hence moving local mental dispositions towards the coexistence side of the continuum. For more details about a series of studies conducted by the author regarding human dimensions of wildlife management in Japan as well as how human dimensions approaches would be effective in different cultural settings, see Sakurai in press.

9.5 RECOMMENDATIONS AND FUTURE DIRECTIONS

- Having wildlife conflicts could provide opportunities where various stakeholders (e.g. local residents and urban students, residents of different generations) collaborate and contribute to community development.
- Letting local residents realize that wildlife could be utilized for community development could potentially become one step towards changing residents' mindset from conflict to tolerance/coexistence.
- Having wildlife around could generate economic or ecological benefit, but also cultural and social benefits such as providing good social relations (interactions and collaboration among people), that could increase positive human well-being.
- Future research should focus on measuring people's tolerance and perceptions towards wildlife and how these change (or not) after programme implementation. Researchers could also select and explore criteria for measuring the conflict-to-coexistence continuum and how each community or case study falls within this framework.

9.6 ACKNOWLEDGEMENTS

This chapter was written based on and modified from Sakurai et al. (2013a) and Ichikawa and Sakurai (2012). This study was funded by JSPS KAKENHI [grant number 25338], the Supporting Organization for Research of Agricultural and Life Science in Tokyo and a Research Grant of Oriental White Stork of Toyooka City in Hyogo.

9.7 References

Abrams, D., Ando, K. & Hinkle, S. (1998). Psychological attachment to the group: Cross-cultural differences in organizational identification and subjective norms as predictors of workers' turnover intentions. *Personality & Social Psychology Bulletin*, 24(10), 1027–39.

Agricultural Production Bureau of Ministry of Agriculture, Forestry and Fisheries. (2009). *Manual for Preventing Wildlife Damage: Capturing Boars, Deer, Macaques, and Crow.* Tokyo: Agricultural Production Bureau of Ministry of Agriculture, Forestry and Fisheries (in Japanese).

Bruskotter, J. T. & Wilson, R. S. (2014). Determining where the wild things will be: Using psychological theory to find tolerance for large carnivores. *Conservation Letters*, 7(3), 158–65.

Cabinet Office Japan. (2014). Annual report on the aging society, Tokyo, Japan. Available from www8.cao.go.jp/kourei/english/annualreport/2014/2014pdf_e.html (accessed July 2015).

Child, B. (1996). The practice and principles of community-based wildlife management in Zimbabwe: The CAMPFIRE programme. *Biodiversity & Conservation*, 5, 369–98.

Conover, M. (2002). *Resolving Human–Wildlife Conflicts: The Science of Wildlife Damage Management.* Boca Raton, FL: Lewis Publishers.

Decker, D. J., Brown, T. L. & Siemer, W. F. (2001). *Human Dimensions of Wildlife Management in North America.* Bethesda, MD: The Wildlife Society.

Division of Rural Development of Tochigi Prefecture. (2011). Agricultural damage by wildlife of 2010 (in Japanese). Available from www.pref.tochigi.lg.jp/g02/houdou/choujyuu-higai.html (accessed July 2012).

Frank, B. (2016). Human–wildlife conflicts and the need to include tolerance and coexistence: An introductory comment. *Society & Natural Resources*, 29, 738–43.

Gandiwa, E., Heitonig, I. M. A., Lokhorst, A. M., Prins, H. H. T. & Leeuwis, C. (2013). CAMPFIRE and human–wildlife conflicts in local communities bordering Northern Gonarezhou National Park, Zimbabwe. *Ecology & Society*, 18(4), art. 7.

Gore, M. L., Knuth, B. A., Curtis, P. D. & Shanahan, J. E. (2006). Stakeholder perceptions of risk associated with human–black bear conflicts in New York's Adirondack Park campground: Implications for theory and practice. *Wildlife Society Bulletin*, 34, 36–43.

Gudykunst, W. B. & Nishida, T. (1999/2000). The influence of culture and strength of cultural identity on individual values in Japan and the United States. *Intercultural Communication Studies*, IX, 1.

Hashimoto, K., Ko, K., Fujinaga, H. & Lutz, R. (2008). Predictive ability of the Theory of Planned Behavior for mental health outcomes in Japanese vs. Chinese students. *Journal of Health Science*, 30, 23–37 (in Japanese with English abstract).

Hyogo Prefecture. (2009). Management Plan of macaques (in Japanese). Available from http://web.pref.hyogo.jp/hw24/documents/000140432.pdf (accessed February 2012).

Hyogo Prefecture. (2010). Management Plan of boars (revised) (in Japanese). Available from http://web.pref.hyogo.jp/hw24/documents/000164532.pdf (accessed January 2012).

Ichikawa, H. & Sakurai, R. (2012). Evaluation of wildlife damage prevention program involving urban students (in Japanese). Paper for 2011 Research Grant of Oriental White Stork of Toyooka City. Available from www.city.toyooka.lg.jp/www/contents/1214890421676/html/common/other/4f7519ef008.pdf (accessed June 2014).

International Union for Conservation of Nature (IUCN). (2016). IUCN Red List: Asiatic Black Bear. Available from www.iucnredlist.org/species/22824/114252336 (accessed November 2018).

Japan Bear Network. (2010). Community Support Program at Nagano Prefecture (in Japanese). Available from www.japanbear.sakura.ne.jp/ckk/0324/ (accessed November 2018).

Kaji, K., Igota, H. & Suzuki, M. (2013). *Science of Hunting for Wildlife Management in Japan*. Tokyo: Asakura Press (in Japanese).

Karuizawa Web. (2011). Preventing encountering bears and boars; 120 people participated in the fifth year of underbrush cutting (in Japanese). Available from www.karuizawa.co.jp/topics/2011/09/5120/ (accessed January 2017).

Kawai, M. & Hayashi, Y. (2009). *Rebellion of Wildlife*. Tokyo: PHP Science World Research Centre (in Japanese).

Knight, J. (2006). *Waiting for Wolves in Japan: An Anthropological Study of People–Wildlife Relations*. Honolulu: University of Hawaii Press.

Manfredo, M. J. (2008). *Who Cares about Wildlife? Social Science Concepts for Exploring Human–Wildlife Relationships and Conservation Issues*. New York: Springer.

Manfredo, M. J., Vaske, J. J., Brown, P. J., Decker, D. J. & Duke, E. A. (2009). *Wildlife and Society: The Science of Human Dimensions*. Washington, DC: Island Press.

Ministry of Agriculture, Forestry and Fisheries. (2007). Agricultural damage by wildlife in each prefecture in 2007 (in Japanese). Available from www.maff.go.jp/j/seisan/tyozyu/higai/h_zyokyo/h19/pdf/ref_data02.pdf (accessed February 2012).

Ministry of Agriculture, Forestry and Fisheries. (2008). Current situation and challenges regarding abandoned agricultural fields (in Japanese). Available from www.maff.go.jp/j/nousin/tikei/houkiti/ (accessed November 2014).

Ministry of Agriculture, Forestry and Fisheries. (2014a). Change of agricultural damage by wildlife (in Japanese). Available from www.maff.go.jp/j/seisan/tyozyu/higai/h_zyokyo2/h26/pdf/160122-b.pdf (accessed January 2017).

Ministry of Agriculture, Forestry and Fisheries. (2014b). Agricultural damage by wildlife in each prefecture in 2014 (in Japanese). Available from www.maff.go.jp/j/seisan/tyozyu/higai/h_zyokyo/h19/pdf/ref_data02.pdf (accessed November 2018).

Ministry of the Environment. (2007). Methodological research of monitoring Specified Wildlife Conservation and Management Plan: Fukushima Prefecture. Ministry of Environment, Fukushima, Japan (in Japanese).

Ministry of the Environment. (2013). Report regarding management of bears (in Japanese). Available from www.env.go.jp/nature/choju/plan/plan3-report/h24report_kuma.pdf (accessed January 2017).

Ministry of the Environment. (2015). Results of survey of population and habitats of deer and boars for implementing Specified Wildlife Capturing Programme based on revision of wildlife law (in Japanese). Available from www.env.go.jp/press/100922.html (accessed January 2017).

Ministry of the Environment. (2016a). Number of bears captured in 2016 (in Japanese). Available from www.env.go.jp/nature/choju/effort/effort12/capture-qe.pdf (accessed January 2017).

Ministry of the Environment. (2016b). Number of casualties caused by bears in 2016 (in Japanese). Available from www.env.go.jp/nature/choju/effort/effort12/injury-qe.pdf (accessed January 2017).

Mishra, C., Young, J. C., Fiechter, M., Rutherford, B. & Redpath, S. M. (2017). Building partnerships with communities for biodiversity conservation: Lessons from Asian mountains. *Journal of Applied Ecology*, 54(6), 1583–91.

Reed, M. S., Stringer, L. C., Fazey, I., Evely, A. C. & Kruijsen, J. H. J. (2014). Five principles for the practice of knowledge exchange in environmental management. *Journal of Environmental Management*, 146, 337–45.

Riley, S. J. & Decker, D. J. (2000). Risk perception as a factor in wildlife stakeholder acceptance capacity for cougars in Montana. *Human Dimensions of Wildlife*, 5, 50–62.

Sakurai, R. (in press). Human Dimensions of Wildlife Management in Japan: from Asia to the World. Tokyo: Springer.

Sakurai, R., Enari, H., Matsuda, N. & Maruyama, T. (2014a). Testing social-psychological theories to predict residents' behavioral intentions regarding wildlife issues – application of Theory of Planned Behavior and Wildlife Acceptance Capacity Model. *Mammal Study*, 54, 219–30 (in Japanese with English abstract).

Sakurai, R., Jacobson, S. K., Matsuda, N. & Maruyama, T. (2015). Assessing the impact of a wildlife education program on Japanese attitudes and behavioral intentions. *Environmental Education Research*, 21(4), 542–55.

Sakurai, R., Jacobson, S. K. & Ueda, G. (2013b). Public perceptions of risk and government performance regarding bear management in Japan. *Ursus*, 24, 70–82.

Sakurai, R., Jacobson, S. K. & Ueda, G. (2014b). Public perceptions of significant wildlife in Hyogo, Japan. *Human Dimensions of Wildlife*, 19, 88–95.

Sakurai, R., Matsuda, N., Maruyama, T. & Jacobson, S. K. (2013a). Overview of the model district program for reducing human–wildlife conflicts in Tochigi Prefecture. *Wildlife & Human Society*, 1(1), 47–54 (in Japanese with English abstract).

Slagle, K. M., Bruskotter, J. T. & Wilson, R. S. (2012). The role of affect in public support and opposition to wolf management. *Human Dimensions of Wildlife*, 17, 44–57.

Toyooka City Government. (2012). Plan of the welfare of the aged and nursing care insurance at Toyooka City (in Japanese). Available from www.city.toyooka.lg.jp/www/contents/1333020316999/files/ko.pdf (accessed January 2017).

Tsunoda, H. & Enari, H. (2012). Ecological role and social significance of reintroducing wolves in Japan under the shrinking society. In A. Paula & H. F. Crussi, eds., *Wolves: Biology, Behaviour and Conservation*. New York: Nova Science Publishers, pp. 177–98.

Ueda, G., Kanzaki, N. & Koganezawa, M. (2010). Changes in the structure of the Japanese hunter population from 1965 to 2005. *Human Dimensions of Wildlife*, 15, 16–26.

Vaske, J. J., Roemer, J. M. & Taylor, J. G. (2013). Situational and emotional influences on the acceptability of wolf management action in the Greater Yellowstone Ecosystem. *Wildlife Society Bulletin*, 37(1), 122–8.

Waylen, K. A., Fischer, A., McGowan, P. J. K., Thirgood, S. J. & Milner-Gulland, E. J. (2010). Effect of local cultural context on the success of community-based conservation interventions. *Conservation Biology*, 24, 1119–29.

Woodroffe, R., Thirgood, S. & Rabinowitz, A. (2005). *People and Wildlife, Conflict or Coexistence?* Cambridge: Cambridge University Press.

Yamabata, N. (2011). Effect of chasing away by village on the home range and appearances of a macaques group: Verification in 7 area of Mie Prefecture. *Journal of Rural Planning*, 30, 381–6 (in Japanese with English abstract).

Yokoyama, M., Sakata, H., Morimitsu, Y., Fujiki, D. & Muroyama, Y. (2008). Current status of and perspectives on The Specified Wildlife Conservation and Management Plans and monitoring for Japanese black bear in Hyogo Prefecture, Japan. *Mammal Science*, 48, 65–71 (in Japanese with English abstract).

Zajac, R. M., Bruskotter, J. T., Wilson, R. S. & Prange, S. (2012). Learning to live with black bears: A psychological model of acceptance. *Journal of Wildlife Management*, 76, 1331–40.

Towards Tolerance and Coexistence

A Comparative Analysis of the Human–Macaque Interface in Sulawesi, Indonesia, and Florida, United States

ERIN P. RILEY

We are in the Anthropocene – the epoch so named for the tremendous impact humanity has had on the diverse ecosystems and organisms that constitute our planet. One attribute, among many, of life in the Anthropocene is that the ecologies and livelihoods of humans and other animals increasingly intersect. As these intersections expand across space and in scale and depth, scholars, conservation practitioners and everyday people alike are recognizing that some of the outcomes of these interactions constitute serious challenges to long-term sustainable coexistence. One such outcome in particular is that of human–wildlife conflict. Conflict is widely understood to mean having incompatible goals. Accordingly, human–wildlife conflict is defined as when the needs and behaviour of non-human animals and humans impact negatively on each other's needs and goals (Madden 2004a, p. 248). Examples of such conflict include when an elephant (*Loxodonta africana*) tramples croplands (Hoare 1999), when large carnivores prey upon domesticated livestock (Suryawanshi et al. 2013), when macaque monkeys (*Macaca tonkeana*) consume important cash crops (Riley 2007a) and when humans, in turn, retaliate against this behaviour.

At the same time, conflict is not always an inevitable outcome of human–wildlife interactions. A number of scholars have therefore recently argued for a move away from framing human–wildlife interactions using a conflict perspective towards one that emphasizes the potential for tolerance and coexistence (Madden 2004b; Peterson et al. 2010; Frank 2016). The idea behind this shift is that rhetoric matters; the words we use to describe how different organisms intersect can in turn shape how we think about those interactions and our motivations and actions in response to those interactions. Peterson et al. (2010), for example, argue that framing human–wildlife interactions as *human–animal conflict* emphasizes antagonism, thereby encouraging, for example, retaliation. Framing human–wildlife interactions using a conflict perspective can also potentially have a narrowing effect when developing new or choosing among existing solutions (Frank 2016). Accordingly, scholars have argued for an alternative framework for analysing the nature and outcomes of human–wildlife interactions; one that allows space for tolerance and coexistence between humans and wildlife, and recognizes that the human–wildlife interface is better conceptualized in terms of a continuum (Madden 2004b; Peterson et al. 2010; Frank 2016). The *conflict-to-coexistence continuum* (Frank 2016) envisions that the human–wildlife interface may be characterized by negative, neutral and positive attitudes and behaviours at divergent, as well as concurrent, times and places.

In this chapter, I use this analytical tool to examine the human–macaque interface. In a number of areas of the world, humans and members of the primate genus *Macaca* live in close proximity and interact with one another across diverse contexts. Here I explore how the conflict-to-coexistence continuum framework plays out when comparing two distinct settings where humans and macaques interface: the rural highlands of Central Sulawesi, Indonesia, where people and Tonkean macaques (*Macaca tonkeana*) share a long history of indigenous cohabitation, and the river banks of the Silver River, Florida, USA, where a feral, introduced population of rhesus macaques (*Macaca mulatta*) lives. Notably, these two locations stand in stark contrast to one another in a multitude of important ways (e.g. geographic location, socio-cultural and economic context, and the 'naturalness' and root of the conflict). However, by exploring the continuum at each site, commonalities begin to emerge, whereby both sites show elements that suggest that coexistence deriving from tolerance and compromise is possible.

10.1 THE LINDU HIGHLANDS, CENTRAL SULAWESI, INDONESIA: SHARING SPACE WITH OTHER PRIMATES, AND OTHER HUMANS

10.1.1 Research Context

The Indonesian island of Sulawesi is, from an ecological standpoint, both peculiar and remarkable: it comprises a mix of Asian and Australasian flora and fauna with high levels of endemism, and yet is considered species depauperate. The primates of Sulawesi are no exception. While there are only three primate genera on the island (*Homo, Macaca, Tarsius*), Sulawesi is home to seven endemic macaque taxa (Fooden 1969) and approximately sixteen taxa of Eastern tarsiers (Groves & Shekelle 2010).

Sulawesi's current human population of more than 17 million people comprises eighty different ethnic groups that are divided in relation to geography, subsistence, language and religion (BadanPusatStatistik Indonesia 2000). While traditional forms of subsistence include swidden (or slash and burn) agriculture and fishing (Davis 1976), today, many communities practise wet-rice agriculture and plantation agriculture of cash crops, including coffee (*Coffea* spp.), cacao (*Theobroma cacao*), palm oil (*Elaeis guineensis*) and cloves (*Syzygium aromaticum*). The emergence of government-supported enterprises, including commercial logging, the transmigration programme and development of cash crop industries, has resulted in major land transformations in the last five decades (Whitten et al. 2002). These forms of land conversion combined with a growing human population size have increased the likelihood of human–macaque interactions, particularly in rural areas at the forest–agriculture edge. Many primates, including the Sulawesi macaques, can flexibly adapt to such alterations in their habitat by incorporating agricultural crops into their dietary repertoire (Supriatna et al. 1992; Riley 2007a; Priston et al. 2012), a behaviour referred to as *crop-raiding*, or more recently, *crop-foraging* or *crop-feeding* (Hockings et al. 2015; Hill 2017; Zak & Riley 2017; Figure 10.1).

Figure 10.1 Camera trap photo capturing two moor macaques (*Macaca maura*) feeding on maize. *A black and white version of this figure will appear in some formats. For the colour version, please refer to the plate section.*

From 2002 to 2004, as part of a broader study on the ethnoprima-tology[1] of *Macaca tonkeana* (Riley 2005), I examined crop-feeding behaviour by Tonkean macaques and villagers' attitudes towards the macaques, nature and protected area conservation in the Lake Lindu highland plain in Lore Lindu National Park, Central Sulawesi, Indonesia. Lore Lindu National Park, comprising a total area of 217,982 hectares, was established in 1993 from two existing reserves and is designated as a United Nations Educational, Scientific and Cultural Organization (UNESCO) Man and the Biosphere Reserve. It provides habitat for a majority of Sulawesi's endemic mammals, including one endemic macaque taxon, *Macaca tonkeana*. The Tonkean macaque is a medium-sized, primarily frugivorous primate that exists in multimale-multifemale social groups (Riley 2007b). This primate shows considerable ecological flexibility in its ability to use multiple forest strata, including a considerable amount of time on the ground, and to persist in heavily altered habitats by feeding on remaining wild foods as well as cultivated foods (Riley 2007a, 2007b, 2008). Lore Lindu National Park also provides watershed protection for two major river catchment systems, the Lariang and the Gumbasa-Palu rivers.

The Lindu highland plain, situated at an altitude of approximately 1,000 metres, is one of two enclaves that are allowed to exist within the park because they have a long history of occupation and are major rice-growing areas. During the designation of the National Park, the granting of enclave status meant that the Lindu people were able to maintain their current agricultural fields and continue to engage in small-scale forest production collection. Three of the four villages in Lindu are primarily occupied by the indigenous *To Lindu* (KailiTado'); however, over the last 60 years, members of other Kaili ethnic groups as well as Bugis people originally from South Sulawesi have migrated to the area in search of available land for agriculture and to benefit from the lucrative tilapia (*Oreochromis niloticus*) fishing industry at the lake (Acciaioli 2000). Although wet-rice agriculture (*sawah*) predominates in Lindu, and is practised by both indigenous Lindu and migrants, since the 1980s, tree cash crops, such as coffee and cacao, have also become important. In Indonesia, cacao is the fourth most important agricultural export, after rubber, palm oil and coconut, supplying more than 12 per cent of the

---

[1] Ethnoprimatology is the study of the ecological and cultural interconnections between human and non-human primates and the implications of those interconnections for conservation and sustainable coexistence (Riley 2013b).

world's cacao; 75 per cent of Indonesia's supply is produced in Sulawesi (Indonesia-Investments.com n.d.). Cacao is a perennial crop that is typically cultivated under forest shade, and is one of the primary cash crops fed on by macaques and other wildlife (Supriatna et al. 1992; Riley 2007a).

10.1.2 Research Approach

To explore the human–macaque interface in Lore Lindu National Park, I adopted an *expanded* community ecology perspective, viewing humans and Tonkean macaques as members of a shared ecological community (Riley 2005). I also employed a mixed-methods approach incorporating primatological, ecological and ethnographic research methods. Primatological and ecological research methods included quantitative measurements of cacao crop loss caused by Tonkean macaques and other wildlife, namely squirrels (*Prosciurillus* sp.) and forest rats (*Taeromys* sp.) (Riley 2007a). Cacao crop loss was measured bimonthly by enumerating the number of cacao pods on each tree and the number of pods consumed by the macaques (see Riley 2007a for more details). The ethnographic component included semi-structured interviews with farmers, focusing on their perceptions of the frequency of crop-feeding behaviour and its impact on their livelihoods (Riley 2007a). I also conducted semi-structured interviews with a broader sample of villagers to assess human forest resource use, knowledge and perceptions of nature and protected area conservation, and human–macaque folklore (Riley 2007a, 2010, 2013a).

10.1.3 Results

Regarding crop-feeding by Tonkean macaques, I found that quantitative measurements of crop loss did not match with farmers' perceptions. For example, while villagers listed the macaques as the most destructive cacao crop-feeding animal, with a number of farmers reporting a 75 per cent loss of harvest, the ecological surveys revealed that a significantly greater number of cacao pods were consumed by forest rats (Riley 2007a). In fact, the percentage of cacao pods consumed by macaques across all gardens only ranged between 0 per cent and 6.4 per cent (Riley 2007a). Nonetheless, farmers' frustration with the macaques was palpable in their words and tone: 'Because the macaques raid the cacao gardens, it would be better to cut all the cacao trees down and plant

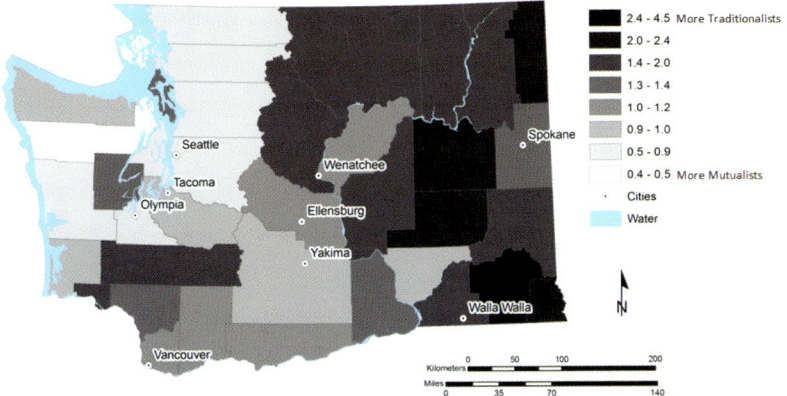

Figure 2.3a Ratio of Traditionalists to Mutualists, from a 2009 survey of Washington residents by county (Dietsch et al. 2011). Ratio represents the number of Traditionalists for every one Mutualist, whereby a number greater than one signifies that there are more Traditionalists than Mutualists in the county, and a number less than one indicates that there are more Mutualists than Traditionalists. *A black and white version of this figure will appear in some formats.*

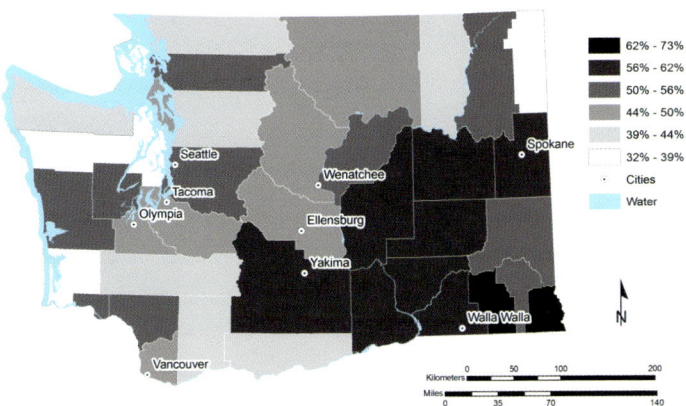

Figure 2.3b Percentage of residents accepting of lethal removal by the state wildlife agency of a nuisance black bear, from a 2009 survey of Washington residents by county (Dietsch et al. 2011). The nuisance situation was described as one where, for example, the bear is getting into rubbish or pet food containers near the resident's home. Darker colours denote a higher percentage of people who find lethal removal of a nuisance black bear to be acceptable. *A black and white version of this figure will appear in some formats.*

Figure 9.2 Landscape of one of the districts where the Jyugai Ranger Programme was implemented. *A black and white version of this figure will appear in some formats.*

Figure 9.3 Activities of cutting brush. *A black and white version of this figure will appear in some formats.*

Figure 10.1 Camera trap photo capturing two moor macaques (*Macaca maura*) feeding on maize. *A black and white version of this figure will appear in some formats.*

Figure 10.2 One of the Silver River macaque groups (*Macaca mulatta*) at the river's edge.
Photo: Erin Riley. *A black and white version of this figure will appear in some formats.*

Figure 10.3 Kayakers provisioning the Silver River macaques (*Macaca mulatta*). Photo: Tiffany Wade. *A black and white version of this figure will appear in some formats.*

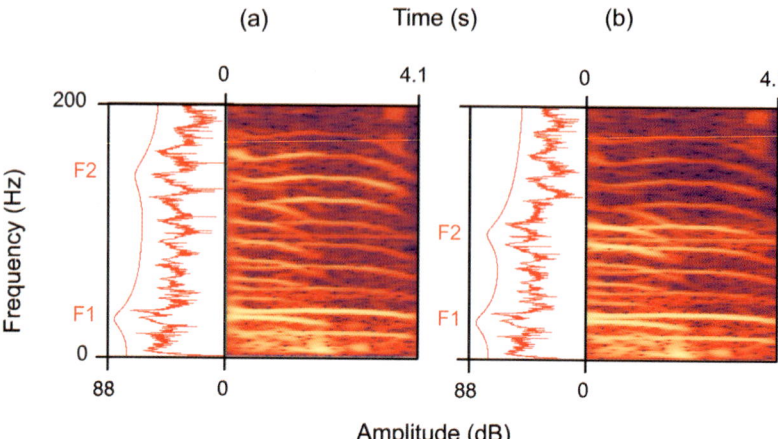

Figure 11.1 Frequency spectrograms of elephant rumbles in response to (a) the bee playback stimulus showing two formants (peaks, F1 and F2) that indicate where the stress in the vocal chords emphasizes the communication information in the elephant rumble, (b) representative spectrogram of a typical elephant rumble from those recorded in response to control white noise playback. Note the different position of F2.

Reproduced from King et al. (2010). *A black and white version of this figure will appear in some formats.*

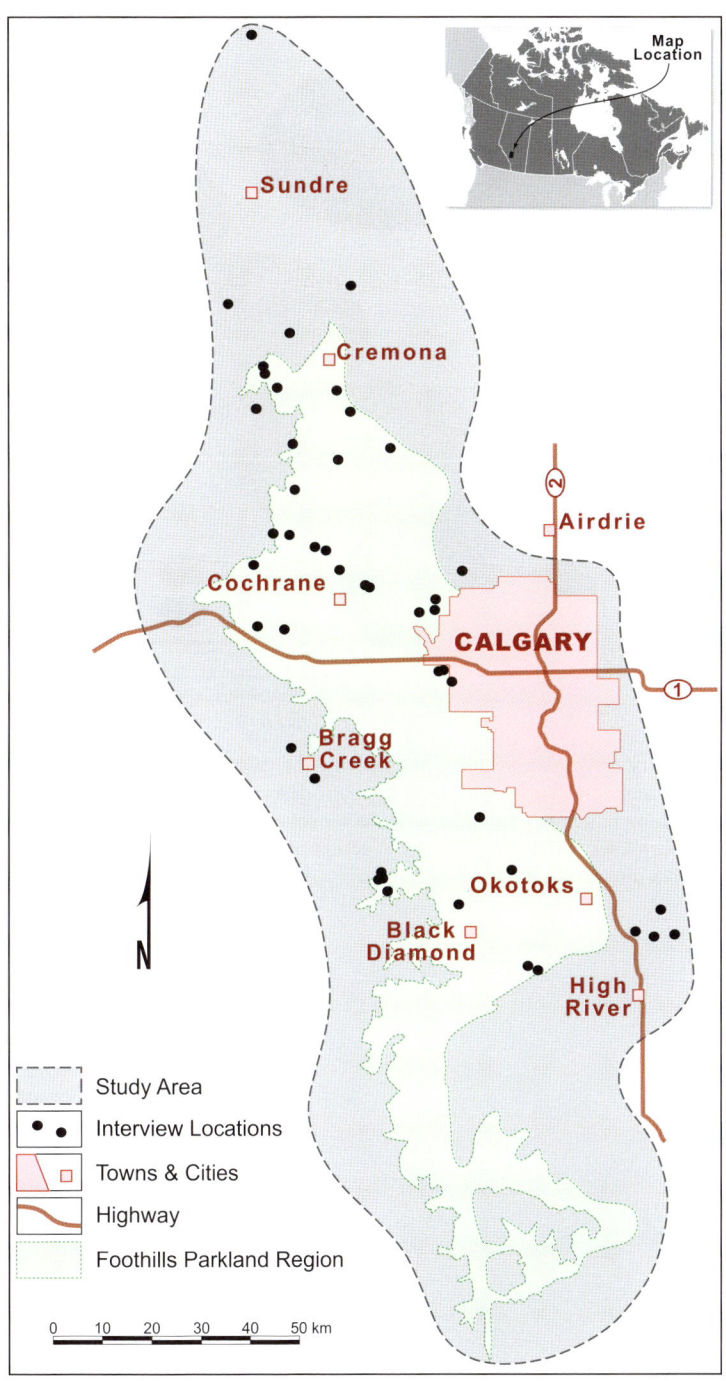

Figure 15.1 Foothills Coyote Initiative study area. *A black and white version of this figure will appear in some formats.*

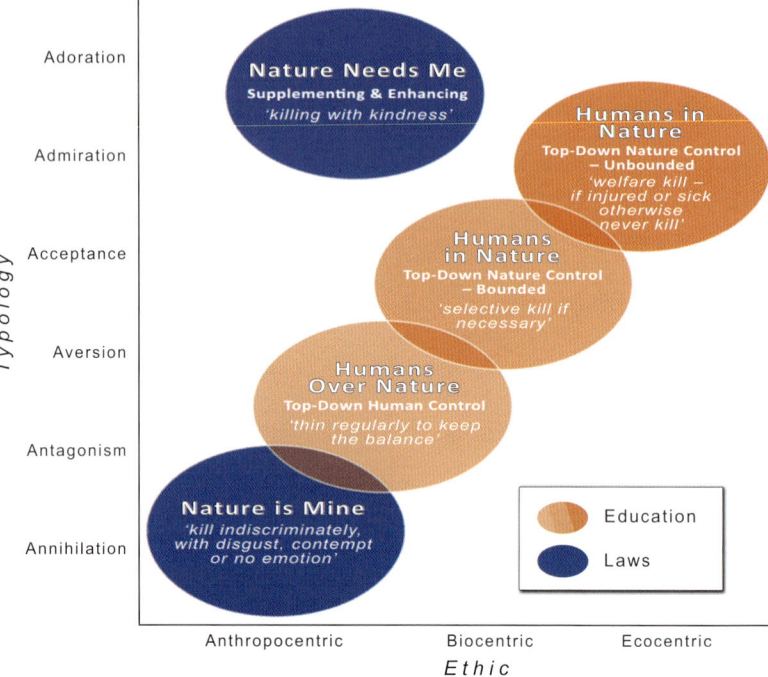

Figure 15.2 Worldviews spectrum showing intersection of behavioural typology, y-axis, and ethic, x-axis, with brief thematic description. Orange represents worldviews open to education, with darker colours more likely to use education. Purple shows worldviews unlikely to accept science knowledge mobilization. *A black and white version of this figure will appear in some formats.*

coffee', and 'I get only ¼ of my harvest, because the other ¾ is taken by the monkeys' (Riley 2007a, p. 477). The fact that the macaques do feed on their crops, albeit less than other wildlife, and the tone of the sentiments expressed, suggest there is considerable human–macaque conflict in the Lindu highlands.

However, I found that despite villagers' negative perceptions towards the macaques and their crop-feeding behaviour, there appeared to be little to no retaliation against them, at least not by the farmers who are indigenous to the area. Instead, tolerance prevailed. The ethnographic research component revealed the existence of human–macaque folklore that explains this seeming paradox. For the indigenous Lindu, not only are Tonkean macaques viewed as guardians of *adat*,[2] but they play a role in a specific folktale that embodies the respect *and* disdain the Lindu have for them (Riley 2010, 2013a, p. 193). The tale tells the story of how a father left behind his young daughter to guard their house and gardens from raiding monkeys while he went to fish at the lake. According to the tale, as soon as the father left the monkeys began to raid their crops, to which the daughter responded by yelling and throwing coals at them. The monkeys, in turn, responded by kidnapping her. Upon hearing the commotion and his daughter's screaming from the lake, the father chased after the monkeys, and fought to rescue his daughter, leaving alive only a male and pregnant female. In the end, the man successfully rescues his daughter, but he also keeps some macaques alive, and then they go their separate ways. The moral of the story then is that the Lindu must not act negatively towards the macaques, or bad things will transpire. This tale was only recounted by a few elders; however, its message remains salient for many Lindu and has resulted in a taboo against harming macaques they encounter in their shared ecological space: 'If monkeys come to our gardens we can only shoo them away . . . you can't be harsh because otherwise they will get angry' (Riley 2010, p. 238).

Tolerance towards the monkeys may also stem from the fact that many in Lindu believe that the forests also belong to the macaques: 'I pity them [macaques] as they, too, need to find food' (Riley 2007a, p. 482). On the other hand, few respondents who were migrants to the Lindu highlands conveyed any sense of cultural connectivity with macaques, and often admitted shooting at them when encountered in their gardens. Migrants and indigenous Lindu also differed in their

---

[2] *Adat* is typically defined as the customary laws (e.g. local traditions and natural resource management practices) developed by the members of various sub-ethnic groups.

perceptions about the forest and its conservation (Riley 2013a). For example, the Lindu speak to the value of the forest in both utilitarian (i.e. for the services and products it provides) and aesthetic terms (i.e. for future generations to appreciate). Many migrants, on the other hand, do not perceive value in its persistence, but rather in what the land *could be utilized for*; that is, further agricultural development.

10.2 THE SILVER RIVER MACAQUES OF FLORIDA: INVASIVE AND EXOTIC, YET BELOVED AND THRIVING

10.2.1 Research Context

Non-human primates have been extinct from the United States for millions of years, but a population of rhesus macaques (*Macaca mulatta*) has been living in the riparian woodlands of north central Florida since they were first introduced by humans nearly 80 years ago. The location of this introduced population is the Silver River, an 11-km watercourse that flows east through Silver Springs State Park (SSSP), Ocala, Florida (Figure 10.2). Although there are no written records on the macaques' initial introduction, a set of local folktales point to their origin. For example, one story suggests that the current population descends from the *extras* that were released after the completion of the original Tarzan films that were filmed on site in the 1930s and 1940s. Another tale, and notably the more widely accepted one, is that they were released in the 1930s by a tour-boat operator who was eager to enliven his *jungle cruises* along the Silver River. The tour-boat operator apparently released them

Figure 10.2 One of the Silver River macaque groups (*Macaca mulatta*) at the river's edge.
Photo: Erin Riley. *A black and white version of this figure will appear in some formats. For the colour version, please refer to the plate section.*

on an island situated in the river, expecting them to remain in place for the enjoyment of his tour patrons. What he did not know, however, was that rhesus macaques are capable swimmers, and so, the released individuals soon escaped from the island (Wolfe & Peters 1987). Eventually tour-boat operators began provisioning them at the river's edge to facilitate viewings for tourists (Maples et al. 1976).

In the 1970s, the scale of provisioning increased dramatically when the management of the attraction at the river's headsprings established two feeding sites where commercial monkey chow and fruit were provided twice daily (Sarris 1980; Peters 1983). This large-scale provisioning is believed to have contributed to considerable growth in the population from a founding group of approximately 10 individuals to over 300 individuals in the mid-1980s (Wolfe 2002). Although heavy provisioning apparently ceased after 1984 (Wolfe & Peters 1987), since that time boaters who navigate the river and encounter the macaques along the river's edge continue to provision them (Figure 10.3).

Figure 10.3 Kayakers provisioning the Silver River macaques (*Macaca mulatta*). Photo: Tiffany Wade. *A black and white version of this figure will appear in some formats. For the colour version, please refer to the plate section.*

Over the years, this population of introduced monkeys has not been without controversy. In the 1980s, the Florida Game and Fresh Water Fish Commission (FGFWFC) was concerned about the population's growing size and potential impact on native ecological communities, and potential for disease transmission,[3] particularly given that park visitors and boaters had ample opportunity to encounter the macaques (Riley 2017). It consequently initiated efforts to control the population, including a sterilization programme and trapping and removal (Montague et al. 1994, p. 11).[4] This decision was not well-received by local residents of Marion County, who generally favoured the continued existence of the macaques along the river; so much so that a group known as the *Friends of the Silver River Monkeys* was formed and successfully secured tens of thousands of signatures from local residents in protest of the monkeys' removal from public lands (Riley 2017). Nonetheless, trapping ensued with approximately 500 monkeys removed between 1984 and 1993 (Wolfe 2002). From 1998 to 2012, a single trapper removed approximately 627 individuals from Silver River and more than 200 from the Cross Florida Greenway along the Ocklawaha River (Sally Lieb, personal communication). In October 2013, the Department of Environmental Protection (DEP) and the Florida Park Service (FPS) permanently banned trapping on state park lands and the Cross Florida Greenway (Pittman 2013). At the end of our study in May 2013, we determined that the Silver River macaque population comprised four social groups totalling approximately 118 individuals (Riley & Wade 2016).

10.2.2 Research Approach

The rhesus macaque's native range includes tropical, subtropical and temperate habitats across Central, South and South East Asia. This wide geographical range and ability to thrive in a diverse array of habitat types, including human-dominated environments, is due to their considerable behavioural flexibility (Southwick et al. 1996). In the case of the Silver River rhesus macaques, this flexibility has enabled them to

---

[3] In 1992, it was determined that the Herpes B virus (*Cercopithecine herpesvirus 1*; BV) was present in the Silver River macaque population. Serious disease due to Herpes B virus is rare in macaques, but when transmitted to humans, it can be fatal if not treated promptly (Elmore & Eberle 2008).

[4] Between 1986 and 1990, the University of Florida Veterinary College sterilized twenty females by tubal ligation (Montague et al. 1994).

survive in a new environment. My colleagues and I were interested in understanding how they are able to do so; in other words, we aimed to examine how they have adjusted ecologically to life in the riparian woodlands of Florida. In addition, because an important element of this population's adjustment is its daily interface with people, we also examined the nature of human–macaque encounters.

10.2.3 Results

We found that the Silver River rhesus macaques have adapted to the comparatively fruit-poor environment of the SSSP by predominately consuming leaves and other vegetative parts (Riley & Wade 2016). They also have incorporated locally available plants native to North America into their dietary repertoire, such as sedges and other semi-aquatic plants (Riley & Wade 2016).

Our observations of the ways people and the macaques interface along the river indicate that both species are highly attracted to one another. Eighty-per cent of the boats (n = 608) we observed were scored as 'encounters', meaning that the people directed their attention to or moved in the direction of the macaques (Riley & Wade 2016). The macaques, in turn, are attracted to the river's edge because of the wild resources the floodplain swamp habitat provides, but also because of opportunities to receive provisions from passing boaters. We found, however, that not all boaters interacted with the macaques in the same way – variation we attribute to differences in use values associated with particular watercraft types. For example, people in motorized boats, such as pontoon boats and motor boats, more frequently provisioned the macaques, whereas kayakers and canoers tended to navigate the river more quietly and rarely attempted to interact with the macaques, choosing instead to simply observe. Interestingly, we noted that the macaques themselves have learned that people in different boats do different things: while they were more likely to ignore kayaks and canoes, motorized boats triggered movement to the river's edge for monitoring for provisions (Riley, unpublished data).

Going into the study, we expected to observe high rates of provisioning – a prediction based on the park's perception of the issue. However, our observations indicate that the majority of encounters solely involve mutual observation, with only a small percentage of boats engaging in provisioning (11.5%, n = 65 of 566 for which we collected data on provisioning boats; Riley & Wade 2016), and little direct contact between boaters and the macaques. These elements of the human–macaque interface along the Silver River stand in stark contrast to human–macaque interactions in previous decades,[5] as well as other sites where human–macaque interactions characteristically involve heavy provisioning and frequent direct contact that result in bites and increased opportunities for disease transmission (e.g. Bali, Indonesia; Engel et al. 2002; Fuentes & Gamerl 2005). These findings suggest that, currently, there is little conflict between park visitors and the macaques. Rather, park visitors are excited about their presence, and hence tolerate coexistence along the river.

10.3 DISCUSSION

The floodplains of Florida and the rural highlands of Sulawesi represent disparate settings where humans and macaques interface. They differ in terms of geographical location, the longevity and 'naturalness' of the interface, and socio-cultural and economic context. A comparative analysis using the conflict-to-coexistence continuum framework reveals further divergences but also parallels between the two sites. In the Lindu highlands of Central Sulawesi conflict is real for farmers who endure daily raids by neighbouring Tonkean macaque groups. In contrast, the human–macaque interface along the Silver River is predominately one of mutual benefit, at least between the macaques and park visitors: the macaques receive provisions from some visitors and park visitors enjoy an enhanced river experience. This site illustrates how *human–wildlife conflict* is not *just* wildlife impacting people. The fact that this primate population has been culled over the years indicates that *humans* have negatively impacted rhesus macaques' *needs and goals*. There also is conflict between different groups of people (e.g. state and park officials versus local residents) *about* the presence of the macaques; thereby supporting recent arguments that the label of *human–wildlife conflict* masks 'the complex and dynamic nature of the underlying conflict that stems from differential values, needs, priorities, and power relations between the human groups concerned' (Hill 2015, p. 3), or *human–human conflict* (see also Peterson et al. 2010; Redpath et al. 2013; Madden & McQuinn 2014). At the Silver River, macaques are deemed invasive, exotic animals unworthy of protection by wildlife and park

---

[5] Based on reports from local informants and as evidenced by vintage postcards illustrating boats full of visitors situated at the river's edge with people directly handing food to the macaques.

officials, yet they have achieved a special place in the hearts of many boaters who frequent the Silver River (Riley et al. unpublished data).

Deeper analysis of broader issues concerning access to resources and the impact of protected area conservation on local livelihoods in Lindu, however, reveals that the human–macaque conflict in Lindu may also involve, and possibly stem from, underlying human–human conflict. For example, when asked about what they perceived to be threats to the National Park, a number of the Lindu respondents envisioned the migrants *themselves* as threats (Riley 2005). This perception likely stems from their fear of the over-exploitation of forest resources and the disregard of Lindu *adat* by migrants: 'They [migrants] go into the forest and make more gardens' and 'Transmigrants don't know the rules of Lindu *adat* about the forest . . . they only feel that they can benefit from it' (Riley 2013a, p. 192). Migrants (who, as I noted above, expressed no qualms about retaliating against crop-feeding macaques), in turn, expressed frustration about the National Park limiting their ability to expand their agricultural land; a frustration that is likely heightened when the macaques feed within their existing crop lands.

Shifting our perspective towards the coexistence end of the continuum allows for additional commonalities between the two sites to emerge. My research in Lindu suggests that respect for nature is linked to place and generated from direct, long-term interactions with it. Such respect can, in turn, result in local support for the long-term protection of the forest via protected area conservation efforts. For the Lindu, who regularly enter, experience and utilize the forest, encounters with wildlife, such as the macaques, result in experiences from which human–primate folklore can derive. And, importantly, this folklore has helped foster a culture of tolerance of the macaques despite their destructive behaviour. Moreover, when I presented the research results back to the farmers, many were surprised by the mismatch between their perceptions and the ecological surveys, leading some to conclude that the extent of the crop losses to macaques was tolerable relative to their total harvest (Riley 2007a). A sense of empathy also pervaded people's narratives: a number of the respondents from Lindu noted that the monkeys need to make a living too ('I pity them [macaques] as they, too, need to find food'; Riley 2007a, p. 482) and that the forest belonged to the monkeys before people planted their crops: 'Because people cut down fruiting trees, monkeys go to the cacao trees' (Riley 2013a, p. 196).

Although the Silver River is not a site where humans and macaques share a long history of ecological sympatry, people who flock to the river and who know about the macaques conveyed a similar sense of empathy. For example, some park visitors expressed a level of concern about the monkeys' well-being, claiming *they look hungry*. These visitors presume provisioning is helpful because they are at least aware that rhesus macaques are non-native, but are unaware that the monkeys can subsist on local plant foods. Conversations we had with boaters revealed that park visitors want to be educated about the macaques, their origin and why feeding them, an act people presume is helpful, can be harmful to monkeys and humans alike.

10.4 CONCLUSION

Living in the Anthropocene means that humans and wildlife are increasingly *sharing space* (Lee 2010). To achieve sustainable coexistence requires that we fully understand the conflict that ensues from the sharing of space, as well as the factors that lead to and promote tolerance. The *conflict-to-coexistence* continuum is a useful analytical tool to advance this approach. Analysis of the human–macaque interface in Lindu and along the Silver River demonstrates that conflict can have many layers; it can simultaneously exist between people and primates and between different groups of people. In Lindu, understanding the human perspective towards human–macaque conflict meant appreciating the multiplicity of the *local community*. In areas of high in-migration, such as Lore Lindu National Park, a resulting diversity in backgrounds (i.e. ethnicities, lifestyles) can translate into differing views, concerns and interests with regard to wildlife and forest use and protection (Ostrom 1999). Because some locals may have no qualms about eliminating crop-feeding macaques, the ability of macaques and humans to coexist at locales such as this may ultimately require specific types of conservation initiatives. For example, initiatives that encourage an appreciation of the biological, morphological and cognitive continuity between humans and other primates as a way to engender more respect and tolerance, and ones that encourage a convergence of values to achieve the mutual satisfaction of human, primate and ecosystem needs (Riley 2010; Malone et al. 2014). In other words, tolerance by all who live at the forest-farm edge may require more than empathy; they may need to perceive the macaques as imbuing utilitarian value. This could be accomplished by highlighting the critical ecological role macaques play in forest regeneration via seed dispersal (Sengupta et al. 2014), particularly for trees that are economically and culturally important to people.

On the Silver River, there is more than just empathy and tolerance; the predominant sentiment among park visitors is deep affiliation for the macaques. While this sentiment by no means captures the perspective of state and park officials of this invasive, exotic animal, the fact that a ban on trapping was put into place suggests that all relevant stakeholders are tolerating coexistence, at least for now. Tolerance rather than conflict towards invasive species is further discussed in Chapter 13 by Harvey and Mazzotti, showcasing how the concepts of conflict and coexistence can be inverted based on the species considered.

10.5 RECOMMENDATIONS AND FUTURE DIRECTIONS

- Because culture is not static, and hence taboos can change, future research in the Lindu highlands could examine current perspectives on and reactions to macaque crop-feeding behaviour. This research could focus on both indigenous and migrant community members to not only assess shifting perceptions but intra-community variation in those perceptions.
- A gap in knowledge is about the impact the Silver River rhesus macaques may be having on the vegetation structure of Florida's riverine floodplains; the predominance of certain key species (e.g. ash trees; *Fraxinus* spp.) in their diet could mean a negative ecological impact. However, the macaques could also be playing an important role in seed dispersal.
- To fully understand the flexibility and invasion success of rhesus macaques in Florida, and the potential for future conflict, future research should focus on examining the overall size and distribution of populations beyond the Silver River. There have been reports of group sightings many kilometres from the Silver River, but systematic observations have not yet been conducted.
- An interdisciplinary, mixed-methods approach that addresses both sides of the conflict – human and non-human – will more fully expose the interface, and hence, better inform efforts to build tolerance and compromise.
- Because perceptions regarding conflict matter, efforts should be made to not only access these perceptions but also to determine whether perceptions align with other measures of the conflict (e.g. data on crop-feeding from direct observations (see Spagnoletti et al. 2017) or remote sensor cameras (see Zak & Riley 2017). These

results, in turn, should be presented back to the communities to foster engagement and tolerance.

- Efforts to revitalize traditional folklore that protects wildlife that interface with people and that highlight the utilitarian value of wildlife may help promote tolerance and sustainable coexistence (e.g. through conservation education programmes with the youth).

10.6 ACKNOWLEDGEMENTS

I am grateful to Beatrice Frank, Jenny Glikman and Silvio Marchini for inviting me to contribute to this edited volume. The research reported here was only possible with permission from LIPI and BTNLL (Indonesia) and Sally Lieb (SSSP, Florida) and Alice Bard (DEP, Florida); funding from the National Science Foundation, Wenner Gren Foundation, Wildlife Conservation Society, the American Society of Primatologists, National Geographic Society / Waitt Institute and San Diego State University; and the research assistance provided by Manto, James, Papa Denis, Pias, Pak Asdi and Tiffany Wade. I am also grateful to Robert Gottschalk, 'Captain Tom' O'Lenick and Mickey Summers for sharing their knowledge of the Silver River macaques and their experiences navigating the Silver River and Ocklawaha waterways.

10.7 References

Acciaoli, G. (2000). Kinship and debt: The social organization of Bugis migration and fish marketing at Lake Lindu, Central Sulawesi. *Biijdragen tot de taal-, landen volkenkunde*, 156(3), 589–617.

BadanPusatStatistik Indonesia. (2000). *Biro Pusat Statistik: 2000 Population Census*. Bogor: BPS.

Davis, G. (1976). *Parigi: A Social History of the Balinese Movement to Central Sulawesi, 1907–1974*. PhD Thesis, Stanford, CA: Stanford University.

Elmore, D. & Eberle, R. (2008). Monkey B virus (Cercopithecine herpesvirus I). *Comparative Medicine*, 58(1), 11–21.

Engel, G., Jones-Engel, L., Schillaci, M., Suaryana, K. G., Putra, A., Fuentes, A. & Henkel, R. (2002). Human exposure to herpes virus B-seropositive macaques, Bali, Indonesia. *Emerging Infectious Diseases*, 8(8), 789–95.

Fooden, J. (1969). *Taxonomy and Evolution of the Monkeys of Celebes*. Basel: Karger.

Frank, B. (2016). Human–wildlife conflicts, the need to include tolerance and coexistence: An introductory comment. *Society & Natural Resources*, 29(6), 738–43.

Fuentes, A. & Gamerl, S. (2005). Disproportionate participation by age/sex class in aggressive interactions between long-tailed macaques (*Macaca fascicularis*) and human tourists at Padangtegal Monkey Forest, Bali, Indonesia. *American Journal of Primatology*, 66(2), 197–204.

Groves, C. & Shekelle, M. (2010). The genera and species of Tarsiidae. *International Journal of Primatology*, 31(6), 1071–82.

Hill, C. M. (2015). Perspectives of 'conflict' at the wildlife–agriculture boundary: 10 years on. *Human Dimensions of Wildlife*, 20(4), 296–301.

Hill, C. M. (2017). Primate crop feeding behavior, crop protection, and conservation. *International Journal of Primatology*, 38(2), 385–400.

Hoare, R. E. (1999). Determinants of human–elephant conflict in a land-use mosaic. *Journal of Applied Ecology*, 36(5), 689–700.

Hockings, K. J., McLennan, M. R., Carvalho, S., Ancrenaz, M., Bobe, R., Byrne, R. W., Dunbar, R. I. M., Matsuzawa, T., McGrew, W. C., Williamson, E. A., Wilson, M. L., Wood, B., Wrangham, R. W. & Hill, C. M. (2015). Apes in the Anthropocene: Flexibility and survival. *Trends in Ecology & Evolution*, 30(4), 215–22.

Indonesia-Investments.com. (n.d.). *Cacao*. Available from www.indonesia-investments.com/business/commodities/cocoa/item241 (accessed November 2018).

Lee, P. C. (2010). Sharing space: Can ethnoprimatology contribute to the survival of nonhuman primates in human-dominated globalized landscapes? *American Journal of Primatology*, 72(10), 925–31.

Madden, F. (2004a). Creating coexistence between humans and wildlife: Global perspectives on local efforts to address human–wildlife conflict. *Human Dimensions of Wildlife*, 9(4), 247–57.

Madden, F. (2004b). Can traditions of tolerance help minimize conflict? An exploration of cultural factors supporting human–wildlife coexistence. *Policy Matters*, 13, 234–41.

Madden, F. & McQuinn, B. (2014). Conservation's blind spot: The case for conflict transformation in wildlife conservation. *Biological Conservation*, 178, 97–106.

Malone, N., Wade, A. H., Fuentes, A., Riley, E. P., Remis, M. & Robinson, C. J. (2014). Ethnoprimatology: Critical interdisciplinarity and multispecies approaches in anthropology. *Critique of Anthropology*, 34(1), 8–29.

Maples, W. R., Brown, A. B. & Hitchens, P. M. (1976). Introduced monkey populations at Silver Springs, Florida. *The Florida Anthropologist*, 29(4), 133–6.

Montague, C. L., Colwell, S. V., Percival, H. F. & Gottgens, J. F. (1994). *Issues and Options Related to Management of Silver River Rhesus Macaques*. United States Biological Survey Technical Report No. 49. Gainesville, FL: Florida Cooperative Fish and Wildlife Research Unit.

Ostrom, E. (1999). Self-governance and forest resources. CIFOR Occasional Papers No. 20.

Peters, E. H. (1983). *Vocal Communication in an Introduced Colony of Feral Rhesus Monkeys* (Macaca mulatta). PhD Dissertation, Gainesville, FL: University of Florida.

Peterson, M. N., Birckhead, J. L., Leong, K., Peterson, M. J. & Peterson, T. R. (2010). Rearticulating the myth of human–wildlife conflict. *Conservation Letters*, 3(2), 74–82.

Pittman, C. (2013). State park will no longer allow trapper to catch wild monkeys for labs. *Tampa Bay Times*, 25 October.

Priston, N. E., Wyper, R. M. & Lee, P. C. (2012). Buton macaques (*Macaca ochreata brunnescens*): Crops, conflict, and behaviour on farms. *American Journal of Primatology*, 74(1), 29–36.

Redpath, S. M., Young, J., Evely, A., Adams, W. M., Sutherland, W. J., White-house, A., Amar, A., Lambert, R. A., Linnell, J. D. & Watt, A. (2013). Understanding and managing conservation conflicts. *Trends in Ecology & Evolution*, 28(2), 100–9.

Riley, E. P. (2005). *Ethnoprimatology of* Macaca tonkeana: *The Interface of Primate Ecology, Human Ecology, and Conservation in Lore Lindu National Park, Sulawesi, Indonesia*. PhD Dissertation, Athens, GA: University of Georgia.

Riley, E. P. (2007a). The human–macaque interface: Conservation implications of current and future overlap and conflict in Lore Lindu National park, Sulawesi, Indonesia. *American Anthropologist*, 109(3), 473–84.

Riley, E. P. (2007b). Flexibility in diet and activity patterns of *Macaca tonkeana* in response to anthropogenic habitat alteration. *International Journal of Primatology*, 28(1), 107–33.

Riley, E. P. (2008). Ranging patterns and habitat use of Sulawesi tonkean macaques (*Macaca tonkeana*) in a human-modified habitat. *American Journal of Primatology*, 70(7), 670–9.

Riley, E. P. (2010). The importance of human–macaque folklore for conservation in Lore Lindu National Park, Sulawesi, Indonesia. *Oryx*, 44(2), 235–40.

Riley, E. (2013a). The human–macaque interface in the Sulawesi Highlands. In R. Corbey & A. Lanjouw, eds., *The Politics of Species: Reshaping Our Relationships with Other Animals*. Cambridge: Cambridge University Press, pp. 189–96.

Riley, E. P. (2013b). Contemporary primatology in anthropology: Beyond the epistemological abyss. *American Anthropologist*, 115(3), 411–22.

Riley, E. P. (2017). Silver River macaques of Florida. In A. Fuentes, ed., *International Encyclopedia of Primatology*. Hoboken, NJ: Wiley Liss, pp. 123–9.

Riley, E. P. & Wade, T. W. (2016). Adapting to Florida's riverine woodlands: The population status and feeding ecology of the Silver River rhesus macaques and their interface with humans. *Primates*, 57(2), 195–210.

Sarris, E. (1980). Aspects and implications of supplemental foraging among provisioned monkeys. *Florida Scientist*, 43(3), 164–8.

Sengupta, A., McConkey, K. R. & Radhakrishna, S. (2014). Seed dispersal by rhesus macaques *Macaca mulatta* in northern India. *American Journal of Primatology*, 76(12), 1175–84.

Southwick, C. H., Yongzu, Z., Haisheng, J., Zhenhe, L. & Wenyuan, Q. (1996). Population ecology of rhesus macaques in tropical and temperate habitats in China. In J. A. Fa & D. G. Lindburg, eds., *Evolution and Ecology of Macaque Societies*. Cambridge: Cambridge University Press, pp. 95–105.

Spagnoletti, N., Cardoso, T. C. M., Fragaszy, D. & Izar, P. (2017). Coexistence between humans and capuchins (*Sapajus libidinosus*): Comparing observational data with farmers' perceptions of crop losses. *International Journal of Primatology*, 38(2), 243–62.

Supriatna, J., Froehlich, J. W., Erwin, J. M. & Southwick, C. H. (1992). Population, habitat and conservation status of *Macaca maurus*, *Macaca tonkeana* and their putative hybrids. *Tropical Biodiversity*, 1(1), 31–48.

Suryawanshi, K. R., Bhatnagar, Y. V., Redpath, S. & Mishra, C. (2013). People, predators and perceptions: Patterns of livestock depredation by snow leopards and wolves. *Journal of Applied Ecology*, 50(3), 550–60.

Whitten, T., Henderson, G. S. & Mustafa, M. (2002). *The Ecology of Sulawesi*. Singapore: Periplus.

Wolfe, L. D. (2002). Rhesus macaques: A comparative study of two sites, Jaipur, India, and Silver Springs, Florida. In A. Fuentes & L. D. Wolfe, eds., *Primates Face to Face: The Conservation Implications of Human–Nonhuman Primate Interconnections*. Cambridge: Cambridge University Press, pp. 310–30.

Wolfe, L. D. & Peters, E. H. (1987). History of the free-ranging rhesus monkeys (*Macaca mulatta*) of Silver Springs. *Florida Scientist*, 50(4), 234–45.

Zak, A. A. & Riley, E. P. (2017). Comparing the use of camera traps and farmer reports to study crop feeding behavior of moor macaques (*Macaca maura*). *International Journal of Primatology*, 38(2), 224–42.

Elephants and Bees

Using Beehive Fences to Increase Human–Elephant Coexistence for Small-Scale Farmers in Kenya

LUCY E. KING

Elephants are so wrapped up in the modern history of East Africa that it is almost inconceivable to describe an era of politics or social development without some element of the pachyderms rising up to take their part in the story. In 1925 elephants roamed over 87 per cent of East Africa and the predominantly hunter-gatherer/pastoral people were described by historians as *living in a sea of elephants* (Parker & Graham 1989). Development and the advent of cultivation during the twentieth century reduced the elephant range to just 27 per cent by 1975 and the relationship changed from one of mild interactions to one of intense competition for resources with 'elephants occupying fragmenting, diminishing islands in a sea of people' (Parker & Graham 1989, p. 296). The parallel escalation of the ivory trade during the latter half of the twentieth century saw a catastrophic decline in the elephant populations in East Africa, with Kenya's number of elephants most notably crashing from around 129,570 in 1973 to 15,279 by 1989 (Douglas-Hamilton 1989). This intense poaching of elephants and reduction in their numbers from large swathes of the Kenyan bush from 1973 onwards corresponded with an expansion of agricultural range by a rapidly increasing human population that boomed from 12 million in 1972 to just under 23 million by 1990. Bush and rangeland areas previously thought impossible for farming activities were becoming increasingly free of elephant presence. The proceeding influx of aid for agricultural and social development in the 1980s and 1990s led to an escalation of farming activities and investment in the country, including areas around the boundaries of national parks and elephant reserves, turning many potential dispersal areas for elephants into farming zones.

History changed course in 1989 with the global ivory trade ban signed into full effect by the world's governments at the Convention on International Trade in Endangered Species of Wild Fauna and Flora (CITES) conference in Switzerland. All international trade in elephant ivory was banned and the price of ivory plummeted, effectively halting the killing of elephants (at least in their savannah range in Africa) overnight. Under this new non-utilization regime, Kenya's elephant populations were once again able to slowly recover.

Although unlikely ever to reach pre-crisis figures, Kenya's elephant population has increased in the last 25 years with between 23,000 and 30,000 now roaming over 22 per cent of the country (Thouless et al. 2016). Despite a new poaching outbreak between 2009 and 2012, the population weathered this dip, and old migration routes are being explored again. It is likely that older matriarchs and experienced bulls who survived the population crash of the 1970s and 1980s still retain the memory of old pastures and dry-season water holes. These elephants are thought to retain, and are passing on, a treasured social history of how to survive and thrive within the landscape and this migratory behaviour in places is bringing them directly into conflict with Kenya's modern-day farmers and settlers (Douglas-Hamilton et al. 2005). With the World Bank estimating 49.7 million people were living in Kenya in 2017, there are now more challenges for this smaller wild elephant population that still has the potential to grow even though its range and habitats are fragmented and disrupted by human development.

11.1 WILDLIFE CONFLICT FOR SMALL-SCALE FARMERS

The juncture of where farms and elephants collide has created an entire industry of research, innovation and political debate. On the farm side, we are witnessing expansion and permanency developing within nomadic and rural settlements, usually hard-working people, unreservedly believing that the land is theirs even if the title deeds remain a political promise. These settlers are rarely poachers; they are typically God-fearing farmers squeezing a living out of semi-arid land that was previously deemed unworthy for agriculture due to unreliable rains and rampant wildlife. What little funds they have are ploughed back into the soil twice a year at the sign of heavy clouds and scratchy harvests remain mostly subsistence in nature. Foraging birds, monkeys, squirrels, porcupines, antelopes and rogue livestock all create constant challenges to the successful harvesting of these small-scale crop plots (Newmark et al. 1994; King 2010). The closer the farms are to the edge of the wildlife refuges, the smaller the farms typically are as costs for crop protection and the needs for multiple animal deterrent methods escalate.

Elephants are typically not as frequently farm visitors as the smaller rodent, ungulate, primate or avian creatures, but they are perceived to be the most frightening crop-raiders with the greatest impact by small-scale farmers (King 2010). Family members often rotate crop-guarding duties through the day and night, using simple methods such as fire, barking dogs, shouting and banging on iron sheets to keep the elephants at bay. Children are often brought into these active deterrent efforts and a lack of sleep can lower school performance and attendance. Some farmers, repeatedly affected, try to erect fences, build stone walls or dig ditches to keep the elephants out but the effort required to maintain such front-line defences usually wears out and elephants soon habituate to these passive methods (Graham & Ochieng 2008; Hoare 2012). Farmers typically resort to active night-time patrolling and quickly become intolerant of these night raiders. Hence, attitudes towards elephants and any local wildlife managers are usually negative and strained, and fatalities on both sides are a sad reality (Dickman 2010; Sakellariadis 2015).

Although these new farms are typically located on old elephant dispersal areas, the geo-politics of modern-day Kenya means that it is unlikely established farming communities will be moved (Litoroh et al. 2012). As the conflict has increased at the farmer–elephant interface, there has been a momentum within the scientific and wildlife management community to find viable methods to convert this conflict juncture into projects aimed at helping communities move towards better coexistence (Dublin & Hoare 2004). Managers are now trying to integrate socio-economic projects with ethical ways to live with elephants through the use of sustainable elephant deterrents (Osborn & Parker 2003). The continuum stretches from essential hard boundaries (for example electric fences around a developed town) down to farmer-managed active deterrents to keep elephants out of small patches of farmland that are located deep within wildlife dispersal areas (see for example McCormick and Keyser 2016).

One such innovative idea developed on this continuum from conflict towards participatory coexistence, has been the development of chili-based elephant deterrents in Zimbabwe, an idea based on the irritation factor of the active ingredient *Capsicum oleoresin* to elephant nasal passages (Osborn 2002). The chili can be managed by farmers and applied in several formats around the farm, either as dried chili-dung bricks that are lit and smoked slowly throughout the evening releasing a potent chili smoke; or as an oil-infused substance that cloth rags can be soaked in and hung around the outside of the farm creating a smelly and painful barrier to any exploratory elephant trunk (Karidozo & Osborn 2015). Chili *bombs* have also been invented using dried chili powder packed into condoms with a firecracker tied into the end (McCormick & Keyser 2016). Once lit and thrown towards a raiding elephant the explosion of the firecracker splits the condom and sprays the unsuspecting elephant with irritating chili. The breadth of innovation around this idea stems from a behavioural understanding that utilizing an element of pain to deter elephants means the elephants are less likely to habituate. Despite proven successes, there still remains the challenge of rural farmers accessing the material or hand-growing enough chili plants to keep elephants at bay for an entire crop-season (Graham & Ochieng 2008).

11.2 HONEY BEES AS A NATURAL DETERRENT FOR ELEPHANTS

The concept of using a natural stimulant to deter elephants is one that we have been exploring at Save the Elephants since 2000 (www .savetheelephants.org), a scientific research organization founded in 1993 by one of the forefathers of elephant behavioural research, Dr Iain Douglas-Hamilton. Fundamental to our ethos is first to understand the behaviour, space and resource needs of elephants before using that knowledge to come up with management tools for their conservation. Specific to human–elephant conflict mitigation ideas, we are working on designing practical and socially appropriate farm-based deterrent methods that do not breach ethical boundaries for these highly intelligent and social animals.

The notion that elephants do not like honey bees and will avoid foraging on trees hosting active beehives was identified in 2000 by Professor Fritz Vollrath from Oxford University and the present chairman of Save the Elephants. Whilst working on Mpala Ranch in Laikipia, northern Kenya, Vollrath interviewed a number of beekeepers who revealed frequent observations of elephants running away from bee swarms (Vollrath & Douglas-Hamilton 2002a). This local knowledge had triggered much discussion amongst the local ranchers on whether or not bees could be used as 'guardians' to protect trees from the destruction caused by elephant foraging. This was tested scientifically by Vollrath and Douglas-Hamilton (2002b) and revealed that under experimental conditions, acacia trees with beehives (either occupied or empty) did indeed have the effect of preventing elephant-foraging damage to the trees. I joined the Save the Elephants team in 2006 to continue with this research question and in my interviews with local farmers and herders, they revealed local folklore that elephants *do not like bees* and will *run away* and *won't go anywhere near them* (King 2010).

Professor Vollrath and Dr Douglas-Hamilton gave me the challenge of further exploring this interaction between African honey bees *Apis mellifera scutellata* and elephants for my masters and then doctoral thesis with Oxford University. I was at first stalled by the seemingly enormous challenge of how to get the two species near each other in order to study their behavioural interaction in any kind of detail. Particularly as the starting hypothesis was that they would avoid each other at all costs, and secondly as both species were quite capable of killing me if I got an experiment wrong. Exploring literature on the extensive audio and play-back research by leading Kenyan elephant scientists who used the method to tease apart an understanding of elephant behaviour in response to vocalisation stimuli (Poole et al. 1988; Poole 1999; McComb et al. 2001; McComb et al. 2003), I decided to use playbacks as a safer method to understand how elephants might respond to honey bees. In 2006 I launched The Elephants and Bees Project (www.elephantsandbees .com) with our first set of field experiments based out of the Save the Elephants Research Centre in Samburu.

11.3 BEE PLAYBACK EXPERIMENTS

Finding an occupied honey bee colony in Samburu National Reserve was not a challenge as they are frequently observed in hollows of acacia trees and the one we found was handily at shoulder height. My Ndorobo field assistant, Lucas, helped me rig up a platform for my directional microphone and HD mini-disc player at the entrance of the hive and by throwing a stone into the hive entrance we made a recording of angry, disturbed honey bees. The eruption of bees out of the tree trunk was quite dramatic to observe and very quickly we could imagine elephants running away from such a stinging, noisy, furious, swarming spectacle. Cutting and duplicating 30 seconds of the most aggressive bee sounds with free PRAAT software gave us a consistent 4-minute bee playback recording to take into the field to play to elephants. We built a fake tree trunk out of reeds and an old vegetable rack and cut a hole out of the middle to sink a wireless speaker into the cavity about a metre above the ground to help recreate a *real* bee colony in a tree. We were able to quietly drop this fake tree trunk and camouflaged speaker within 10–20 metres of elephant families resting under trees in the middle of the day. Slowly driving away, we placed ourselves at a 45-degree angle away from the speaker and usually 40–60 metres away from the elephants. This method ensured that when we pressed play on the recording (using a wireless transmitter) the erupting bee sound would not be associated with the car or our human smells and that we would be closely observing the reaction of elephants to the sound of a disturbed honey bee colony.

This methodology worked surprisingly well. Our very first playback experiment with the fake-bee-sound-emitting tree trunk was beyond dramatic. The Rivers Family, a migrant elephant family well known to the Save the Elephants research team, were resting under a leafy *Kigalia africana* tree near the Ewaso N'giro river in Samburu. A young infant was sleeping flat on the soft sand and seven adult females and four calves were all snoozing and flapping their ears in the shade. We filmed this peaceful scene for our two-minute control period, taking notes of any alert behaviour, headshaking, dusting or movements. Once the bee stimulus was triggered, we filmed as the elephants stood stock still for a moment then ears came out, heads came up and started turning from side to side, trunks started to twist and smell the air and a gentle foot nudged the calves on the ground to get up. One female seemed particularly agitated and within 40 seconds of hearing the onset of the bee sound, she lifted her head and moved quickly away from the shade of the tree and the bee sounds. All the females followed her while the largest female, and presumed matriarch, paused the longest at the back of the group looking with alarm and annoyance at the direction of the speaker. Once the whole group had left the tree, they picked up speed and started to run at pace creating a cloud of dust as they fled away from the bee sound. Forgetting my filming protocol, I can hear myself whispering 'oh wow!' over the footage and that very first video remains one of our most watched research clips that has gone around the world on TV and social media as a classic example of how elephants respond to the sound of disturbed wild honey bees (see link for video: https://player.vimeo.com/video/72656689?title=0&byline=0&portrait=0&color=f5ce40). We repeated that bee playback experiment multiple times with seventeen different Samburu elephant families, including an entire series of fifteen control experiments where we played the sound of a rushing waterfall which contained all the frequencies emitted by honey bees but with the frequencies absorbed into the *white noise* of the waterfall. A total of sixteen out of the seventeen families ran or moved away on average 64 m to the bee sounds compared to just eight out of the fifteen control families who moved away on average just 20 m. Additionally, the speed of retreat was faster to bee sounds, with eight out of the sixteen moving away within just 10 seconds of bee sound onset compared to the control families where no elephants moved within 10 seconds and only four families had moved within the first 80 seconds, a significantly slower response. We were also able to play both bee and the control sound of white noise to nine of the same family groups, varying the playback order and leaving at least a seven-day gap between playbacks. There was a significant difference between sounds for both speed of retreat (latency of response) and distance moved from the same elephants. We also observed increased dusting and headshaking in those elephants hearing bee sounds, suggesting that the elephants understood what the sound of bees meant and they were trying to ensure that there were no stinging insects around the eyes or face (King et al. 2007).

11.4 BEE ALARM CALLS

Our experiments opened a window into our understanding of how elephants might respond to a wild colony of real African honey bees and during the weeks of fieldwork I also observed another behavioural phenomenon. Often I would not realize that there were other elephants hidden in the bushes nearby who would emerge during the playbacks and join the elephants in their retreat from the bee sounds. I found this particularly intriguing. Were they alarmed by the distant bee sounds and coming to join up with relatives to retreat together as a *safety in numbers* mode or were they actively communicating with each other about this potential threat?

I decided to repeat the experiment with some high-quality microphones that could capture any infrasonic sounds that the elephants might be emitting during their response to the bee sounds. Recording infrasonic sounds from wild elephants in the bush is not easy; it requires expensive equipment that needs consistent power and a mechanism to download and visualize acoustic data that are out of our human range of hearing. Looking for a skilled collaborator for this project, we contacted Dr Joseph Soltis, an expert bioacoustician working with animal behaviour at Disney's Animal Kingdom in Florida. He agreed to join our team in Kenya to repeat the playback experiments with me but this time with an array of three infrasonic Earthworks QTC micro-phones and Marantz PDM recorders placed around the elephants during the playback period. Each microphone was capable of picking up any low-frequency vocalizations (down to 3 Hz) that captured vital acoustic data well below the hearing of most human ears.

Our duplication of the bee playback experiment worked with similar results. Ninety-three per cent of the elephants moved away on average 72 m from the bee playback (compared to just 46 per cent of the control groups who moved just 32 m) and at the same time we got 217 recordings of elephants using rumbles to communicate with each other during the experiments. A total of 160 calls were in response to bee playbacks whilst elephants listening to the control of white noise emitted only 57 calls. Additionally the vocalizations to bee sounds continued and increased during the two-minute post-stimulus recording phase. There was a significant difference both in the rate of vocalizations and in the structure of the clearest 120 low-frequency calls between our bee and control groups with the bee response call showing a significantly higher second formant in the call structure (King et al. 2010; Figure 11.1).

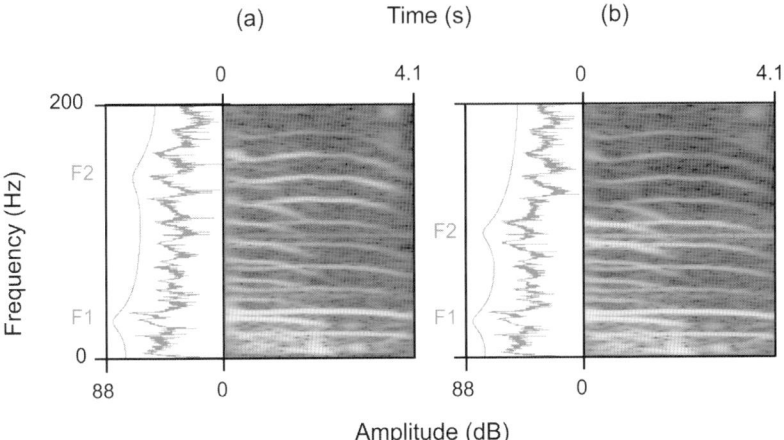

Figure 11.1 Frequency spectrograms of elephant rumbles in response to (a) the bee playback stimulus showing two formants (peaks, F1 and F2) that indicate where the stress in the vocal chords emphasizes the communication information in the elephant rumble, (b) representative spectrogram of a typical elephant rumble from those recorded in response to control white noise playback. Note the different position of F2.
Reproduced from King et al. (2010). *A black and white version of this figure will appear in some formats. For the colour version, please refer to the plate section.*

The excitement of this discovery amongst our team cannot be over-stated. However, we held onto these results for a further year, deciding to first play back the infrasonic calls we had identified to other elephants to see if there was any response to these *bee alarm calls*. This time we had to haul around a giant FBT MAXX 4A speaker and battery pack with capabilities of playing frequency components down to 20 Hz to enable the output of these special calls. Incredibly, the elephants listening to the playback of these low-frequency bee alarm calls also moved away and shook their heads as if there were bees in the area (King et al. 2010).

We did not stop there. The following year Joseph Soltis and I rounded off the set of experiments with a series of playbacks using Samburu warrior voices in an attempt to compare the bee alarm call we had identified with the alarm call from elephants to a known and recognized threat in the area, that of humans. The data were clear. The alarm call structure from elephants in response to humans had a significant difference in the mean fundamental frequency and in the first formant of the call structure, which were elevated compared to calls emitted in response to bee stimuli (Soltis et al. 2014). Furthermore, there was no headshaking or dusting behaviour in response to Samburu warrior voices. We have learnt that our Samburu elephants have a way to referentially communicate about these two specific threats to each other and that they can distinguish between a warning call about bees and a warning call about humans.

11.5 ELEPHANTS AND LIVE BEES

This understanding of how elephants might react to disturbed honey bees was supplemented by a one-off unique observation we had of an elephant group resting under a tree with a live beehive hung several metres above their heads. The elephants appeared to be snoozing com-fortably, unalarmed by the hive but on closer observation, we saw that in the heat of the day, the bees were very docile. Only one or two lazy bees were flying in and out of the hive and, clearly, there was no angry buzzing like our playback recording going on. We watched this peaceful scene for some time thinking how to *disturb* the beehive in order to witness a real interaction between elephants and wild honey bees. We decided to throw a stone at the hive. Possibly not the most sophisticated scientific method, but there was no other way to access the hive safely with the elephant family underneath. My assistant hit the hive first time and miraculously the stone zinged off into the bushes without bouncing off an elephant on its way down. For three seconds nothing happened. Then suddenly the hive erupted with angry bees flying out to see what had disturbed their midday rest. The elephants picked up their heads in fright, just like the playback experiments we were conducting. Within 30 seconds, the elephants had fled the scene and we watched in awe as they shot away across the savannah in a dust storm. Admiration turned to self-protection as the bees turned on us and we drove away in the opposite direction to out-pace them (King 2010).

These extensive and detailed field observations of the pure behavioural response of elephants to honey bees gave weight and momentum to my first design of a beehive fence. Fully appreciating that a bee colony needed to be disturbed to generate the angry bee sounds and stinging defence behaviour that appeared to be enough to send elephants running, I worked for some time on a beehive fence design. We then needed a willing community of farmers who were suffering from elephant raids to test the pilot idea for us.

11.6 BEEHIVE FENCE PILOT TRIAL IN LAIKIPIA

A farming community in Ex-Erok in Laikipia was ideally situated for our project. The farmers had been integrated into some elephant deterrent programmes a few years before and they were also active beekeepers. The farmers listened to my descriptions of elephants running from bee sounds, looked at my beehive fence drawings with some scepticism, but ended up agreeing wholeheartedly to help me with the experiment. Together we built the first beehive fence around the plot of an active farmer named Ephrahim Maina. Maina was already a beekeeper and demonstrated a motivation that can often be found amongst rural Kenyan farmers who have to innovate to survive in these semi-arid areas. The 90-m beehive fence took two days to build using traditional log beehives and strong wooden posts that held up a thatched roofing system that would keep the interlinked hives cool (see Text Box 11.1). It seemed like a simple concept and it was certainly inexpensive to build. The well-used, traditional hives cost around $8 each to buy off a Thara-kan beekeeper from Meru and the wire and few nails and posts that I bought in Nanyuki meant the whole experimental fence cost just under $300. The hives smelt of bees and honey comb but remained empty of live bee colonies for the six-week pilot experiment. Nevertheless, the beehive fence-protected plot had only thirty-eight elephants enter the farm compared to the next door control farm without a beehive fence

Text Box 11.1

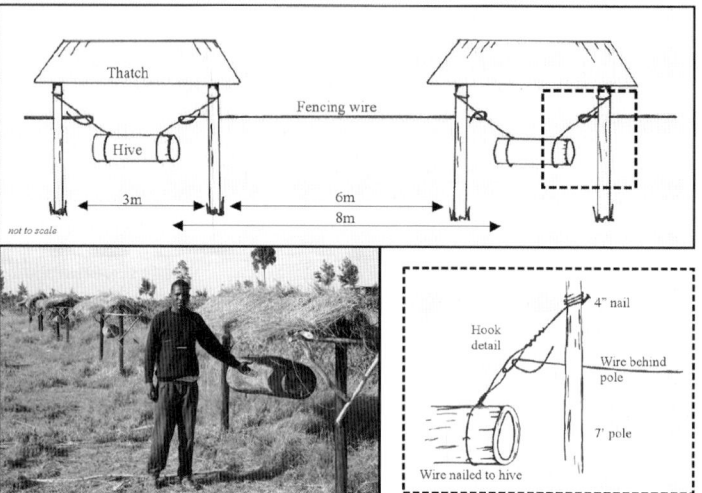

Figure 11.2 Our first 90-m beehive fence was built with interlinking traditional log beehives and grass-thatched shade roofs.

Our first beehive fence using 9 traditional log beehives was approximately 90 m in length. The fence was constructed with beehives hung under small thatched roofs. Huts were spaced 6 m apart, allowing the hives to be spaced 8 m apart. An elephant walking between huts would be less than 4 m from the nearest hive, the minimum distance elephants in the study area approached solitary beehives. The beehives swung freely, suspended by tightly secured fencing wire to the top of the seven-foot poles. Each hive was linked to the other with strong, taut, fencing wire that hooked to the centre of the permanent wire of each hive and was, crucially, behind the upright poles on the crop side of the fence. An intruding elephant trying to enter the field will avoid the complex solid structure of the huts and will be channelled between them. As the elephant tries to push through the thigh-high wire, it causes the attached beehives to swing violently, thereby disturbing and releasing the bees to irritate or sting the elephant. However, if forced, the interlinking wire will break away before the beehive is pulled down. This also prevents elephants being trapped inside the farm as they can break out without damaging the hives.

Reproduced from King et al. 2009 with permission from John Wiley and Sons. © 2009 Blackwell Publishing Ltd.

which had eighty-seven elephants successfully raid during the same time period. The beehive fence farmer was also able to harvest most of his crops compared to 90 per cent of the control farm being damaged by elephants (King et al. 2009).

11.7 BEEHIVE FENCE TRIALS IN NORTHERN KENYA

With construction of the Laikipia electric wildlife fence in 2008 and the closure of the elephants' route into the Ex-Erok community, we moved to a new farm–elephant conflict site south-east of Buffalo Springs Reserve to continue testing the beehive fence concept. Here we worked with a fascinating community of Ngisonyoka Turkana settlers who were farming in a stretch of land between the Ngare Mara and the Ngare Nite rivers. The elders had moved to the site in the late 1970s after intense inter-tribal conflict in Turkana land and the community had grown to sixty-two families by 2008. Simple thatched huts were stationed along a rocky plateau with farming sites positioned on either side of the ridge in two areas called Chumviyere and Etorro. With low annual rainfall, wildlife conflict and no ability for irrigation, farming was a hit-and-miss affair but this community had adapted by grazing livestock around and inside the national reserves and tapping into other natural resources – charcoal, bushmeat and ivory. The community was being monitored under the Monitoring the Illegal Killing of Elephants (MIKE) pro-gramme and in the four years prior to our project nine elephants had been poached within an 8-km radius (Douglas-Hamilton et al. 2010). Elephants were seemingly regarded as both a pest and as another resource for income and with the level of crop-raiding that was taking place, it had become a serious danger zone for the Samburu elephants and the community appeared far from the peaceful end of a coexistence continuum.

Teaming up with Wilson Lelukumani, Save the Elephants' energetic MIKE field officer, we used a rapid rural appraisal technique to conduct our participatory trial with the community. This technique using calendars, mapping and a series of group, gender-specific and individual interviews with the members of both Chumviyere and Etorro helped to incorporate the knowledge and opinions of these rural farmers into our own farm field trial design (King 2010). Wilson was Turkana himself and his insight and language skills enabled us to understand this previously ostracized community in great depth. Although everyone gained income from their farms and occasional paid work, cutting down trees for charcoal was the activity that generated the highest income. Elephants were unanimously regarded as the worst crop-raiders closely followed by monkeys, porcupines, squirrels, birds and antelope. All farmers had to remain awake at night to guard their crops against elephants and hence daytime sleeping was a necessity that took them away from productive work during the day and rendered the farms vulnerable to daytime crop-raiders like baboons and birds (King 2010).

One of our participatory group activities was the creation of an annual calendar for crop-planting and harvesting events overlaid with the time of year that destructive elephant crop-raiding events occurred. These qualitative data were both accurate and helpful in establishing the beehive fence trial to fit in with the first crop-growing season. After explaining how the fence worked and earning some enthusiasm for participating in the project, 81 per cent of the sixteen farmers interviewed stated, very politely, that the beehive fence would not be successful at keeping elephants out of their farms. Their honesty was encouraging and set off a theme to our two-year trial in the area when we frequently had very frank and open discussions about the project, the design of the fence and the improvements that could be made. The farmers became fully engaged with the entire process of the project and during the construction process two farmers, Peter and Lobenyo, reflected on the effort required to build the thatched roofs to cover the hives from the sun and came to me with a concept to speed up the build. This kind of engagement and innovation was highly encouraging and we quickly adopted Peter and Lobenyo's new free-swinging roof design that not only sped up construction but also required fewer materials and was easier to maintain.

A total of thirty-four front-line farms were selected for the trial, seventeen of which were protected by conjoined beehive fences containing 171 Kenyan Top Bar beehives and seventeen farms were established as control farms without a beehive fence. All farms had established thorn bush barriers around the outside and these were left in place as they helped deter other crop-raiders such as baboons and antelope. The first planting season of the study in 2009 was affected by a devastating regional drought but in our second season the rains and crops came back to the study site and we recorded forty-five farm visit events by elephants to the study farms. Of these farm visits, thirty-two resulted in elephants approaching the thorn bush and successfully breaking the thorn barriers and entering the farms. No elephants were deterred by the thorn bush barriers. In contrast, only one elephant broke a beehive fence to gain access to the farm and both hives on either side of the break point were unoccupied with bees at the time. Of the remaining thirteen elephant visits, eight elephants approached right up to the beehive fence, walked along it for some time and then backed away into the bush. In five more events we recorded elephants walking along the line of the beehive fence until they got to the *end* where the thorn bush barriers started again and they broke through into the farms at this juncture. However, on six occasions the elephants entered the farm from the other side, directly coming through the houses and entering the farms from behind the barriers. In six of these cases elephants were able to successfully break out of the beehive fence when being chased by the farmer (King et al. 2011).

These data were highly encouraging. Not only did the beehive fences provide significant protection to the farms from elephants coming directly at them from the bush but also it proved to the farmers that thorn bush barriers were no deterrent at all for crop-raiding elephants. Furthermore, 98 of the 171 hives were occupied at least once during the trial providing the farmers with a total of 106 kg of honey at the end of the second year. This provided an approximate income of $290 for the farmers to share. This income was less than expected due to the earlier drought conditions and a spate of honey badger attacks, but was nevertheless regarded as significant for the farmers involved in the trial. It provided an additional motivation to maintain the beehive fences between crop-growing seasons so that the bees could continue to generate the farmers an income. Out of the sixteen farmers interviewed in 2008 before the trial, only one individual believed the beehive fences would be *very successful* at deterring elephants. Yet we were encouraged to record that the remaining fifteen farmers had modified their opinion two years later from *not successful* to *very successful* by the end of the trial (King 2010).

11.8 BEEHIVE FENCE TRIALS IN TAITA-TAVETA

During a project delay caused by the drought of 2009, I was invited to a high human–elephant conflict site in Taita-Taveta, southern Kenya, to build two trial beehive fences for the people of Mwakoma village. The community of around 150 households was incredibly welcoming from the first day of my arrival and was situated in a unique location on the side of Sagalla hill, only 15 minutes from Voi town. The village looked over Tsavo East National Park, home to 6,214 elephants (Ngene et al.

2013) and it was from here that elephants would travel out of the park boundary, cross the Mombasa highway and into the farms to crop-raid during the night. The community was starting to build a new primary school as the children no longer felt safe walking the 3 km to the neighbouring village school as raiding elephants could be seen roaming around early in the mornings and would sometimes block the main village road. One farmer had tragically been killed the year before by an elephant in his farm, and I was shown two huts that had been entirely knocked down by elephants attempting to access the grain stored inside.

We held a large community meeting at the fledgling primary school to discuss the issue of elephants and how the farmers were suffering from repeat crop-raiding events. Emotions were high and there were plenty of stories shared about the level of conflict. Indeed, interviews our team conducted in this community revealed that *fear, anger and helplessness* were predominant emotions expressed amongst the farmers in relation to the intensity of elephant crop-raiding that was going on (Sakellariadis 2015). At this initial meeting, the community selected two families who had the worst affected one-acre farms and we set out to build each of them a trial beehive fence using Kenyan Top Bar Hives (KTBH) constructed during a workshop a village elder, Hesron Nzumu, helped to organize at the school. Throughout the whole process, the community were willing, interested, positive and helpful and it was a pleasure to come back over the next ten months to visit the farms with Nzumu and collect the simple data sheets that the farmers had agreed to fill out. The data showed that fifty-two elephants had visited the two farms in the ten-month study period and only one elephant had broken the beehive fence.

The community were hugely encouraged by the trial and when I returned to Sagalla in June 2012 I was pleased to see both pilot beehive fences still being used three years on. During a second participatory gathering at Mwakoma Primary School, the community democratically created a list of the ten worst affected farms by selecting farmers who were perceived as being active and would be able to manage the bee-keeping side of the project. With collaborative support from the local Kenya Wildlife Service (KWS) wardens in Tsavo East National Park and funding from the Disney Conservation Fund (DCF) and The Rufford Foundation, we set about constructing a further six beehive fences with a new design aimed at both testing a new beehive and reducing the cost of the fence by 50 per cent.

We constructed four beehive fences with the more expensive Lang-stroth beehives. These larger, modern hives house the queen bee and her egg-laying activities in a ten-bar *brood chamber* separated from the honey super boxes with a neat queen excluder wire mesh made of local coffee wire. The worker bees are small enough to easily pass through the mesh ensuring they can feed the brood, clean the hive and make honey at the same time. The design is ingenious and provides a beekeeper with an entire box of pure, brood-free honey combs that ensures easy extraction without disturbing the queen. The Langstroth beehives were adapted by attaching small wooden blocks on the outside through which we passed fencing wire to attach the hives to the fence. We built a further two beehive fences with KTBH hives to join our two pilot KTBH beehive fence farms already in action (see Text Box 11.2).

During the trials in both Ngare Mara and Sagalla, our hive occupation had rarely peaked over 50 per cent so incorporated into the new project was an idea to remove half of the beehives from the fence design and replace them with two-dimensional 'dummy' beehives (Text Box 11.2). The added advantage to the reduced cost was that this design ensured that the hives were now 20 m apart from each other, which reduced fighting between bee colonies and helped with increasing occupation rates for the beehive fences. In terms of affecting elephant deterrence intensity, we hypothesized that should an elephant still try to push between a hive and a dummy, there would still be live bees on at least one side of the raider (Text Box 11.2).

A further development was an adaptation invented by Nzumu who was by now both a project officer and also one of our beehive fence farmers. His idea was to use live posts for hanging the hives instead of buying wooden posts from the hardware store. We coppiced or cut *Commiphora* tree trunks and left them for 3–4 days for the sap to run and then replanted them as posts for the beehives. The posts quickly sprouted roots and shoots providing stability to them and shade branches for the beehives. Within a year or two of being planted, the posts create a leafy canopy and the interlinking branches avoid the need for replacing the thatched shade roof for the hives. The farmers quickly adopted Nzumu's idea as termites ate their 'dead' posts and the live *Commiphora* posts were easily cut for free from the surrounding bush.

By September 2012 we had our eight beehive fences built and ready for the upcoming crop season and natural swarms had already occupied a dozen of the 114 hives. Each of the eight plots had conjoining control plots on the same farm of equal size and the farmers were trained in data collection using simple maps of their farms where any elephant movements around the fence could be drawn on by pencil. Our new

Text Box 11.2

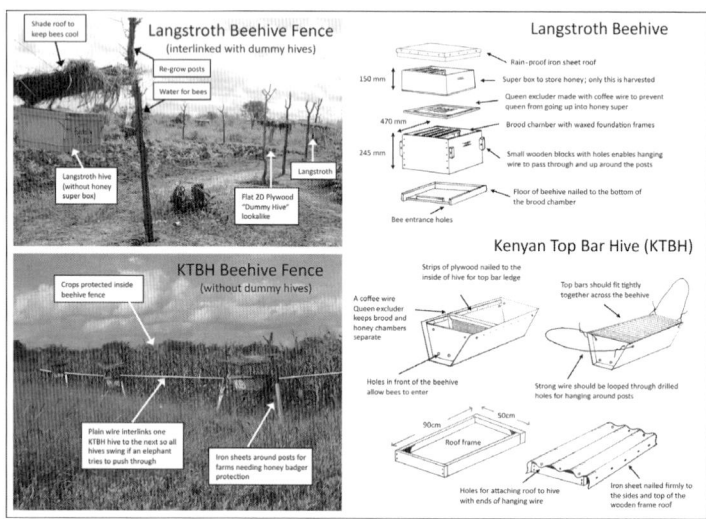

Figure 11.3 Construction of Langstroth beehives and Kenyan Top Bar Hives (KTBH) used in elephant-deterrent beehive fences and setup and components of the fences.

The waxed combs of the Langstroth hives help attract bees and provide ample storage for honey in the super box above, separated from the brood chamber by a horizontal queen excluder. The KTBH hives have a vertical queen excluder separating the front and back chambers for brood and honey respectively. Real hives and dummy hives (two-dimensional plywood) hang alternately along plain fencing wire attached to regrowing *Commiphora* sp. posts (King et al. 2017). These 2-D dummy hives were made out of 9 mm plywood and by cutting twelve hive-like shapes out of one plywood sheet, costing just $3 each. Instead of a $50 beehive being hung every 10 m, we alternated hive–dummy–hive–dummy, etc. around each one-acre farm. This meant that we only needed twelve real beehives and twelve 2-D dummy hives to protect one acre. Iron sheets tied onto the posts help deter honey badger attacks, as their claws cannot grip onto the iron.

Reproduced from King et al. 2017, © 2017 Society for Conservation Biology.

community project officer, Emmanuel Mwambingu, came and checked the sheets once a raiding event was called in. Emmanuel and Nzumu were both given a week's intensive beekeeping training at Baraka Agricultural College, which provided our team with two experienced community members to assist the farmers with bee-related questions and hive management. During the first 18 months of the project, elephant visits appeared to be slow and steady and data trickled in. Early data showed promising results in deterring the few elephants that reached the farms from the park and by 2013 demand for beehive fences from the rest of the community was so intense that we constructed a further two DCF-funded Langstroth fences in February 2013, giving us a sample of ten farmers and 131 beehives to work with.

11.9 THE ELEPHANTS AND BEES RESEARCH CENTRE

Support for the project from both KWS park wardens and the community appeared genuine and was cemented by the sub-chief of Mwakoma village who approached us in November 2013 with a unique offer of an acre of land to construct an office so that we could expand the project. KWS and the Education Secretary for Voi County presided over the colourful opening ceremony of the Elephants and Bees Research Centre in March 2014. Once we had a serious base for our project's activities, the output boomed. We set up a rain gauge, established better beehive monitoring systems, trained a third community project officer for daily GPS monitoring of elephant movements, constructed a honey processing room and attracted teams of Kenyan and international volunteers to help with monitoring, maintenance and honey harvesting events. The boost in capacity for data collection and management also coincided with the initiation of the new Standard Gauge Railway in early 2014, a flagship development project for Kenya. This railway construction began along the boundary of the project site between Sagalla and Tsavo East National Park generating considerable disturbance and affecting the natural movement of elephants in and out of the park. Some elephants from the park were likely deterred from crossing the construction line, but ones that did make it over got stuck on the Sagallan side of the highway and caused a noticeable increase in elephant crop-raid events in our farms (King et al. 2017).

KWS was aware of the increasing conflict occurring in the Sagalla community and donated thirty International Fund for Animal Welfare (IFAW)-funded beehives to the project in 2014, enabling us to construct a further three beehive fences in Mwakoma village. The farmers were pleased to receive these beehives from KWS and much discussion resulted where our farmers perceived this to be a good gift from the park administration for helping them to coexist with elephants that roamed outside the park boundaries.

As news of the success of beehive fences started to spread, demand for them grew rapidly and the next-door village of Mwambiti was keen to get involved. In late 2014 we completed our rapid rural assessment in Mwambiti and began constructing a further eleven beehive fences in this village during 2015/16. At the time of writing we have twenty-five beehive fences in action in the two villages but uptake and demand is outstripping our funding and capacity and even with two more Kenyan project officers and volunteers assisting us with monitoring, research and support, we are still under capacity for the demand generated by the community for this new income-generating elephant mitigation method. More communities are now coming on board to test the beehive fence idea around the Tsavo Conservation Area.

11.10 RESULTS OVERVIEW FROM SAGALLA BEEHIVE FENCE STUDY

Data from the first 43 months of the initial ten Mwakoma village farms show that the beehive fences have deterred 80 per cent of the 253 elephants that approached the farms and attempted to access the crops inside. Of the 43 elephants that did manage to break a beehive fence to access a field of crops, 34 elephants (71 per cent) came during the driest time on the project when there had been less than 30 mm of rain in the previous 28 days and the elephants were likely looking for supplementary food (Figure 11.4).

Additionally, in 70 per cent of the successful beehive fence breakage events (n = 18), the group size of elephants that accessed the farms was mostly 1–2 elephants with just eight events where groups of 3–4 elephants broke through. This very low figure meant that farmers were able to successfully chase the elephants out of the farms without too much damage being done. None of the larger groups' sizes of 5–10 elephants broke a beehive fence (Figure 11.5).

During 2014–15, the six Langstroth farmers had a mean occupation rate in their sixty-eight hives of 68 per cent, resulting in harvesting 206 kg of raw Elephant-Friendly Honey™ (Figure 11.6). This contrasted to the four KTBH farmers who only managed an average occupation rate of 32 per cent in their sixty-three KTBH hives and only harvested 22 kg of honey between them. The Langstroth hives were more successful in almost all aspects of the project – they were quicker to attract and retain

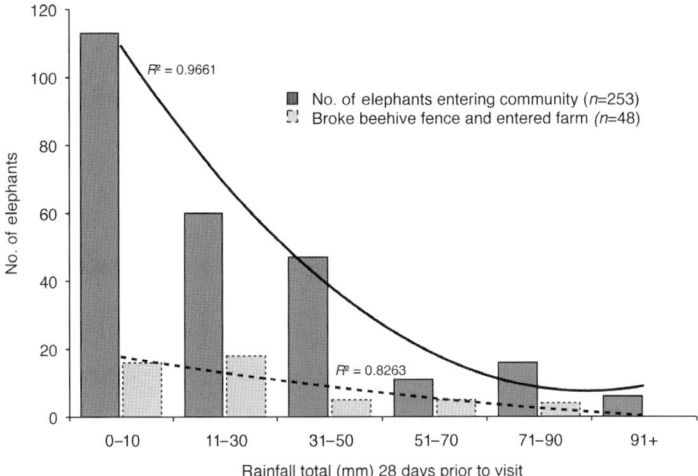

Figure II. 4 Frequency of elephants visiting farms and breaking through beehive fences relative to biannual rainfall periods.
Reproduced from King et al. (2017).

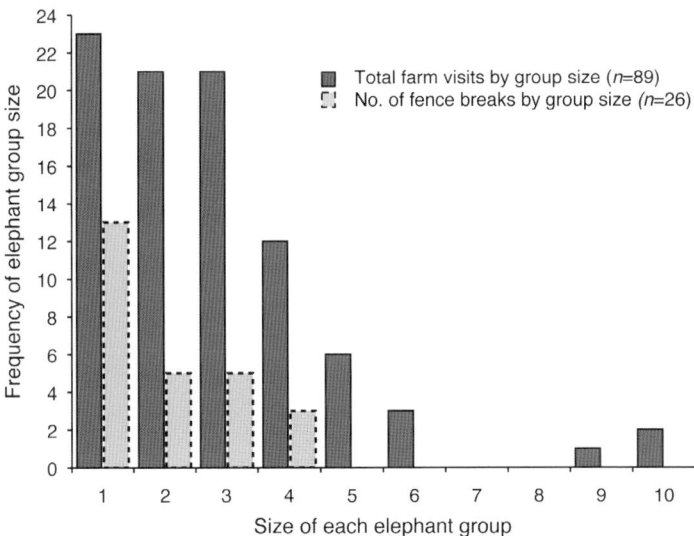

Figure II.5 Elephant visits to farms and number of incidences of elephants breaking through beehive fences by group size.
Reproduced from King et al. (2017).

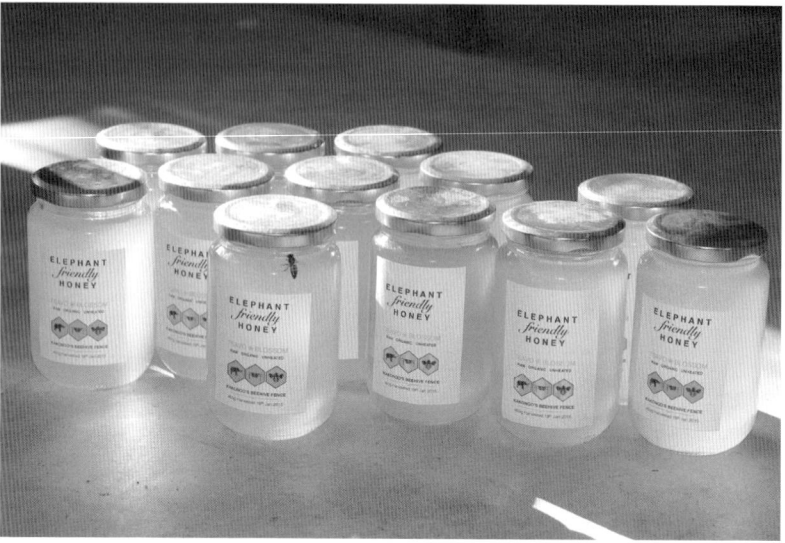

Figure 11.6 Elephant-Friendly Honey™

bees during the dry season, were easy to check and harvest and they produced more honey per hive than the KTBH hives. Notably, there were only seven events in the 43-month period where breakages of the beehive fence occurred between hives that were fully occupied with bees (King et al. 2017). Despite costing around $20 more per hive to purchase we recommend Langstroth hives for beehive fences as without live bees in the hives, the fences lose their *natural electricity* and may eventually lose their deterrence effect against elephant raiders.

Honey production from the beehive fences created a circulatory motivation for the farmers to maintain the beehive fences all year round. The potential income from the hives of $3 per super box bar of honey (one Langstroth bar produced on average 850 g of honey) was significant income to subsistence farmers who might be earning only $20–$100 a month in cash from other sources. With ten bars in each super box and twelve hives per farm, the potential income from a 68 per cent occupied beehive fence could be $245 per harvest season. In reality we saw significantly less income being generated from the hives than hoped ($684 income generated for all ten farmers) and this was largely due to unpredictable rainfall and honey badger attacks on fully laden beehives, which caused repeated frustration for the farmers. We bolstered all beehive fences by wiring strips of 70 cm iron sheets up each of the fence posts that greatly helped to reduce honey badger raiding success but this predator still remains a remarkable challenge for the beehive fence concept. Despite the lower than hoped-for honey income, the beehive fence still remains the only elephant deterrent fence that actually generates the farmer an income throughout the year and the income certainly outweighs the minimal maintenance costs.

Housing several bee colonies around a farm also has significant pollination benefits for the types of crops being grown in the subsistence, pesticide-free farms in Sagalla. Encouraging beehive fence farmers to grow more honey bee pollinated crops is certainly another way to help boost yields and profitability from farms. We are now encouraging our farmers to grow bee-pollinated crops that are less palatable to elephants. One of these trial crops is sunflowers which are quick growing, provide ample bee nectar and pollen and the seeds can be either dried and eaten, or pressed to provide cooking oil. Farmers testing it for us in Sagalla have reported that elephants 'don't eat it' and planting this as a buffer crop behind the beehive fences to hide the maize and more tasty crops could be an added boost to the effectiveness of the beehive fence.

11.11 COPING BETTER WITH ELEPHANTS

Human perceptions and attitudes towards wildlife are notoriously difficult to measure in a quantifiable way as there is often subjectivity and variability between interviewees and locations. A total of forty-three farmers from both our Ngare Mara and Sagalla beehive fence project sites were interviewed in 2013. Sakellariadis' survey of our participating farmers revealed a significant difference between the perception of farmers with and without beehive fences on whether or not they felt able to 'deal with the problem of elephants on their own' (Figure 11.7).

Although the survey sample size was small and the Sagalla beehive fence farmers had only been participating in the project for one year at the time of the survey, we felt encouraged by these results that beehive fences were one tool that farmers were not only actively able to manage themselves, but that they appear to positively influence the perception that farmers have in their own ability to deal with crop-raiding incidents (Sakellariadis 2015). On the continuum to coexistence, these farmers appear to be halfway there.

Perception of 43 farmers (with and without beehive fences) on their ability to deal with the problem of elephants on their own

(1=very able to cope - 5=not able to cope at all)

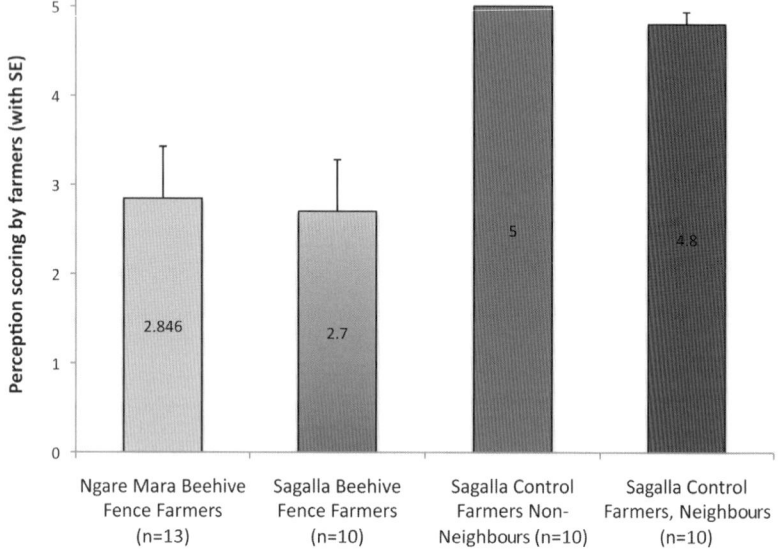

Figure 11.7 Comparison of confidence about coping capacity with elephants between farmers with and without beehive fences in Ngare Mara and Sagalla. Data extracted from Sakellariadis (2015).

11.12 CONCLUSIONS

Farmers who live with elephants need to start thinking differently about what they grow and where and how to implement different deterrent methods to their farms to have the most effective boundary protection they can without breaking ethical limits. Beehives may not be appropriate to have around the house due to an increased risk of bee stings, but they are effective at protecting the majority of a farm boundary where it is harder to see approaching elephants in the dark. Perhaps a ditch, fire-burning stations or bright flashing lights can be used nearer the house, leaving the bees to defend the more distant area of the farm. Boosting the effectiveness of beehive fences from 80 per cent to closer to 100 per cent could, for example, be reached by tying chili-oil-soaked rags to the wire interlinking the beehives to double the negative physical reaction from approaching elephants. Creating this toolbox of options for combining or rotating farm-based deterrents needs more research so the best combination of options for different geographical, climatic and cultural conditions can be found. Without doubt, we believe that beehive fences should now be included in this toolbox of options for small-scale farmers living with elephants. The triple benefits of both reduced elephant crop-raids, pollination services and honey production creates an income-generating fence that can help farmers to become more tolerant of their elephant neighbours. With the jigsaw of quantitative and qualitative data generated by the Elephants and Bees Project over the past eleven years, we believe beehive fences do positively contribute to participating farmers being shifted more towards the coexistence end of the conflict-to-coexistence continuum.

11.13 RECOMMENDATIONS AND FUTURE RESEARCH

- Beehive fences are now being tested in fourteen countries across Africa and Asia. Whether Asian elephants will react the same to Asian bees is part of an ongoing research project by the Elephants and Bees team but more active test sites are needed.
- Research into combining active and passive deterrent methods to keep elephants out of farmland is greatly needed as well as proactive political will to define agreed wildlife dispersal areas from agricultural zones.
- Beehive fences are effective at keeping up to 80 per cent of elephants out of farms and our data show that Langstroth beehives are the most effective hive to use both for attracting and retaining bees and for valuable honey yields.
- To download the research papers and our free Beehive Fence Construction Manual visit www.elephantsandbees.com.

11.14 References

Dickman, A. J. (2010). Complexities of conflict: The importance of considering social factors for effectively resolving human–wildlife conflict. *Animal Conservation*, 13, 458–66.

Douglas-Hamilton, I. (1989). Overview of status and trends of the African elephant. In S Cobb, ed., *The Ivory Trade and Future of the African Elephant*. Oxford: Ivory Trade Review Group.

Douglas-Hamilton, I., Krink, T. & Vollrath, F. (2005). Movement and corridors of African elephants in relation to protected areas. *Naturwissenschaften*, 92, 158–63.

Douglas-Hamilton, I., Wittemyer, G. & Ihwagi, F. (2010). Levels of illegal killing of elephants in the Laikipia-Samburu MIKE site. CITES document CoP15 Inf.40 (Rev.1).

Dublin, H. T. & Hoare, R. E. (2004). Searching for solutions: The evolution of an integrated approach to understanding and mitigating human–elephant conflict in Africa. *Human Dimensions of Wildlife*, 9, 271–8.

Graham, M. D. & Ochieng, T. (2008). Uptake and performance of farm-based measures for reducing crop-raiding by elephants *Loxodonta africana* among small holder farms in Laikipia District, Kenya. *Oryx*, 42, 76–82.

Hoare, R. (2012). Lessons from 15 years of human–elephant conflict mitigation: Management considerations involving biological, physical and governance issues in Africa. *Pachyderm*, 51, 60–74.

Karidozo, M. & Osborn, F. (2015). Community based conflict mitigation trials: Results of field tests of chilli as an elephant deterrent. *Journal of Biodiversity and Endangered Species*, 3, 1.

King, L. E. (2010). *The Interaction between the African Elephant and the African Honeybee and its Potential Application as an Elephant Deterrent.* PhD Thesis. Oxford: University of Oxford.

King, L. E., Douglas-Hamilton, I. & Vollrath, F. (2007). African elephants run from the sound of disturbed bees. *Current Biology*, 17, 832–3.

King, L. E., Douglas-Hamilton, I. & Vollrath, F. (2011). Beehive fences as effective deterrents for crop-raiding elephants: Field trials in northern Kenya. *African Journal of Ecology*, 49, 431–9.

King, L. E., Lala, F., Nzumu, H., Mwambingu, E. & Douglas-Hamilton, I. (2017). Beehive fences as a multi-dimensional conflict mitigation tool for farmers co-existing with elephants. *Conservation Biology*, 31, 743–52.

King, L. E., Lawrence, A., Douglas-Hamilton, I. & Vollrath, F. (2009). Beehive fence deters crop-raiding elephants. *African Journal of Ecology*, 47, 131–7.

King, L. E., Soltis, J., Douglas-Hamilton, I., Savage, A. & Vollrath, F. (2010). Bee threat elicits alarm call in African elephants. *PLoS One*, 5(4), e10346.

Litoroh, M., Omondi, P., Kock, R. & Amin, R. (2012). *Conservation and Management Strategy for the Elephant in Kenya.* Nairobi: Kenya Wildlife Service.

McComb, K., Moss, C., Durant, S., Baker, L. & Sayialel, S. (2001). Matriarchs as repositories of social knowledge in African elephants. *Science*, 292, 491–4.

McComb, K., Reby, D., Baker, L., Moss, C. & Sayialel, S. (2003). Long-distance communication of social identity in African elephants. *Animal Behaviour*, 65, 317–29.

McCormick, S. & Keyser, C. (2016). Planning for resilience in East Africa through policy adaptation, research and economic development (PRE-PARED) project. *Human–Elephant Conflict Prevention Toolkit and Best Practices.* Africa First International, USAID and Tetra Tech ARD, USA.

Newmark, W. D., Manyanza, D., Gamassa, D.-G. M. & Sariko, H. I. (1994). The conflict between wildlife and local people living adjacent to protected areas in Tanzania: Human density as a predictor. *Conservation Biology*, 8, 249–55.

Ngene, S., Njumbi, S., Nzisa, M., Kimitei, K., Mukeka, J., Muya, S., Ihwagi, F. & Omondi, P. (2013). Status and trends of the elephant population in the Tsavo-Mkomazi ecosystem. *Pachyderm*, 53, January–June, 38–50.

Osborn, F. V. (2002). *Capsicum oleoresin* as an elephant repellent: Field trials in the communal lands of Zimbabwe. *Journal of Wildlife Management*, 66, 674–7.

Osborn, F. V. & Parker, G. (2003). Towards an integrated approach for reducing the conflict between elephants and people: A review of current research. *Oryx*, 37, 80–4.

Parker, I. S. C. & Graham, A. D. (1989). Elephant decline (part I) downward trends in African elephant distribution and numbers. *International Journal of Environmental Studies*, 34, 287–305.

Poole, J. H. (1999). Signals and assessment in African elephants: Evidence from playback experiments. *Animal Behaviour*, 58, 185–93.

Poole, J. H., Payne, K., Langbauer, W. R. & Moss, C. J. (1988). The social contexts of some very low frequency calls of African elephants. *Behavioural Ecology & Sociobiology*, 22, 385–92.

Sakellariadis, A. (2015). *Shifting the Dynamic Equilibrium among Perceptions, Attitudes, and Behaviors*. MS Thesis. New Haven, CT: Yale School of Forestry and Environmental Studies.

Soltis, J., King, L. E., Douglas-Hamilton, I., Vollrath, F. & Savage, A. (2014). African elephant alarm calls distinguish between threats from humans and bees. *PLoS One*, 9(2), e89403.

Thouless, C. R., Dublin, H. T., Blanc, J. J., Skinner, D. P., Daniel, T. E., Taylor, R. D., Maisels, F., Frederick H. L. & Bouché, P. (2016). African Elephant Status Report 2016: An update from the African Elephant Database. Occasional Paper Series of the IUCN Species Survival Commission, IUCN, Gland, Switzerland: No. 60 IUCN / SSC Africa Elephant Specialist Group.

Vollrath, F. & Douglas-Hamilton, I. (2002a). African bees to control African elephants. *Naturwissenshaften*, 89, 508–11.

Vollrath, F. & Douglas-Hamilton, I. (2002b). Elephants buzz off! *SWARA Magazine*. September–December, pp. 20–1.

12

The Twin Challenges of Preventing Real and Perceived Threats to Human Interests

OMAR OHRENS, FRANCISCO SANTIAGO-ÁVILA
AND ADRIAN TREVES

Humans and other species have historically competed over resources and space, often resulting in interspecies conflict. As one example, humans have hunted or domesticated wild herbivores for protein, which are also consumed by predators. This leads to conflict between predators and humans over food resources, subsequently leading people to retaliatory killing of carnivores, which poses a major threat to predator populations (Woodroffe & Ginsberg 1998; Chapron et al. 2014). Although societies have developed mitigation strategies to reduce such conflicts, the rise of social discord between people who value carnivores and those who do not has sometimes affected the use of mitigation strategies, whether lethal or non-lethal (Treves & Karanth 2003; Treves et al. 2006; Redpath et al. 2013; Treves & Bruskotter 2014; Woodroffe & Redpath 2015). Differences in interests between people can lead to imposed solutions that benefit some people over others, due to power relations or prevailing attitudes (Redpath et al. 2013; Treves et al. 2015). Because of this imbalanced decision-making, a proposed method may not be implemented as planned (Fishbein & Yzer 2003) or may be dismantled later (Karanth & Madhusudan 2002), even when functionally effective.

The non-implementation of solutions to human–carnivore conflict highlights hidden cognitive mechanisms that have been described by social psychologists' theories (e.g. Ajzen's Theory of Planned Behaviour (hereafter TPB); see Text Box 12.1), in which a complex mix of social norms, emotions and external conditions can influence people's decisions and actions (Fishbein & Yzer 2003; Wieckzorek Hudenko 2012; Amit & Jacobson 2017; Schlüter et al. 2017). The cognitive dimensions of human behaviour interact with both individual appraisals of

Text Box 12.1 Theory of Planned Behaviour (adapted from Ajzen 1991)

This theoretical framework describes how intentions to perform certain behaviours are predicted by cognitive variables such as attitudes towards the behaviour (i.e. evaluation of the behaviour in question), subjective norms (i.e. social pressure to perform the behaviour) and perceived behavioural control (i.e. self-efficacy or perceived capacity to perform the behaviour).

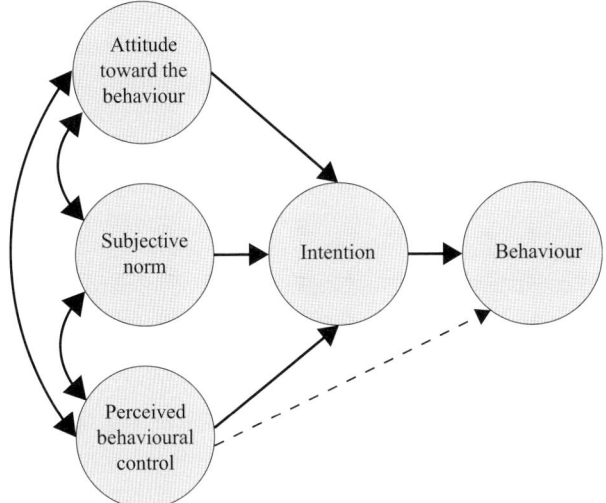

Figure 12.1 Theory of Planned Behaviour.
Adapted from Ajzen 1991.

effectiveness – which does not necessarily correlate with functional effectiveness – and uncertainty about effectiveness to alter implementation. We use the terms effective (*powerful in effect; producing a notable effect*, www.oed.com) and effectiveness because these allow us to address both the potential of individual actors to achieve coexistence and the efficacy of technical devices to attain that goal. We do not use the term efficacy as it is more limited (*not used as an attribute of personal agents*, www.oed.com) and avoid efficient because of its potential for confusion with feasible (*capable of being done, accomplished or carried out; possible, practicable*, www.oed.com). We also focus on evidence for effectiveness of an intervention from actual experimental trials under working conditions (not under laboratory conditions), not idealized claims of effectiveness that have not yet been realized through real-world testing.

Scientific research has shown that numerous methods of intervention can promote coexistence between people and carnivores (Inskip & Zimmermann 2009; Treves et al. 2009; McManus et al. 2015). However, few have been scientifically evaluated along multiple criteria of effectiveness, cost-efficiency, environmental consequences, social acceptability (Shivik et al. 2003; Breitenmoser et al. 2005; Inskip & Zimmermann 2009; Treves et al. 2009; Zarco-González & Monroy-Vilchis 2014; McManus et al. 2015) and adequacy of implementation. Here, we lay out an integrative framework for understanding the implementation of interventions for coexistence and conflict, which includes both the effect in preventing future damages (functional effectiveness, 'FE' hereafter) and the individual human perceptions of effectiveness of an intervention (perceived effectiveness, 'PE' hereafter). In some cases, conflicting perceptions of effectiveness and FE can lead to negative outcomes for wildlife or property owners, where the goals of conservation and coexistence with wild animals may be jeopardized. We expose the cause-and-effect logic underlying decisions to intervene or not, where both explicit and hidden mechanisms are considered. By understanding better how FE and PE relate, we believe the field can avoid a sterile debate claiming that *people are irrational* on the one hand or that *technical experts have no common sense* on the other hand. Avoiding such misunderstandings may improve intervention design and implementation, conservation and coexistence efforts, policy, conflict resolution and scientific analysis of human–wildlife conflict and coexistence (HWCC).

12.1 THE THEORY BEHIND FE AND PE

FE in our context of HWCC measures *whether the intervention reduces future attacks by wildlife* (Treves et al. 2016). Because empirical measurement of wildlife damage and its attribution to wildlife is a technical skill with a measurable rate of errors (e.g. Plumer et al. 2018), FE differs markedly from human opinion of the effectiveness of an intervention, to which we return below. Nevertheless, FE is difficult to evaluate rigorously. Biomedical sciences have pioneered in experiments yielding strong inference about the FE of interventions. For instance, randomized control trials (*gold standard* hereafter; see Text Box 12.2), are the most robust methods to estimate the effectiveness of an intervention (Grimshaw et al. 2000; Mukherjee 2010). Avoiding biases at several stages and reducing the effect of confounding variables are indispensable advantages of this method. For instance, there have been four

---

Text Box 12.2 Definition of Gold, Silver and Platinum-Standard Experiments (see Treves et al. 2016)

Gold Standard

Random assignment of treatments and controls, without detectable biases in sampling, treatment, measurement or reporting. It produces the strongest inference and evidence of effectiveness of an intervention. Examples of this were reported in Treves et al. (2016).

Silver Standard

Non-random assignment of treatments. Includes quasi-experimental designs with haphazard assignment of treatments, such as case-control or Before-After Control-Impact (hereafter BACI) designs. Produces weaker inference because of potential pre-existing differences between treatment and control replicates, and because of confounding temporal effects coincident with the treatments.

Platinum Standard

A gold-standard experiment in which 'blinding' prevents intervenors from influencing measurers and vice versa, and other recommendations from Ioannidis (2005) are put in place by researchers, such as registered reports in which the methods are peer-reviewed before the experiment begins.

---

recent reviews on the FE of methods to reduce carnivore predation on livestock, which revealed diverse interpretations and standards of evidence (Miller et al. 2016; Treves et al. 2016; Eklund et al. 2017; van Eeden et al. 2018a, 2018b). One of the main results of all four reviews was the high variability in the effectiveness of interventions. Moreover, all four reviews concurred that strong inference was scarce because of a lack of experimental controls. Because there has been little consensus until now on standards of evidence for FE (van Eeden et al. 2018b), at least one of the above reviews used measures of PE (did the livestock owner report satisfaction or perceive reduction in losses of livestock?).

Because strong inference depends on careful experiments that oppose hypotheses (Platt 1964), Treves et al. (2016) emphasized that only a handful of studies in North America and Europe had ever produced strong inference about interventions to prevent predation on livestock. Although their goal was to review studies that fulfilled the gold-standard criteria, only two tests of non-lethal method met that standard between 1973 and 2016 and zero for lethal methods of intervention. Therefore, they had to relax the criteria to include silver-standard studies (a total of ten studies under this criteria) (see Text

Box 12.2 for definition). Furthermore, a 2018 re-evaluation of one of the tests of lethal methods led to its removal from the list of functionally effective methods (Santiago-Ávila et al. 2018a), given concerns related to their identification of study subjects (potential sampling bias) and the construction of their dependent variable (potential measurement bias). In summary, we highlight the importance of implementing rigorous and robust designs that measure functional effectiveness with strong inference. This will prevent implementation of ineffective interventions that would lead to wasted resources and harm to animals (wild and domestic) and, therefore, not promote coexistence. We also conclude that after more than 40 years of studies with weak inference or flawed designs, societies seeking evidence-based policy on wildlife control may find little certainty. That can lead to choices of interventions based solely on PE. In the next section, we define PE so future research will maintain a clear separation between FE and PE.

PE in the context of HWCC measures *individual-perceived reduction in damages of an intervention*. For example, most readers would accept that two individuals could perceive the same effect differently to each other and, neither PE may be identical to the scientific measurement of a functional effect. The logical inference in both cases is that PE relies on subjective cues that can be accurate or not. Human brains and senses are not scientific tools for unbiased measurement. For instance, several studies have demonstrated the influence that factors like experience, context, cognition and perceptual biases (e.g. preconceived ideas about something) have on filtering individual observations (Starr 1969; Kellert 1985; Slovic 1987; Finucane et al. 2000; Wieczorek Hudenko 2012). In this section, we attempt to explain more precisely the conditions under which FE and PE do and do not overlap, and the role that overlap plays in fostering or hindering coexistence with others, especially non-human others.

12.1.1 PE Components and Development of Framework

Differences of perception between two persons relates both to physical constraints on perceptual abilities (e.g. sensory and motor constraints) and to psychological factors that influence appraisals (Starr 1969; Slovic 1987). The field of psychology has a long history of investigating appraisals and two major conclusions have emerged. Human brains make rapid appraisals on the order of milliseconds, using more ancient regions of the brain such as the amygdala (Whalen et al. 1998; Morris et al. 1999).

Rapid appraisals (e.g. emotions – fear of snakes) often have high survival value and are difficult to modulate by the slower, cortical regions of the brain (Öhman & Mineka 2001; Barrett 2006; Lindquist et al. 2012). Fast appraisals captured by the amygdala may even go unnoticed by the perceiver, who simply may not be aware of the stimulus (i.e. unconscious pathway) (Esteves & Öhman 1993; Whalen et al. 1998). Human brains also make slower appraisals on the order of tenths of seconds, using more recently evolved regions of the brain such as the frontal cortex (Ajzen 1991; Treves & Pizzagalli 2002; Kahnemann 2003). For instance, when humans face obstacles or threats, their preferred solutions reflect both the rapid-affective (as simple as like or dislike) and slower-cognitive responses (should I like or dislike this?), which may integrate numerous criteria that reflect both the characteristics of the obstacle or threat, and the perceiver's own attributes including experiences and perceived social norms (e.g. how others perceive the situation and what they expect from the subject) (Kahnemann 2003; Wieczorek Hudenko 2012). The way the different appraisals replace each other or integrate is not yet well understood generally and largely unknown for HWCC. In summary, there is a mixed route of decision-making relevant to behaviour based on a rapid, automatic pathway (e.g. affective) combined with a slower, reasoned one (e.g. conscious) (Kahnemann 2003; see Chapter 4).

Building on the above research into cognition and behaviour, investigators of HWCC decision-making suggest that both cognitive (rational) and affective (emotional) components are relevant and important in understanding human behaviour. This is significant given that emotions (e.g. fear) will most likely predominate during these interactions and, therefore, would influence human behaviour (Johansson & Karlsson 2011; Wieczorek Hudenko 2012; Frank et al. 2015; Sponarski et al. 2015). Thus, in our treatment of PE, we restrict ourselves to referring simplistically and similarly to a mixture of affect (rapid responses) and cognition (slower responses) rather than the exclusive use of one or the other.

Observers or non-evaluators may disagree with scientific measurement of FE and will, therefore, behave differently from evaluators. For instance, confirmation bias can be understood loosely as a tendency to ignore information that conflicts with pre-existing beliefs, and to focus on information that conforms to a person's beliefs (Dunwoody 2007; Wieczorek Hudenko 2012). Related but sometimes acting separately, humans may change their perceptions, and behaviours that follow from those perceptions, if the bearer of the new message is familiar and trusted *versus* unfamiliar or untrusted (Dunwoody 2007; Powell et al. 2007). For instance, trust and familiarity have been addressed through research on social norms. Addressing HWCC explicitly, Heberlein (2012) described norms as behavioural regularities and as being closely related to one's role in a social group. Social norms can trump attitudes when it comes to shaping behaviours and expectations (Kinzig et al. 2013). Further, norms of acceptable behaviour and those enforced by social pressure can govern over alternate rules or motivations (e.g. laws or mechanistic explanations for behaviour such as income needs), as in the case of illegal behaviours (Jones et al. 2008; Marchini & Macdonald 2012). For example, social norms strongly influenced the intention to kill jaguars in Brazil more so than retaliation due to livestock predation. People's intention to kill carnivores was driven by the thought that their social peers killed carnivores (Marchini & Macdonald 2012). Furthermore, the decision to act may depend on the individual's perceived behavioural control over that action or the phenomenon being perceived (Ajzen 1991; Fishbein & Yzer 2003; Amit & Jacobson 2017). Discriminating the two cognitive mechanisms (social norms or behavioural control) may be very difficult because of the hidden nature of cognitive processing that precedes action or inaction. Finally, perceptions might change following an intervention event or before, during and after an intervention took place. For example, a farmer may ask himself questions like: (1) *Will this intervention reduce damages or threats?* (before implementation), (2) *Is this reducing damages?* (during), (3) *Do I like the outcome?, Were there any unexpected consequences?* (immediately after); and (4) *Would I try it again?* (longer after; see PE in Figure 12.2). Some authors (Ajzen 1991; Fishbein & Yzer 2003) predict that events may produce changes in intentions or in perceptions of behavioural control, with the effect that the original measures of these variables no longer permit accurate prediction of behaviour.

Here, we build and expand on the TPB (Ajzen 1991) as well as more recent work on behaviour change in the literature on human–environment interactions (Fishbein & Yzer 2003; Wieczorek Hudenko 2012; Amit & Jacobson 2017; Schlüter et al. 2017) to offer a schematic figure both to illustrate the complexity of human cognition as it relates to PE, and as a heuristic tool for partitioning the process of PE into more manageable components for analysis, as discussed previously (Figure 12.2). For instance, Amit and Jacobson (2017) described an expanded model adapted from Ajzen's TPB (1991) applied to human–

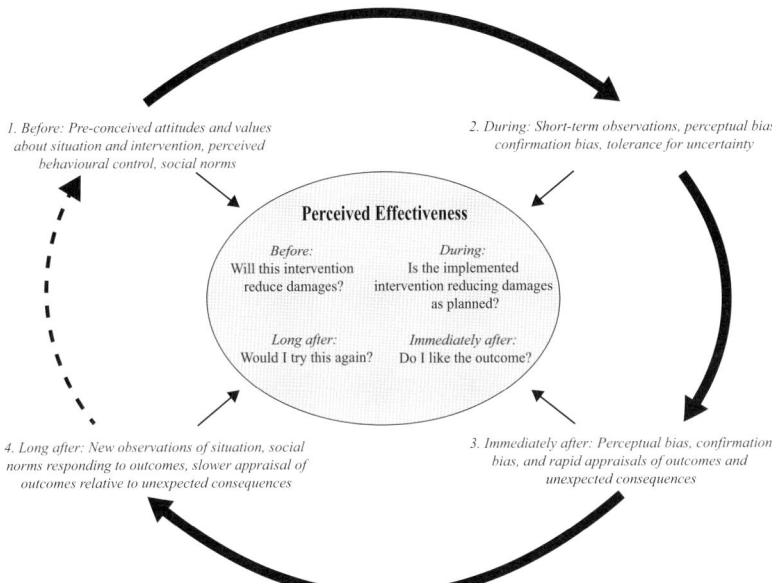

Figure 12.2 Perceived effectiveness framework adapted from social-psychological decision-making theories. In this adapted framework, human cognition variables are laid out chronologically from the upper left running clockwise from pre-implementation of an intervention to long-term post-implementation. The dashed arrow indicates the possibility of re-starting the process adaptively if the implementers are not satisfied.

carnivore conflict mitigation strategies. This expanded model included additional factors such as emotions and situational variables (i.e. livestock mortality rates by carnivores, income from livestock production and size of the property) that may influence farmers' decision-making behaviour related to the adoption of an intervention or not. Here we simplify intervention choice or implementation down to the most important causal variables so that we can integrate FE and PE. Integration of both will help us to identify and understand the circumstances when they do or do not align and, therefore, focus on where and how we should put our efforts on interventions aimed at coexistence.

12.1.2 Integrative Framework: Theory of Relationship between Functional and Perceived Effectiveness

So far, we have described the theory behind FE and PE independently. Now, we want to integrate the two concepts to propose a hypothesis. Our

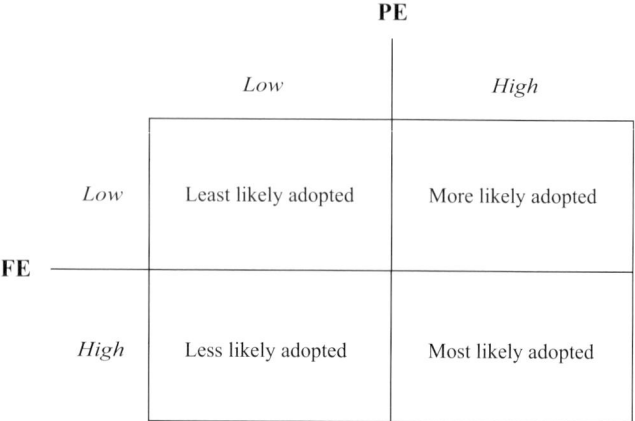

Figure 12.3 Hypothesis that integrates concepts of perceived effectiveness (PE) and functional effectiveness (FE).

hypothesis is that a scientifically proven functionally effective intervention (high FE) is more likely to be adopted if PE $\geq$ FE, than if PE $<$ FE. Alternatively, an ineffective intervention (low FE) is more likely to be adopted if PE $>$ FE, than if PE is low (Figure 12.3).

This hypothesis highlights two cases of important conservation and coexistence concern. We predict that (1) non-adoption of a functionally effective intervention (high FE and low PE, lower left in Figure 12.3) leads to political conflicts between researchers and stakeholders in addition to adoption of another intervention method, which might in turn lead to (2) the adoption of an ineffective intervention (low FE and high PE, upper right in Figure 12.3). We predict outcome (2) leads to wasted resources and harm to animals without improving coexistence. In both cases, our goal is to predict the factors that are influencing the decisions and suggest outcomes for coexistence.

Here, we propose three cognitive processes that may influence PE and the decision to implement an intervention: (1) uncertainty about FE, (2) ecological and social side effects and outside interest groups influences (e.g. social norms) and (3) ability to implement (e.g. feasability, behavioural control[1]). These cognitive processes do not act separately, presenting levels of overlap and correlation between them (Ajzen 1991; Fishbein & Yzer 2003; Amit & Jacobson 2017). Nevertheless, all of the cognitive processes underlying PE might contribute to the decision to

---

[1] The degree to which an individual perceives a behaviour is under their control.

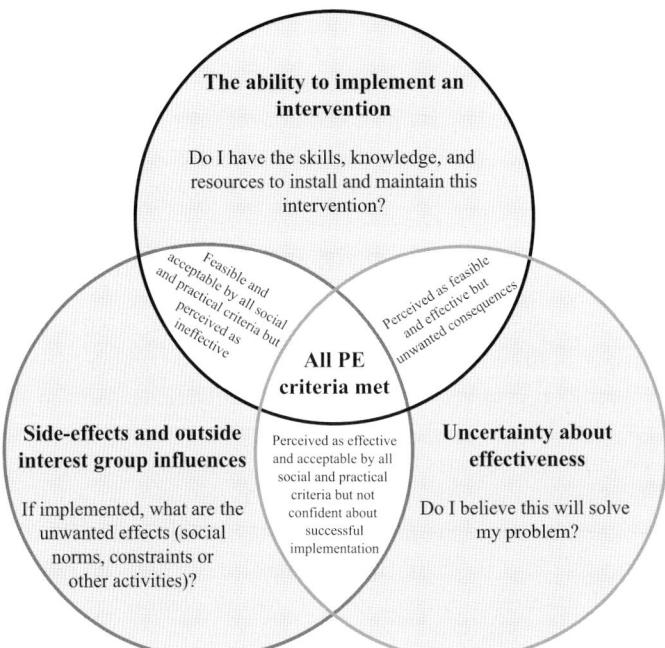

The ability to implement an intervention

Do I have the skills, knowledge, and resources to install and maintain this intervention?

Feasible and acceptable and practical by all social criteria but perceived as ineffective

Perceived as feasible and effective but unwanted consequences

All PE criteria met

Side-effects and outside interest group influences

If implemented, what are the unwanted effects (social norms, constraints or other activities)?

Perceived as effective and acceptable by all social and practical criteria but not confident about successful implementation

Uncertainty about effectiveness

Do I believe this will solve my problem?

Figure 12.4 Integrative framework for effectiveness of interventions regarding human–wildlife conflicts. Decision-making variables span different groups and levels, indicated by overlapping circles. The overlapped area indicates the co-occurrence of PE variables. Examples of questions are provided for each variable (circle) which might influence decision-making. The bottom right circle relating to uncertainty is meant to predict that FE (scientific evidence) is still filtered through a cognitive process relating to uncertainty if the FE applies to the subject in question.

act (implement an intervention), which we define as an area where all three processes overlap in Figure 12.4. Because two of the three cognitive processes have nothing to do with FE (social norms and perceived behavioural control), we predict in many instances FE $\neq$ PE. We predict that FE is more likely to equal PE and that appropriate action would follow when a trusted messenger demonstrates the intervention or testifies to its usefulness (Dunwoody 2007) (reducing uncertainty), when unintended side effects are minimized or eliminated, and when resource or technical aid is provided to improve perceived control over the intervention.

A framework should guide the testing of a hypothesis, if the predictions are articulated properly and measured appropriately. Our integrative framework helps to explain why an implementer may decline or dismantle an intervention that shows evidence of FE, because FE does not address side effects, social norms or feasibility. Likewise, our framework helps to explain why technical experts often disapprove of the actual methods in use by lay persons. For instance, the implementation was feasible and accepted by social norms but the technical expert may see a design flaw that precludes FE. Our framework would improve future coexistence if it exposes mismatches between PE and FE so that intervention designers and implementers can include persuasive interventions if needed.

Below we explore some cases in which PE $\neq$ FE yet FE is high. Our first example addresses a non-lethal intervention in which social norms are favourable and uncertainty is low but individuals seem to rate the feasibility (perceived control) as low. A proven intervention such as livestock guarding dogs (LGDs hereafter) can reduce livestock losses in a variety of situations (Gehring et al. 2010; Treves et al. 2016), but many livestock owners express concerns about their ability to raise, maintain and train such dogs or share the belief that these dogs do not work on large, open pastures despite evidence to the contrary (Espuno et al. 2004). If we are correct that other components of PE are moderate to highly conducive to adoption but a perceived lack of behavioural control or ability to implement an LGD is widespread (issues and assistance with proper training and caring for guarding dogs), then adoption might be promoted by training and demonstration projects with owners.

Other methods of intervention seem to be perceived as feasible (high perceived behavioural control) yet are not adopted widely. For example, in Sweden subsidized fencing to protect livestock has not led to widespread installation by farmers (Frank & Eklund 2017), although individuals accepted the help initially and this intervention has substantial evidence of FE (Karlsson & Sjöström 2011; Ängsteg et al. 2014). According to our hypotheses and its predictive framework, some component of PE must be low or missing. We predict that a social norm exists against the subsidized fencing or that after installation farmers are discovering side effects or infeasible aspects.

It is tempting for scholars to assume that when PE $\neq$ FE, the layperson needs more information (the information-deficit hypothesis). Our framework suggests instead that other important cognitive processes may be blocking adoption and maintenance of the implemented method. Uncertainty and novelty of methods can dampen adoption. For example, differences between sites where FE experiments take place and the actual site of implementation could elevate uncertainty. Small differences in livestock husbandry, carnivore species or landscapes can raise doubts about FE in even the most willing adopter. Even after implementation, a person might abandon the method if outcomes are not as promised. Slower appraisals that arise from unexpected outcomes, as well as dynamic social norms might lead to dismantling or defection of implementers. Moreover, the farmer may oppose the general view of effectiveness due to disagreement over conservation goals and might, therefore, dismiss and contest research (Redpath et al. 2013; Woodroffe & Redpath 2015). Therefore, the presentation of information and its acceptance by various audiences is best understood by studying the communication process and participants, more than by the content of the communication.

It is widely believed that owners of domestic animals should be engaged actively in decision-making to help build trust and meet PE criteria. For example, participant engagement approaches have been described as helpful when promoting adoption of interventions (Treves et al. 2006, 2009; Reed 2008; Woodroffe & Redpath 2015). Nevertheless, it does not necessarily follow that participants must be engaged in groups to decide on each other's interventions. That might amplify social norm imposition (peer pressure) that could drive PE further from FE. The ideal scenario for coexistence in HWCC is for both people and wildlife to be protected with FE interventions that meet PE criteria. The ideal scenario would be researchers measuring the cognitive components of PE before attempting an intervention. What we are proposing here has not yet been fully tested but promising projects are under way (see Text Box 12.3 below).

12.2 CASE STUDIES ON PERCEIVED EFFECTIVENESS OF METHODS TO REDUCE DAMAGES TO LIVESTOCK

We reviewed various case studies regarding PE of interventions with the goal of comparing them with the proposed integrative framework, and then give guidance on how to design a study to measure these components. We selected three studies where we addressed at least one of the cognitive processes or components described in our integrative framework (see Text Box 12.3).

Our three examples have highlighted discrepancies between PE and FE, but do not serve to test our hypothesis rigorously. We lack a study of FE combined with measures of PE at the same site that are both focused on the same intervention, regardless of how many subjects benefited

Text Box 12.3 Case Studies Illustrating Cognitive Processes or Components of the Integrative Framework

Case Study 1: Integrating Proposed Framework to Improve Coexistence between Pumas and People in Chile

Research began by measuring attitudes among Aymara indigenous people in northern Chile towards pumas and perceptions of methods to protect livestock. These baseline data revealed low perceived behavioural control (owners felt they needed help to implement any intervention), that non-lethal methods were viewed as an option by respondents (i.e. permissive social norms) (Ohrens et al. 2016). Furthermore, researchers had very weak evidence about FE of any method for the predator, the livestock or the region. Subsequently, authors offered help to intervenors (owners) by attempting a participatory intervention planning workshop (see methods in Treves et al. 2009) to select a non-lethal method of their preference (Ohrens et al. in press). This participatory process (i.e. local engagement) might have helped to overcome PE about what would be effective and what would not, given equal uncertainty among methods. Additionally, researchers attempted to improve participants' perceptions of behavioural control. Only one of twelve participants in the experiment abandoned the project midway, the remaining eleven accepted the placebo control in a cross-over (reverse-treatment) design (Ohrens et al. in press), and after the end of the experiment all eleven requested to keep the light deterrent device they had tested. Although this example attempted to integrate several criteria of PE, it did not measure social norms explicitly and does not yet demonstrate long-term adoption.

Case Study 2: Lethal Interventions against Jaguars in Brazil

The second study, done in Amazonia and Pantanal, Brazil (Marchini & Macdonald 2012), measured social norms regarding lethal control of jaguars. To gather specific variables that could help to predict behaviour and intentions to use lethal methods, the authors followed the TPB (Ajzen 1991) and separated social norms into several components (e.g. descriptive norm, social identity) to measure cognitive aspects of coexistence or illegal killing of jaguars. The authors concluded that peer group pressures and other social norms (cultural beliefs about men and jaguars) were important predictors of the intention to kill jaguars, independently from wealth or economic losses, which did not predict that intention well. Apparently, respondents believed that killing jaguars would save cattle despite lack of evidence of FE (low uncertainty about the method), and that belief was amplified by social norms. Nevertheless, farmers who expressed an intention to kill jaguars reported substantial variation in their ability to do so (Marchini & Macdonald 2012). In sum, implementation (illegally killing a jaguar) was predicted strongly by behavioural control and the expected positive social benefits of doing so. In such a situation, measuring FE or intervening to raise uncertainty about the effectiveness of killing jaguars to protect cattle may be irrelevant. Conservationists aiming at coexistence should address the social norm affecting those individuals who intended to kill jaguars or report the ability of those individuals to act on their beliefs.

Case Study 3: Perceived Effectiveness of Interventions in South Africa

We combined two studies that similarly presented measurements on uncertainty of effectiveness, and retention of interventions over time. The first study, from McManus et al. (2015), applied a pseudo-control design to measure the effect of lethal interventions compared to subsequent non-lethal ones. The authors found that livestock losses and related costs declined after implementing a variety of different non-lethal methods. Therefore, FE of non-lethal methods was concluded to be higher than FE of lethal methods. Follow-up interviews revealed that six of the eleven farmers continued the effective non-lethal methods twelve months after the team stopped measuring livestock losses. However, after thirty-six months only four of eleven farmers continued the effective non-lethal interventions. The reasons that seven farmers abandoned the non-lethal methods included unexpected outcomes (dog that may have killed livestock was shot by neighbour), ability to implement (farmer found it easier to implement lethal method) and uncertainty of effectiveness (lethal method perceived more effective). We infer that FE was not sufficient to assure long-term adoption of a non-lethal method. Several components of PE resurfaced over time and a lower FE method supplanted the method with higher FE (McManus et al. 2015).

The second study conducted by Rust et al. (2013) applied a quasi-experimental design (before-and-after), without controls, to measure attitudes of farmers to the performance of LGDs in protecting livestock from cheetahs as well as costs associated with their implementation. Researchers documented that LGDs were perceived as cost-effective in reducing livestock predation by carnivores. Mean perceived annual predation for the total participating farms (n = 70) were reduced by 33 to 100% after LGD placement. The authors reported that from a total of 97 LGDs, 22% (n = 21 dogs) were removed from farms. Reasons for dog removal were mostly reported to be related to the farmer's perception of the dog's behaviour and capacity (uncertainty of effectiveness) followed by a few cases that were related to the owner's capacity to implement dog training or husbandry properly (ability to implement). Again, a FE method in the short term proved to have longer-term problems in a minority of cases or at least the PE of the method diminished over time.

from the intervention (i.e. a continuous measure of FE). With a sufficient sample of respondents, such a study could test our hypothesis by correlating PE to each PE component and to individual experiences of FE across subjects.

Alternately, we would need a study across many sites that compares aggregated PE measures for each site to the binary variable of FE (i.e. was it effective at that site or not?). Under those conditions, the intervention does not need to be the same across sites because site-specific PE and FE are being compared to each other (within-subject correlation).

Such a study would provide a more general test of our hypothesis, but would lack the specificity to reveal clearly which component of PE was responsible for any observed mismatch because different biophysical, socio-political and intervention designs would cloud the interpretation of results. Regardless, either type of study would help to advance research on preventing HWCC. We expect coexistence would be promoted as a result.

12.3 GUIDELINES TO MEASURE PERCEIVED EFFECTIVENESS OF INTERVENTIONS

For this purpose, we present guidelines and steps in designing and conducting research regarding our PE criteria. We will focus on the intent of coexistence interventions, and how they affect PE, and each of its components. For example, we need to: (1) use the integrative framework to target and focus on components that have not been addressed in former studies conducted in the same locations (e.g. define research questions), (2) select robust designs to reduce all sorts of biases (e.g. design of studies), (3) develop methods to target research questions (e.g. questionnaires, appropriate framing and design of questionnaires) (see Marchini & Macdonald 2012; St John et al. 2014) and (4) consider temporality within study design (e.g. before, during, after and follow-up measurements) (see McManus et al. 2015) (see summary in Table 12.1).

12.3.1 Study Design for PE

We propose to apply questionnaires to farmers randomly selected within a study area, a common method in social sciences, to measure our proposed components (Newing et al. 2011). The focus of questionnaires may depend on the amount and type of existing information that is related to our framework and available at the site. However, for our purposes we will target all components described earlier (Figure 12.2 and 12.4). We recommend that questionnaires follow the timescale presented in our PE framework; with questions that target information before, during, immediately and long after implementation of interventions. At the same time, we suggest following the construct of our proposed integrative framework to design questions that measure each component. For example, questions can be in the form of statements for each variable within components, using Likert scale answers (from

Table 12.1 *Guidelines to measure perceived effectiveness of interventions*

Timing relative to implementation	Response variable	Predictors		
		Uncertainty of effectiveness	Social norms (measured within participants and outside interest groups)	Ability to implement
Before	Intention to implement?	Measure the participant's appraisal of future effectiveness	Measure the likely gain or loss of social status if they implement (based on perceptions relative to others) Measure side effects from outside interest groups as perceived by participant and associates	Measure anticipated feasibility (cost, skill, time, side effects other than social ones)
During	Maintain implementation?	Measure the participant's appraisal of ongoing effectiveness	Measure the actual gain or loss of social status as perceived by participant, associates and outside interest groups (based on perceptions relative to others)	Measure ongoing actualized feasibility (cost, skill, time, side effects other than social ones)
Shortly after	Appraisal of outcomes?	Measure the participant's conclusion about effectiveness	Measure the actual gain or loss of social status as perceived by participant, associates and outside interest groups (based on perceptions relative to others)	Measure final, actualized feasibility (cost, skill, time, side effects other than social ones) and the cost-benefits of outcomes
Long after	Adopt and promote with others?	Measure the participant's willingness to continue use or communicate outcomes to others	Measure the actual gain or loss of social status as perceived by participant, associates and outside interest groups (based on perceptions relative to others)	Measure long-term side effects and the costs and the cost-benefits of outcomes

strongly agree to strongly disagree) (Newing et al. 2011). This is a commonly used method to measure latent constructs such as attitudes and behaviours. Here are some examples of statements for each component: (1) *I am confident about continuing to use the intervention* (ability to implement), (2) *I feel social pressure to use a specific intervention* (side effects and outside group influences), (3) *The intervention has been very effective in reducing attacks on livestock* (uncertainty of effectiveness) (see Marchini & Macdonald 2012; St John et al. 2014). To test, we can use a general linear model (GLM) between integrative framework variables as predictors (e.g. ability to implement, uncertainty of effectiveness, social norms) and the binary result of the intention or not (0 or 1) to implement the proposed intervention as response variable.

12.4 CONCLUSIONS: TYING BACK TO COEXISTENCE

Interventions aim at promoting coexistence by reducing negative interactions between wildlife and humans. Under our integrative framework, we hypothesize that the successful adoption of proven effective interventions is more likely if functional and perceived effectiveness align (PE $\geq$ FE and FE is high), which in the long term should promote and foster coexistence. Similarly, Heberlein (2012) argued that to approach environmental problems successfully, more than one of his proposed fixes (e.g. technical, cognitive and structural) need to be addressed. Analogously, our framework is proposing to address both technical (i.e. technical solution to reduce livestock losses – FE) and structural-cognitive fixes (indirect solution that attempts to address people's attitudes and behaviours towards wildlife – PE, also see Treves et al. 2006, 2009) to improve coexistence. We recommend interdisciplinary measurement of both human cognition and behaviour as well as experimental tests of FE. By promoting PE and FE alignment, we fall to the right side of the conflict-to-coexistence continuum, aimed at improving positive attitudes/behaviours towards wildlife (Frank 2016).

Our framework (Figure 12.4) predicts that political conflicts will arise in two different ways when FE $\neq$ PE. When PE $>$ FE and FE is low, technical experts will object to the implementation of an ineffective intervention, and the political conflicts and disputes that follow will be on trust in science, as well as legitimacy of unscientific decisions, among others. If opposing interest groups are involved, the interest group that either ideologically prefers the intervention or prefers science-based decision-making will take sides. When PE $<$ FE and FE is high (case study 3), we predict technical experts will find themselves trying to persuade laypeople to implement something they are resistant to try. If technical experts fail, then the likely outcome would be the case where a lower FE method is implemented (PE > FE, FE is low).

Without evidence for high FE, PE tends to sway decisions and will determine which intervention is implemented. Confirming that FE is high before implementing an intervention is especially important if decision-makers perceive that non-human animals do not deserve moral consideration. If an intervention has low FE and is implemented nonetheless, non-human animals – wild and domestic – are likely to suffer. Moreover, our inability to deliberate fairly with non-humans and the power asymmetry between parties may undermine coexistence between humans and non-humans (Favre 1979; Hutchins & Wemmer 1986). Within this social and structural context, the implementation of interventions with PE > FE that can be harmful or lethal to non-human animals (e.g. lethal methods, translocation) should be viewed most sceptically by youth and future generations and by current adults concerned with ethics, legitimacy and precautionary principles. Here, emerging fields such as compassionate conservation and practices such as predator-friendly farming can help in providing principles and guidance on the implementation of socially acceptable interventions that promote animal well-being (Ramp & Bekoff 2015; Wallach et al. 2015; Johnson & Wallach 2016). By emphasizing coexistence with individual non-humans, these fields promote the moral standing of non-humans and attempt to equitably consider their interests when deciding to intervene in their lives (Santiago-Ávila et al. 2018b).

12.5 RECOMMENDATIONS AND FUTURE DIRECTIONS

- Strengthen the rigour of science for understanding adoption and maintenance of interventions for coexistence.
- Collect both ecological (FE) and social-psychological (PE) variables when evaluating an intervention aimed to reduce conflict. This would enable a more balanced interdisciplinary understanding of social–ecological systems, such as human–wildlife interactions.
- Test hypotheses of particular interventions in a rigorously designed study. This would help in better design and implementation of interventions to reduce conflicts (see guidelines in Table 12.1).
- Address current gaps in the use of gold-standard designs to evaluate both FE and PE of methods and their implications for carnivore and wildlife conservation in general.

• Address current gaps in knowledge on possible unexpected effects of non-lethal interventions on predators and other wildlife (e.g. disruption of behaviour and social organization).

12.6 References

Ajzen, I. (1991). The theory of planned behavior. *Organizational Behavior & Human Decision Processes*, 50, 179–211.

Amit, R. & Jacobson, S. K. (2017). Understanding rancher coexistence with jaguars and pumas: A typology for conservation practice. *Biodiversity & Conservation*, 26, 1353–74.

Ängsteg, I., Ängsteg, R., Levin, M., Karlsson, J., Eklund, A. & Råsberg, A. (2014). *Stängsling mot stora rovdjur*. Sweden: Viltskadecenter, SLU.

Barrett, L. F. (2006). Solving the emotion paradox: Categorization and the experience of emotion. *Personality & Social Psychology Review*, 10, 20–46.

Breitenmoser, U., Angst, C., Landary, J.-M., Breitenmoser-Wursten, C., Linnell, J. D. C. & Weber, J.-M. (2005). Non-lethal techniques for reducing depredation. In R. Woodroffe, S. Thirgood & A. Rabinowitz, eds., *People & Wildlife: Conflict or Coexistence?* Cambridge: Cambridge University Press, pp. 49–71.

Chapron, G., Kaczensky, P., Linnell, J. D. C., von Arx, M., Huber, D., Andrén, H., López-Bao, J. V., Adamec, M., Álvares, F., Anders, O., Balčiauskas, L., Balys, V., Bedő, P., Bego, F., Blanco, J. C., Breitenmoser, U., Brøseth, H., Bufka, L., Bunikyte, R., Ciucci, P., Dutsov, A., Engleder, T., Fuxjäger, C., Groff, C., Holmala, K., Hoxha, B., Iliopoulos, Y., Ionescu, O., Jeremić, J., Jerina, K., Kluth, G., Knauer, F., Kojola, I., Kos, I., Krofel, M., Kubala, J., Kunovac, S., Kusak, J., Kutal, M., Liberg, O., Majić, A., Männil, P., Manz, R., Marboutin, E., Marucco, F., Melovski, D., Mersini, K., Mertzanis, Y., Mysłajek, R. W., Nowak, S., Odden, J., Ozolins, J., Palomero, G., Paunović, M., Persson, J., Potočnik, H., Quenette, P.-Y., Rauer, G., Reinhardt, I., Rigg, R., Ryser, A., Salvatori, V., Skrbinšek, T., Stojanov, A., Swenson, J. E., Szemethy, L., Trajçe, A., Tsingarska-Sedefcheva, E., Váňa, M., Veeroja, R., Wabakken, P., Wölfl, M., Wölfl, S., Zimmermann, F., Zlatanova, D. & Boitani, L. (2014). Recovery of large carnivores in Europe's modern human-dominated landscapes. *Science*, 346, 1517–19.

Dunwoody, S. (2007). The challenge of trying to make a difference using media messages. In S. C. Moser & L. Dilling, eds., *Creating a Climate for Change*. Cambridge: Cambridge University Press, pp. 89–104.

Eklund, A., López-Bao, J. V., Tourani, M., Chapron, G. & Frank, J. (2017). Limited evidence on the effectiveness of interventions to reduce livestock predation by large carnivores. *Scientific Reports*, 7, 2097.

Espuno, N., Lequette, B., Poulle, M. L., Migot, P. & Lebreton, J. D. (2004). Heterogeneous response to preventive sheep husbandry during wolf recolonization of the French Alps. *Wildlife Society Bulletin*, 32, 1195–1208.

Esteves, F. & Öhman, A. (1993). Masking the face: Recognition of emotional facial expressions as a function of the parameters of backward masking. *Scandinavian Journal of Psychology*, 34, 1–18.

Favre, D. S. (1979). Wildlife rights: The ever-widening circle. *Environmental Law,* 9, 241–81.

Finucane, M. L., Slovic, P., Mertz, C. K., Flynn, J. & Satterfield, T. A. (2000). Gender, race, and perceived risk: The 'white male' effect. *Health, Risk & Society,* 2, 159–72.

Fishbein, M. & Yzer, M. C. (2003). Using theory to design effective health behavior interventions. *Communication Theory,* 13, 164–83.

Frank, B. (2016). Human–wildlife conflicts and the need to include tolerance and coexistence: An introductory comment. *Society & Natural Resources,* 29, 738–43.

Frank, J. & Eklund, A. (2017). Poor construction, not time, takes its toll on subsidised fences designed to deter large carnivores. *PLoS ONE,* 12, e0175211.

Frank, J., Johansson, M. & Flykt, A. (2015). Public attitude towards the implementation of management actions aimed at reducing human fear of brown bears and wolves. *Wildlife Biology,* 2, 122–30.

Gehring, T. M., VerCauteren, K. C., Provost, M. L. & Cellar, A. C. (2010). Utility of livestock-protection dogs for deterring wildlife from cattle farms. *Wildlife Research,* 37, 715–21.

Grimshaw, J., Campbell, M., Eccles, M. & Steen, N. (2000). Experimental and quasi-experimental designs for evaluating guideline implementation strategies. *Family Practice,* 17, 11–18.

Heberlein, T. (2012). *Navigating Environmental Attitudes.* Oxford: Oxford University Press.

Hutchins, M. & Wemmer, C. (1986). Wildlife conservation and animal rights: Are they compatible? In M. W. Fox & L. D. Mickley, eds., *Advances in Animal Welfare Science 1986/87.* Washington, DC: The Humane Society of the United States, pp. 111–37.

Inskip, C. & Zimmermann, A. (2009). Human–felid conflict: A review of patterns and priorities worldwide. *Oryx,* 43, 18–34.

Ioannidis, J. P. A. (2005). Why most published research findings are false. *PLoS Medicine,* 2, 696–701.

Johansson, M. & Karlsson, J. (2011). Subjective experience of fear and the cognitive interpretation of large carnivores. *Human Dimensions of Wildlife,* 16, 15–29.

Johnson, C. N. & Wallach, A. D. (2016). The virtuous circle: Predator-friendly farming and ecological restoration in Australia. *Restoration Ecology,* 24, 821–6.

Jones, J. P. G., Andriamarovololona, M. M. & Hockley, N. (2008). The importance of taboos and social norms to conservation in Madagascar. *Conservation Biology,* 22, 976–86.

Kahnemann, D. (2003). A perspective on judgment and choice: Mapping bounded rationality. *American Psychologist,* 58, 697–720.

Karanth, K. U. & Madhusudan, M. D. (2002). Mitigating human–wildlife conflicts in southern Asia. In J. Terborgh, C. P. Van Schaik, M. Rao & L. C. Davenport, eds., *Making Parks Work: Identifying Key Factors to Implementing Parks in the Tropics.* Covelo, CA: Island Press, pp. 250–64.

Karlsson, J. & Sjöström, M. (2011). Subsidized fencing of livestock as a means of increasing tolerance for wolves. *Ecology & Society*, 16, art. 16.

Kellert, S. R. (1985). Public perceptions of predators, particularly the wolf and coyote. *Biological Conservation*, 31, 167–89.

Kinzig, A. P., Ehrlich, P. R., Alston, L. J., Arrow, K., Barrett, S., Buchman, T. G., Daily, G. C., Levin, B., Levin, S., Oppenheimer, M., Ostrom, E. & Saari, D. (2013). Social norms and global environmental challenges: The complex interaction of behaviors, values, and policy. *BioScience*, 63, 164–75.

Lindquist, K. A., Wager, T. D., Kober, H., Bliss-Moreau, E. & Barrett, L. F. (2012). The brain basis of emotion: A meta-analytic review. *Behavioral & Brain Sciences*, 35, 121–202.

Marchini, S. & Macdonald, D. W. (2012). Predicting ranchers' intention to kill jaguars: Case studies in Amazonia. *Biological Conservation*, 147, 213–21.

McManus, J. S., Dickman, A. J., Gaynor, D., Smuts, B. H. & Macdonald, D. W. (2015). Dead or alive? Comparing costs and benefits of lethal and non-lethal human–wildlife conflict mitigation on livestock farms. *Oryx*, 49, 687–95.

Miller, J. R. B., Stoner, K. J., Cejtin, M. R., Meyer, T. K., Middleton, A. D. & Schmitz, O. J. (2016). Effectiveness of contemporary techniques for reducing livestock depredations by large carnivores. *Wildlife Society Bulletin*, 40, 806–15.

Morris, J. S., Öhman, A. & Dolan, R. J. (1999). A subcortical pathway to the right amygdala mediating 'unseen' fear. *Proceedings of the National Academy of Sciences*, 96, 1680–5.

Mukherjee, S. (2010). *The Emperor of All Maladies: A Biography of Cancer*. New York: Scribner.

Newing, H., Eagle, C. M., Puri, R. K. & Watson, C. W. (2011). *Conducting Research in Conservation: Social Science Methods and Practice*. London: Routledge.

Öhman, A. & Mineka, S. (2001). Fears, phobias, and preparedness: Toward an evolved module of fear and fear learning. *Psychological Review*, 108, 483–522.

Ohrens, O., Treves, A. & Bonacic, C. (2016). Relationship between rural depopulation and puma–human conflict in the high Andes of Chile. *Environmental Conservation*, 43, 24–33.

Ohrens, O., Bonacic, C. & Treves, A. (2019). Non-lethal defense of livestock against predators: Flashing lights deter puma attacks in Chile. *Frontiers in Ecology & the Environment*.

Platt, J. R. (1964). Strong inference. *Science*, 146, 347–53.

Plumer, L., Talvi, T. N., Männil, P. & Saarma, U. (2018). Assessing the roles of wolves and dogs in livestock predation and suggestions for mitigating human–wildlife conflict and conservation of wolves. *Conservation Genetics*, 1–8. https://doi.org/10.1007/s10592-017-1045-4

Powell, M., Dunwoody, S., Griffin, R. & Neuwirth, K. (2007). Exploring lay uncertainty about an environmental health risk. *Public Understanding of Science*, 16, 323–43.

Ramp, D. & Bekoff, M. (2015). Compassion as a practical and evolved ethic for conservation. *BioScience*, 65, 323–7.

Redpath, S. M., Young, J., Evely, A., Adams, W. M., Sutherland, W. J., Whitehouse, A., Amar, A., Lambert, R. A., Linnell, J. D. C., Watt, A. & Gutiérrez,

R. J. (2013). Understanding and managing conservation conflicts. *Trends in Ecology & Evolution*, 28, 100–9.

Reed, M. S. (2008). Stakeholder participation for environmental management: A literature review. *Biological Conservation*, 141, 2417–31.

Rust, N. A., Whitehouse-Tedd, K. M. & MacMillan, D. C. (2013). Perceived efficacy of livestock-guarding dogs in South Africa: Implications for cheetah conservation. *Wildlife Society Bulletin*, 37, 690–7.

Santiago-Ávila, F., Cornman, A. M. & Treves, A. (2018a). Killing wolves to prevent predation on livestock may protect one farm but harm neighbours. *PLoS ONE*, 13, e0189729.

Santiago-Ávila, F., Lynn, W. & Treves, A. (2018b). Inappropriate consideration of animal interests in predator management: Towards a comprehensive moral code. In T. Hovardas, ed., *Large Carnivore Conservation and Management: Human Dimensions*. New York: Routledge, pp. 227–51.

Schlüter, M., Baeza, A., Dressler, G., Frank, K., Groeneveld, J., Jager, W., Janssen, M. A., McAllister, R. R. J., Müller, B., Orach, K., Schwarz, N. & Wijermans, N. (2017). A framework for mapping and comparing behavioural theories in models of social-ecological systems. *Ecological Economics*, 131, 21–35.

Shivik, J., Treves, A. & Callahan, P. (2003). Nonlethal techniques for managing predation: Primary and secondary repellents. *Conservation Biology*, 17, 1531–7.

Slovic, P. (1987). Perception of risk. *Science*, 236, 280–5.

Sponarski, C., Vaske, J. & Bath, A. (2015). The role of cognitions and emotions in human–coyote interactions. *Human Dimensions Wildlife*, 20, 238–54.

Starr, C. (1969). Social benefit versus technological risk. *Science*, 165, 1232–8.

St John, F. A. V., Keane, A. M., Jones, J. P. G. & Milner-Gulland, E. J. (2014). Robust study design is as important on the social as it is on the ecological side of applied ecological research. *Journal of Applied Ecology*, 51, 1479–85.

Treves, A. & Bruskotter, J. (2014). Tolerance for predatory wildlife. *Science*, 344, 476–7.

Treves, A., Chapron, G., López-Bao, J. V., Shoemaker, C., Goeckner, A. R. & Bruskotter, J. T. (2015). Predators and the public trust. *Biological Reviews*, 92, 248–70.

Treves, A. & Karanth, K. U. (2003). Human–carnivore conflict and perspectives on carnivore management worldwide. *Conservation Biology*, 17, 1491–9.

Treves, A., Krofel, M. & McManus, J. (2016). Predator control should not be a shot in the dark. *Frontiers in Ecology and the Environment*, 14, 380–8.

Treves, A. & Pizzagalli, D. (2002). Vigilance and perception of social stimuli: Views from ethology and social neuroscience. In M. Bekoff, C. Allen & G. Burghardt, eds., *The Cognitive Animal: Empirical and Theoretical Perspectives on Animal Cognition*. Cambridge, MA: MIT Press, pp. 463–9.

Treves, A., Wallace, R. B., Naughton-Treves, L. & Morales, A. (2006). Co-managing human–wildlife conflicts: A review. *Human Dimensions of Wildlife*, 11, 383–96.

Treves, A., Wallace, R. B. & White, S. (2009). Participatory planning of interventions to mitigate human–wildlife conflicts. *Conservation Biology*, 23, 1577–87.

van Eeden, L. M., Crowther, M. S., Dickman, C. R., Macdonald, D. W., Ripple, W. J., Ritchie, E. G. & Newsome, T. M. (2018a). Managing conflict between large carnivores and livestock. *Conservation Biology*, 32, 26–34.

van Eeden, L. M., Eklund, A., Miller, J. R. B., López-Bao, J. V., Chapron, G., Cejtin, M. R., Crowther, M., Dickman, C., Frank, J., Krofel, M., Macdonald, D. W., McManus, J., Meyer, T. K., Middleton, A. D., Newsome, T., Ripple, W. J., Ritchie, E. G., Schmitz, O. J., Stoner, K. J., Tourani, M. & Treves, A. (2018b). Carnivore conservation needs evidence-based livestock protection. *PLoS Biology*, 16(9), e2005577.

Wallach, A. D., Bekoff, M., Nelson, M. P. & Ramp, D. (2015). Promoting predators and compassionate conservation. *Conservation Biology*, 29, 1481–4.

Whalen, P. J., Rauch, S. L., Etcoff, N. L., McInerney, S. C., Lee, M. B. & Jenike, M. A. (1998). Masked presentations of emotional facial expressions modulate amygdala activity without explicit knowledge. *Journal of Neuroscience*, 18, 411–18.

Wieczorek Hudenko, H. (2012). Exploring the influence of emotion on human decision making in human–wildlife conflict. *Human Dimensions of Wildlife*, 17, 16–28.

Woodroffe, R. & Ginsberg, J. R. (1998). Edge effects and the extinction of populations inside protected areas. *Science*, 280, 2126–8.

Woodroffe, R. & Redpath, S. M. (2015). When the hunter becomes the hunted. *Science*, 348, 1312–14.

Zarco-González, M. M. & Monroy-Vilchis, O. (2014). Effectiveness of low-cost deterrents in decreasing livestock predation by felids: A case in Central Mexico. *Animal Conservation*, 17, 371–8.

13

Conflict and Coexistence with Invasive Wildlife

Examining Attitudes and Behaviours towards Burmese Pythons in Florida

REBECCA G. HARVEY AND FRANK J. MAZZOTTI

Perspectives on human–wildlife interactions are moving in a positive direction. The long-dominant *conflict* framework is under scrutiny for its narrow focus on negative interactions (Frank 2016) and its call for technical solutions that do not address underlying social causes of conflict (Dickman 2010; Redpath et al. 2015). Researchers argue that shifting the focus from conflict towards more positive concepts of coexistence and tolerance will encourage innovative solutions to conservation challenges (Peterson et al. 2010; Fisher 2016). Frank (2016) defines coexistence and tolerance and describes their relationship along a *conflict-to-coexistence continuum* spanning from extremely negative attitudes and behaviours towards a species, through a midpoint of neutrality or passivity, to extremely positive attitudes and behaviours. To develop this continuum as an analytical tool, she calls for social-science research into positive and neutral attitudes and behaviours towards wildlife.

Human–wildlife conflict most often refers to interactions in which large, charismatic species damage crops, kill livestock or threaten or harm people (Fisher 2016; Frank 2016). Most of the literature involves native species of conservation interest (Redpath et al. 2015) for which conflict with human activities implies negative consequences for biodiversity (Young et al. 2010). However, conflict may be defined more broadly to include any situation in which wildlife threatens human interests, including economic interests, environmental resources or human health (White & Ward 2010). By this general definition, invasive alien species are one emerging arena of human–wildlife conflict. An alien species is defined as a species, subspecies or lower taxon, introduced outside its natural past or present distribution, that might survive and subsequently reproduce (Convention on Biological Diversity (CBD)

Secretariat 2002) and an invasive alien species is defined as an alien species whose introduction and/or spread threaten biological diversity (CBD Secretariat 2002). Invasive alien species (hereafter *invasive species*) have proliferated in recent decades through increasing global trade and travel combined with changing environmental and habitat conditions, and are now implicated as a leading cause of species extinctions and population declines worldwide (White et al. 2008; Simberloff et al. 2012). Thus, contra other human–wildlife interactions, *conflict* with invasive animals (i.e. killing them with the ultimate goal of eradication) suggests a positive outcome for biodiversity and is the status quo for many scientists and resource managers.

Some ecologists question the conflict mentality that characterizes invasive species management. They argue that because natural systems are dynamic and permanently altered by humans, definitions of *native* and *alien* may be based on arbitrary distinctions in time and space (Warren 2007) and restoration to a historically arbitrary *rightful* state is unfeasible (Davis et al. 2011). Invasive species are often the visible outcomes of deeper ecological processes like habitat loss and alteration, and may be scapegoated by managers and funding agencies rather than addressing the more complex issues (Thompson 2014). Because eradication is rarely effective or possible, these critics argue that conservationists should adopt a longer-term and 'more conciliatory approach' towards alien species (Thompson 2014, p. 153) that emphasizes their ecological functions rather than their geographic origins (Davis et al. 2011). This perspective has evoked an intense backlash, with some characterizing scepticism of the negative impacts of invasive species as a form of science denialism (Russell & Blackburn 2017). One hundred and forty-one scientists responded to Davis et al. (2011) arguing that invasion biology targets the most harmful alien species, that the severe impact of invaders should not be downplayed and that eradication is indeed possible for some introductions (Simberloff et al. 2011). A survey of invasion biologists found widespread agreement that the native/alien dichotomy is meaningful and current invasion rates are unprecedented, but divergence over how much invasive species threaten biodiversity and how much their field should advocate against invaders (Young & Larson 2011).

Public and stakeholder attitudes towards invasive species management may be as important as those of ecologists in driving decision-making (White et al. 2008; Shine & Doody 2011). Attitudinal research comprises a tiny portion of the invasive species literature, but is becoming more prevalent and integrated into the field of invasion biology (Estévez et al. 2014). Many studies in this burgeoning literature take a conflict perspective (i.e. that invasive species are harmful and must be controlled) and measure public support for controlling invaders as their key dependent variable. Greater support for control has been associated with socio-demographic variables like age, male gender (Bremner & Park 2007) and education (Fischer & van der Wal 2007) and with situational variables such as type of species (e.g. rats over more desirable species like rhododendron and ducks) and control methods (e.g. sterilization and lethal injection over poisoning; Bremner & Park 2007). Support for control has been associated with knowledge about invasive species (Sharp et al. 2011; Huang & Lamm 2016) and under-lying values such as naturalness, balance (Fischer & van der Wal 2007) and degree of *ecocentrism* (Sharp et al. 2011). Other authors emphasize that layperson attitudes can be rich and complex, spanning multiple attitudinal dimensions including both positive and negative elements (e.g. a species can be seen as harmful but beautiful; Schüttler et al. 2011). Public thought processes about invasive species have been likened to those of ecological professionals, undercutting the conventional assumption that the public needs to be educated to accept scientific views (Fischer et al. 2014). However, different stakeholder groups have vastly different perceptions of species impacts and benefits leading to disagreements over management (Garciá-Llorente et al. 2008). Such differences may be heightened when invasive animals evoke strong affection (like feral cats; Wald et al. 2013) or contempt (like cane toads (*Rhinella marina*); Shine & Doody 2011).

This chapter examines two types of attitudes towards a well-known invasive predator in Florida, USA: the Burmese python (*Python molurus bivittatus*). These attitudes, which we term *acceptance of pythons on the landscape* (or *acceptance*) and *appreciation of pythons' intrinsic qualities* (or *appreciation*) are measured across the valence spectrum from negative, through a neutral midpoint, to positive. Our overall goal is to contribute to conceptual development of Frank's (2016) conflict-to-coexistence continuum by analysing relationships between attitudes and the specific behaviour of participating in a python removal competition. Frank (2016) distinguishes tolerance (an attitude) from coexistence (a behaviour) and describes a correspondence between them as they move along the negative-to-positive continuum. She offers a general definition of tolerance, spanning neutral-to-positive attitudes when 'species are per-ceived as beneficial to the personal, spiritual, cultural, economic, social, or political wellbeing of society' (Frank 2016, p. 740). However, attitudes towards wildlife can be parsed into many different dimensions (Kellert 1984, 1996), and attitudes are often inconsistent with behaviours (Wicker 1969; Eagly & Chaiken 1993). Attitudes are internal states, whereas behaviours are observable acts that are strongly influenced by external factors such as situational constraints and social norms (Heberlein 2012). All behaviours are specific, so attitudes need to be measured at a corresponding level of specificity to predict behaviours (Ajzen & Fishbein 1977; Heberlein 2012).

This study examines how two specific attitudes (acceptance and appreciation) relate to participation in the 2016 Python Challenge™, a public competition to remove Burmese pythons from state lands. We surveyed samples of two populations: general Florida residents (hereafter referred to as the *general public*) and Python Challenge™ participants. Behaviourally, it is evident that these two populations occupy different ranges of Frank's conflict-to-coexistence continuum with respect to Burmese pythons. Python Challenge™ participants are on the negative side of the continuum (i.e. actively hunting and killing pythons) whereas members of the general public are likely situated within the passive middle (i.e. uninvolved in python control efforts). However, their respective positions along a negative-to-positive *attitudinal* continuum are unknown. Therefore, the primary objective of this chapter is to describe and compare the two attitudes (acceptance and appreciation) between the inactive general public and active Python Challenge™ participants. We apply Frank's (2016) concept of a continuum by visually depicting each group's attitudes along a negative-to-positive spectrum and examining their correspondence with behaviour. Our second objective is to identify factors (including socio-economic characteristics and knowledge of invasive species) that underlie each group's attitudes towards pythons. We pay particular attention to relationships between knowledge and attitudes to contribute to theoretical discussion of the *deficit model* of public knowledge (i.e. the idea that public attitudes can and should be changed through provision of information (Heberlein 2012; Fischer et al. 2014).

13.1 METHODS

13.1.1 Study Context

Native to Southeast Asia and introduced to the United States via the pet trade, Burmese pythons are well established in Everglades National Park and surrounding natural areas of South Florida (Snow et al. 2007). This invasive predator threatens many native wildlife species with particularly damaging impacts on mammal populations (McCleery et al. 2015). Pythons are notoriously difficult to detect and remove because of their cryptic nature and use of inaccessible habitats (Hunter et al. 2015). Therefore, despite considerable financial investments to develop and test control methods, no management option currently exists to effectively remove pythons at the requisite landscape scale (Reed & Rodda 2011). Burmese pythons in Florida exhibit previously undocumented behaviours, such as homing ability (Pittman et al. 2014) and salinity tolerance (Hart et al. 2012), which may guarantee their persistence well into the future.

The Florida Fish and Wildlife Conservation Commission (FWC) has operated an unpaid python removal programme since 2009, which permits a group of experienced volunteers to remove pythons from three state wildlife management areas. In addition, FWC allows anyone to remove pythons from twenty-two public lands throughout the year without a permit or licence required. In 2013, FWC and partner agencies launched the first public participation incentive programme, the 2013 Python Challenge™ (pythonchallenge.org) in which 1,582 participants collectively removed sixty-eight Burmese pythons in five weeks (12 January–10 February 2013). While this number may appear low, it represents the highest number of pythons removed in any similar time period from 2008 to 2012 (Mazzotti et al. 2016). Recommendations from the 2013 event included clarifying competition rules and providing more opportunities for hands-on training in python detection and capture methods.

The second Python Challenge™ was held in 2016 with the goals of raising awareness of invasive species and engaging the public in Everglades conservation. The 2016 Python Challenge™ included a python removal competition (16 January–14 February) and provided other participation options such as public events and online contests. Lessons from 2013 led to more deliberate communication about the difficulty of detecting pythons and greater opportunities for in-person trainings (in addition to the required online training). FWC's records indicated that 1,066 people registered for the python removal competition, collectively removing 106 snakes.

13.1.2 Survey Administration

We surveyed the Florida general public and registered Python Challenge™ participants after the 2016 Python Challenge™ in late February,

2016. To survey the Florida general public, we purchased an *opt-in* internet panel ($N = 505$) from Qualtrics, LLC. The opt-in internet panel is a non-probability sampling method increasingly used in public opinion research (Baker et al. 2013), which can provide minimally biased results that sometimes outperform traditional probability samples (Vavreck & Rivers 2008). Qualtrics applied quotas, based on 2010 USA Census data, to represent the Florida population according to gender, age, race, ethnicity and Florida region of residence. We chose these four variables (the maximum Qualtrics could feasibly apply) to ensure the general public sample reflected population segments underrepresented among Python Challenge™ participants (women, minorities, north/central Florida residents). Regional quotas were determined based on the proportion of the state population within each of FWC's five administrative regions (Northwest, North Central, Northeast, Southwest and South; myfwc.com). Respondents were also screened to ensure that they were at least eighteen years old and had not participated in either the 2013 or 2016 Python Challenge™. To reduce bias, the survey topic was concealed from respondents until they chose to participate.

The Python Challenge™ participant survey sample was necessarily non-random, consisting only of participants who registered between 5 January and the end of the python removal competition on 14 February (599 participants were approached and 287 responded for a response rate of 47.9 per cent). This is because we had also surveyed a pre-event cross-section of participants, which consisted of those who had registered between 1 October 2015 and 5 January 2016. In this chapter we only analyse post-event data because our goal is to examine variability in attitudes across the two groups of respondents, rather than to evaluate pre-post effects of the Python Challenge™. However, using only post-event data biases our results by only representing later registrants (i.e. those who registered from 5 January–14 February) rather than all Python Challenge™ participants. In preliminary analyses we found that later registrants differed significantly from earlier registrants in terms of age, prior hunting experience, knowledge of invasive species and appreciation of pythons' intrinsic qualities. We consider implications of these biases in the discussion section.

13.1.3 Measurement

Our dependent variables are two composite scales measuring two different types of attitudes towards Burmese pythons. The acceptance scale includes five items (modified from Sharp et al. 2011; Harvey et al. 2015) reflecting the view that Burmese pythons are an acceptable part of the Florida environment, have the right to live in Florida and should not be killed (Table 13.1). This inversely corresponds to Kellert's (1984) *ecologistic* attitude (defined as primary concern for the environment as a system, for interrelationships between wildlife species and natural habitats) and also includes a *moralistic* dimension (i.e. primary concern for the right and wrong treatment of animals; Kellert 1984). Principal component analysis, with oblique Promax rotation, indicated that the items measured the same underlying concept (factor loadings = 0.79–0.88).

The appreciation scale includes three items (factor loadings = 0.71–0.85) measuring appreciation of pythons' intrinsic qualities including their beauty, lack of fearfulness and appeal as pets (Table 13.1). This reflects the inverse of a *negativistic* attitude (primary orientation an active avoidance of animals due to indifference, dislike or fear) combined with a *humanistic* element (i.e. primary interest and strong affection for individual animals, principally pets; Kellert 1984).

Reliability analysis confirmed that internal consistency was high for the acceptance scale (Cronbach's α = 0.90) and moderate for the appreciation scale (Cronbach's α = 0.65; Vaske 2008). We computed composite scales as the mean response of the individual items (from 1 = *strongly disagree* to 5 = *strongly agree*). Distributions did not deviate markedly from normal (skewness values of 0.84 and 0.30, respectively) so we treated the ordinal scales as continuous variables in linear regression analyses (Vaske 2008).

Knowledge about invasive species was measured as the sum of seven items (factor loadings = 0.47–0.78; Cronbach's α = 0.78; range 1 to 10, skewness 0.31). This included six dichotomous (1 = *yes*, 0 = *no*) questions asking if they had ever heard of the following species in Florida: Indo-Pacific lionfish (*Pterois volitans* and *P. miles*), Argentine black and white tegu (*Salvator merianae*), Nile monitor (*Varanus niloticus*), Old World climbing fern (*Lygodium microphyllum*), melaleuca (*Melaleuca quinquenervia*) and water hyacinth (*Eichhornia crassipes*). The seventh item was *How knowledgeable do you feel you are about invasive species in general?* (1 = not knowledgeable, 2 = slightly knowledgeable, 3 = fairly knowledgeable, 4 = highly knowledgeable). Both surveys also contained socio-economic measures including age, gender, ethnicity, race, education and geographic region of residence; and the behavioural measure of prior experience hunting any animals.

Table 13.1 *Indices and items used to measure attitudes towards Burmese pythons in Florida, based on a post-event survey of 2016 Python Challenge™ participants and the Florida general public*

Index and items	Python Challenge™ participants ($N = 287$)	Florida general public ($N = 505$)	t-value	Cronbach alpha
Acceptance of pythons on the landscape	1.65 (0.64)	2.29 (1.01)	−10.76***	0.90
I feel that Burmese pythons have the right to live in Florida.	1.51 (0.77)	2.28 (1.17)	−11.11***	
I accept Burmese pythons as a permanent part of the Everglades ecosystem.	2.02 (1.09)	2.30 (1.19)	−3.33**	
Invasive species have as much right to exist in Florida as native plants and animals.	1.56 (0.81)	2.30 (1.18)	−10.21***	
I feel that it is wrong to kill wildlife, even if it is an invasive species.	1.60 (0.86)	2.40 (1.22)	−10.71***	
Wildlife managers should worry less about getting rid of invasive species and just let nature run its course.	1.54 (0.79)	2.15 (1.11)	−8.86***	
Appreciation of pythons' intrinsic qualities	3.23 (0.81)	2.01 (0.85)	19.52***	0.65
I think Burmese pythons are beautiful animals.	3.81 (1.13)	2.54 (1.24)	14.47***	
I would enjoy having a Burmese python (or other large constrictor snake) as a pet.	1.81 (1.13)	1.51 (0.99)	3.61***	
I would feel scared if I saw a Burmese python while hiking.[b]	1.92 (1.09)	4.03 (1.24)	−24.62***	

[a] Scale ranged from 1 (*strongly disagree*) to 5 (*strongly agree*).
[b] This item was reverse-coded before the scale was computed.
** $p < 0.01$, *** $p < 0.001$.

13.2 DATA ANALYSIS

We downloaded the participant and general public survey data from Qualtrics and merged them into a single data set in IBM SPSS Statistics 22. On the participant survey item non-response was up to 6 per cent, and these cases were treated as missing values and excluded from analyses. Missing values were not an issue on the general public surveys because Qualtrics required respondents to answer all survey questions.

We first compared socio-economic characteristics of the Python Challenge™ participants and general public respondents using χ^2 tests for categorical variables and a ttest for age. We also used t-tests to compare knowledge, acceptance and appreciation scores between the two groups. We created frequency polygons to visually compare Python Challenge™ participants and general public response distributions on the two attitude scales.

We performed two multiple linear regressions, separately for the general public and Python Challenge™ participants, to predict acceptance and appreciation. Observations that included complete data on all of the model variables were included in regression analyses ($n = 505$ for general public, $n = 233$ for Python Challenge™ participants). We included six independent variables (age, gender, Hispanic ethnicity, education level, prior hunting experience and knowledge about invasive species). Women and older people have been found to express greater concern for invasive species impacts (Harvey et al. 2015) and greater support for control programmes (Bremner & Park 2007; Sharp et al. 2011). Women also demonstrate greater fear of snakes (Kaltenborn et al. 2006; Prokop et al. 2009) and higher risk perceptions towards other carnivores (Harvey et al. 2017). Hispanics/Latinos are an important cultural group in Florida, representing 24 per cent of the population (www.census.gov), and were found in a prior study to prefer leaving invasive species alone (Harvey & Mazzotti 2016). Hunters represent an organized and influential stakeholder group often found to have negative attitudes towards carnivores (Ericsson & Heberlein 2003; Naughton-Treves et al. 2003) and to differ from non-hunters in terms of numerous attitudinal dimensions (e.g. utilitarian, naturalistic, humanistic, moralistic; Kellert 1984). Finally, knowledge of invasive species has been associated with greater support for control efforts (Bremner & Park 2007; Sharp et al. 2011; Huang & Lamm 2016) and is considered by many invasion biologists as important to ensure public acceptance of scientific opinions about invasive species (Fischer et al. 2014).

13.3 RESULTS

13.3.1 Respondent Characteristics

As stated, the general public sample was designed to represent the Florida population in terms of gender (50.7% female, 49.3% male), age (mean 49.3 SD 17.9), race (75.8% white, 19.0% black, 5.1% other), ethnicity (25.7% Hispanic, 74.3% non-Hispanic) and region of residence (34.9% South Florida, 65.1% other area of Florida). However, the general public sample was more highly educated than the Florida population according to 2010 USA Census data (42.8% of survey respondents versus 26.4% of Florida residents had a Bachelor's degree; www.census.gov).

Python Challenge™ participants were younger on average than general public respondents (mean 46.2, SD 14.4, $t = -2.61$, $p < 0.01$) and much more likely to be white (95.9%, $\chi^2 = 55.57$, $p < 0.001$), non-Hispanic (94.8%, $\chi^2 = 48.89$, $p < 0.001$) and male (88.1%, $\chi^2 = 113.66$, $p < 0.001$). They were also more highly educated than the general public (52.6% possessed a Bachelor's degree, $\chi^2 = 6.82$, $p < 0.01$) and more likely to reside in South Florida (38.7%) or out of state (14.6%, $\chi^2 = 91.53$, $p < 0.001$). Participants were much more likely to have prior hunting experience (68.5%) compared to the general public (14.1%; $\chi^2 = 235.88$, $p < 0.001$). Mean responses on the invasive species knowledge scale (range 2–10) were 6.57 (SD 2.05) among Python Challenge™ participants and 3.51 (SD 2.15) among the general public ($t = 18.71$, $p < 0.001$).

13.3.2 Attitudes towards Burmese Pythons

Python Challenge™ participants' mean score on the acceptance scale was between *strongly disagree* (1) and *disagree* (2) on the five-point scale (Table 13.1). Participants were more likely to agree with the statement *I accept Burmese pythons as a permanent part of the Everglades ecosystem* (mean 2.02) than with the other four statements composing the scale. General public mean responses fell between values of *disagree* (2) and *neither agree nor disagree* (3). The general public was significantly more likely than participants to accept pythons on the landscape ($t = -15.99$, $p < 0.001$).

On the other hand, Python Challenge™ participants were significantly more likely than the general public to appreciate pythons' intrinsic qualities ($t = 29.43$, $p < 0.001$; Table 13.1). The two groups diverged sharply on the belief that pythons are beautiful animals and on the feeling that they would be scared if they saw a python while hiking.

Python Challenge™ Participants: Florida General Public:

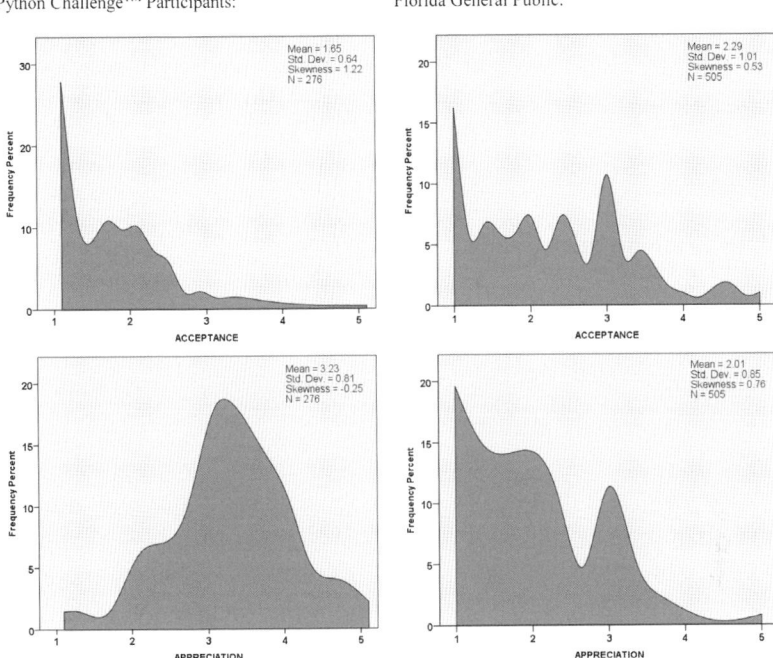

Figure 13.1 Frequency polygons for Python Challenge™ participants' (left) and Florida general public's (right) responses on two measures of attitudes towards Burmese pythons: acceptance of pythons on the landscape (top) and appreciation of pythons' intrinsic qualities (bottom). Responses range from 1 (strong negative attitude) to 5 (strong positive attitude).

Participants were also significantly more likely than the general public to say they would enjoy having a python as a pet, although average responses for both groups were below 2 on this item. Pearson correlation between the two scales was 0.22 ($p < 0.001$) for Python Challenge™ participants and 0.42 ($p < 0.001$) for the general public. Figure 13.1 illustrates how the two attitudes align along a continuum from extremely negative attitudes on the left to extremely positive attitudes on the right.

13.3.3 Factors Predicting Python Challenge™ Participant Attitudes

In the Python Challenge™ participant multiple linear regression models, the highest bivariate correlation (Pearson r) between independent variables was -0.29 ($p < 0.001$, between prior hunting experience

Table 13.2 *Standardized regression coefficient values (β) from multiple linear regression models used to predict Python Challenge™ participants' acceptance of Burmese pythons on the landscape and appreciation of pythons' intrinsic qualities*

Variable	Acceptance	Appreciation
Age	−0.083	−0.061
Female gender	0.011	−0.015
Hispanic/Latino	0.015	0.088
Bachelor's degree or higher	0.050	−0.121
Prior hunting experience	−0.066	−0.007
Knowledge of invasive species	−0.192**	0.334***
N	233	233
F	1.86	6.01***
Adjusted R^2	2.2%	11.5%

$* \ p < 0.05, \ ** \ p < 0.01, \ *** \ p < 0.001.$

and gender), and the lowest tolerance value was 0.89 (for prior hunting experience), indicating that multicollinearity was not a problem (Vaske 2008). The independent variables only explained 2.2% (adjusted R^2) of the variance in acceptance score and the model was not statistically significant ($F(6,226)= 1.86$, $p = 0.088$). The model predicting appreciation was significant ($F(6,226)= 6.01$, $p < 0.001$) and explained 11.5% of the variance. Knowledge of invasive species exhibited a significant negative relationship with acceptance and a significant positive relationship with appreciation. None of the other socio-economic variables were significant predictors of either attitude (Table 13.2).

13.3.4 Factors Predicting General Public Attitudes

Multicollinearity was not a problem in the general public models. The highest bivariate correlation (Pearson r) between independent variables was −0.37 ($p < 0.001$, between gender and age) and the lowest tolerance value was 0.79 (for gender). The independent variables explained 19.6% (adjusted R^2) of the variance in acceptance score and 9.6% of the variance in appreciation score. Both models were statistically significant: acceptance ($F(6,498)= 21.44$, $p < 0.001$) and appreciation ($F(6,498) = 9.88$, $p < 0.001$). Acceptance exhibited a significant positive relationship with Hispanic ethnicity, and significant negative relationships with age, knowledge of invasive species, education and prior hunting experience. Appreciation was positively related to knowledge of invasive species, and negatively related to female gender and age (Table 13.3).

Table 13.3 *Standardized regression coefficient values (β) from multiple linear regression models used to predict the Florida general public's acceptance of Burmese pythons on the landscape and appreciation of pythons' intrinsic qualities*

Variable	Acceptance	Appreciation
Age	−0.376***	−0.179***
Female gender	−0.088	−0.185***
Hispanic/Latino	0.120**	0.083
Bachelor's degree or higher	−0.095*	−0.060
Prior hunting experience	−0.090*	0.038
Knowledge of invasive species	−0.102*	0.231***
N	505	505
F	21.44***	9.88***
Adjusted R^2	19.6%	9.6%

* $p < 0.05$, ** $p < 0.01$, *** $p < 0.001$.

13.4 DISCUSSION

Ecologists are in the midst of a debate over invasive species' threats to biodiversity and the role scientists should play in condemning them. The conventional conflict-oriented view, i.e. that the severe impacts of many invasive alien species justify persistent control efforts (e.g. Simberloff et al. 2011, 2012), is being challenged by a more coexistence-oriented perspective that criticizes the native–alien distinction and accepts alien species as part of dynamic ecosystems (e.g. Davis et al. 2011; Thompson 2014). This tension between conflict and coexistence in regard to invasive species dovetails with a broader paradigm shift from conflict to coexistence in discourses about human–wildlife interactions. This chapter compared two groups' attitudes towards invasive Burmese pythons, in relation to their behaviour and other underlying variables, with the goal of applying and refining Frank's (2016) conflict-to-coexistence continuum.

We examined relationships between a specific behaviour (participating in the 2016 Python Challenge™) and two types of attitudes towards Burmese pythons in Florida. The first attitude, acceptance, reflects an *anti-ecologistic* and moralistic view (Kellert 1984) that invasive pythons have a right to exist in Florida and should not be controlled. This is similar to what Sharp et al. (2011) labeled *absolute ecocentrism*, which they found correlated with a management preference of leaving invasive species alone. Python Challenge™ participants' acceptance scores were significantly lower than those of the general public and were skewed towards the negative side of the spectrum, indicating an unwillingness to coexist with Burmese pythons in Florida. Participants were so consistent in their low acceptance, in fact, that the multiple regression model only explained 2.2 per cent of the variance and was not statistically significant. By this measure, participants' attitude and behaviour towards pythons generally aligned on the negative (conflict) side of the spectrum, seeming to fit Frank's (2016) notion of attitude–behaviour correspondence along the conflict-to-coexistence continuum.

However, the attitude–behaviour relationship is mediated by attitudinal specificity (Heberlein 2012). Correspondence is theoretically strongest when the measured attitude and behaviour match in specificity according to action, target, context and time (Ajzen & Fishbein 1977). Here, the behaviour of interest is hunting and removing (*action*) Burmese pythons (*target*), an invasive species that is scientifically understood to currently threaten the Everglades ecosystem (*context and time*). Our acceptance measure corresponds to the action (e.g. *to kill wildlife*), target (Burmese pythons), context (invasive species in Florida) and, to some degree, time (e.g. *a permanent part of the Everglades ecosystem*). The specificity of this measure likely explains its consistency with behaviour. Similarly, Hayman et al. (2014) found that specific risk beliefs predicted behaviours towards American alligators (*Alligator mississippiensis*) in Florida, whereas general attitude towards alligators did not.

Our findings were different for the second attitude measure, appreciation of pythons' intrinsic qualities. Python Challenge™ participants' mean appreciation score was twice as high as their acceptance score, with more than half scoring above the midpoint of the scale. Many Python Challenge™ participants thus hold contradictory attitudes of liking the animal but strongly disliking its presence in the environment, an ambivalence that has been documented in attitudes towards other invasive wildlife (Schüttler et al. 2011). The reason for this ambivalence is that appreciation and acceptance, although both positive, are very different types of attitudes. The appreciation measure corresponds in its target (Burmese pythons) but does not address the ecological context of pythons as an invasive species nor the action of killing pythons. Rather, it captures a general admiration of the species' beauty and humanistic value combined with a lack of fear. This appreciative attitude was much higher among participants than the general public, illustrating the complexity of a behaviour like participating in the Python Challenge™. What appears on the surface to be human–wildlife conflict (removing Burmese pythons) in fact consists of a combination of at least two behaviours: (1) searching for snakes in South Florida natural areas and (2) attempting to remove and kill those snakes if found. Participants' attraction to Burmese pythons may in fact be one motivation for taking part in the Python Challenge™. Nearly one-third of participants surveyed described themselves as *amateur or expert herpetologists*, and more than half reported that they had previously kept reptiles or amphibians as pets. For these enthusiasts, participating in the Python Challenge™ is likely an extension of their interest in *herping*, i.e. searching for reptiles and amphibians in the field.

As noted, our post-event sample of participants consisted only of later registrants and thus was biased towards younger, less knowledgeable participants with less hunting experience and lower appreciation of pythons' intrinsic qualities. Specifically, the differences between the PRE and POST participant samples on these variables were as follows: mean age (50.0 PRE, 46.2 POST; $t = 3.20$, $p < 0.01$), prior hunting experience (79.8% PRE, 68.5% POST, $\chi^2 = 8.55$, $p < 0.01$), mean knowledge score (6.96 PRE, 6.57 POST, $t = 2.22$, $p < 0.05$), mean appreciation score (3.37 PRE, 3.23 POST, $t = 2.03$, $p < 0.05$). Had we surveyed a representative sample of all participants, increases in both knowledge and appreciation would have resulted in a similar statistical relationship between the two variables. Changes in distributions of age and hunting experience may have strengthened those variables' relationships with attitudes, but likely not enough to alter the regression results that found knowledge to have the strongest effect on both attitudes in models that explained minimal variance overall (particularly for acceptance).

The general public was largely neutral-to-negative on both attitudes, generally corresponding with their behaviours in the passive middle of the continuum. Nearly half of the general public had acceptance scores of *strongly disagree* to *disagree* suggesting that they may support active control of Burmese pythons. However, the other half fell within the midrange of the scale (between *disagree* and *agree*) indicating they do not feel strongly about the issue. If tolerance is understood as beginning at the neutral midpoint of a negative-to-positive continuum (Bruskotter & Fulton 2012; Frank 2016), this finding suggests that much of the general public currently tolerates the presence of Burmese pythons in Florida. Acceptance was highest among people who were younger, less educated and of Hispanic ethnicity. As python management continues into the future, these segments of the public could become weary of long-term appeals for financial and political support. Thompson (2014, p. 218) points out that 'attempted eradication or control is a waste of time if active support (or at least passive cooperation) is required of people who believe the target species to be useful, desirable, attractive or simply harmless'.

General public mean appreciation score was lower than mean acceptance score, indicating that public dislike for pythons may be based more on emotional reactions to the species' intrinsic qualities than on considerations of its ecological role (see Chapter 4). Schüttler et al. (2011) similarly found that negative attitudes towards invasive predators were due to fear more than to the species' invasiveness. Snakes are among the most feared and disliked of all animals (Thorpe & Salkovskis 1997; Pagani et al. 2007), and the prevalence of fear across cultures suggests an evolutionary basis (Öhman 2009). Negativistic attitudes are prevalent even towards endangered species of snakes, undercutting support for their conservation (Knight 2008). We found strong negative effects of gender and age on appreciation, confirming prior findings for fear-based attitudes towards snakes (Kaltenborn et al. 2006; Prokop et al. 2009) and other dangerous carnivores (Harvey et al. 2017). Attitudes rooted in social identities (e.g. youth and masculinity) tend to be strong, emotional and unlikely to change in response to information (Heberlein 2012). When undesirable animals like snakes, toads and rats are also invasive species, they are likely to elicit *extreme opprobrium* (Thompson 2014, p. 204). This finding suggests that public tolerance would be more widespread for a *desirable* invasive species such as a bird or mammal. It also suggests that tolerance of Burmese pythons may decrease if the population becomes more conspicuous (e.g. if it spreads from natural areas into residential areas).

Among both survey groups, knowledge of invasive species was associated with lower acceptance and higher appreciation of pythons. This reflects prior research associating knowledge with higher ecologistic and lower negativistic attitudes towards animals (Kellert & Berry 1981). The positive relationship between knowledge and appreciation is likely not one of direct causality, but rather the effect of an external factor such as overall interest in and experience with reptiles. Attitudes based on direct experience and those with a strong affective (emotional) component rarely change through the provision of knowledge (Heberlein 2012). Morgan and Gramann (1989) demonstrated that affective attitudes towards snakes did not change through information alone, but did change when information was combined with direct contact and modelling (watching a respected person interact with a snake). By comparison, information campaigns have been effective at increasing awareness and knowledge about invasive species (Garciá-Llorente et al. 2008; Sharp et al. 2013) and beliefs that invasive species are an ecological threat (Harvey & Mazzotti 2016). Prokop et al. (2009) found that knowledge influenced *scientistic* attitudes towards snakes (i.e. interest in snake biology) but not negativistic attitudes which were based on fear. Therefore, we may expect that an ecologistic attitude like our acceptance measure would be responsive to educational efforts, particularly among the general public whose attitudes are relatively weak and malleable because they are not based on direct experience (Heberlein 2012). This conclusion may be agreeable to proponents of the deficit model of public knowledge (Fischer et al. 2014) because it suggests that the public can be educated to adopt scientific understandings of invasive species. However, it does not answer the question of whether society should view invasive wildlife through a lens of conflict or coexistence. That is not an ecological question but a value judgement.

13.4.1 Recommendations and Future Directions

- This study applied Frank's (2016) conflict-to-coexistence continuum to an examination of a single behaviour (participation in the 2016 Python Challenge™) in relation to two attitudes towards Burmese pythons. A more holistic application of the continuum would entail examining a range of positive and negative behaviours towards wildlife in relation to attitudes (e.g. Bruskotter et al. 2015).
- The two attitudes examined here each represented a combination of attitude types based on Kellert's (1984) typology. We treated *acceptance of pythons on the landscape* as primarily an ecologistic attitude, although it also included a moralistic component; we treated *appreciation of pythons' intrinsic qualities* as a (reverse) negativistic attitude although it also included a humanistic component. Future research should separate these attitudinal components into individual measures to facilitate interpretation and comparison with existing literature.
- This study identified social variables (age, gender, Hispanic ethnicity, education and hunting experience) that affected attitudes towards Burmese pythons among the Florida general public. Qualitative research would be valuable to explore how social identities, norms and values shape reactions to invasive wildlife, particularly in a multicultural context like South Florida.

- Ongoing registration for the 2016 Python Challenge™ generated research design challenges (and biased samples) because more than half of participants registered after the PRE survey was conducted. Future surveys can produce representative samples by approaching all registered participants, or a random sample, at a single point of time. Alternatively, the same participants could be surveyed before and after the event (using a matched-samples design) if the intent were to examine changes in attitudes or behaviours resulting from the event.

- Discourse on human–wildlife conflict and coexistence can be expanded beyond the conventional focus on charismatic native species, to include a range of situations where wildlife threatens the environment, the economy or human health. The conflict-to-coexistence framework is particularly relevant to alien and invasive species because it dovetails with a current debate among ecologists over the extent to which alien species should be tolerated in new environments.

- Our results indicate that much of the Florida general public currently tolerates the presence of invasive Burmese pythons, when tolerance is defined as neutral-to-positive attitudes towards a species (Frank 2016). This tolerance is fairly remarkable given public aversion to snakes, and is likely due in part to pythons' presence in mainly remote areas of Florida where people do not encounter them. None-theless, this degree of tolerance suggests that many segments of the public may not consider invasive species a worthwhile target of agency control efforts and taxpayer dollars.

- Compared to the uninvolved public, active participants in the 2016 Python Challenge™ were much less tolerant of Burmese pythons in Florida. Although they disliked pythons' presence in the environment, many participants expressed a general appreciation of the animal's intrinsic qualities. These results highlight the multidi-mensionality of attitudes towards wildlife. Practitioners should avoid over-generalizing about *positive* or *negative* attitudes towards a species and rather focus on identifying specific types of attitudes (e.g. natur-alistic, ecologistic, humanistic, moralistic, scientistic, aesthetic, utili-tarian, dominionistic, negativistic; Kellert 1984).

- Likewise, conservationists should guard against the assumption that increasing people's knowledge about wildlife will improve their atti-tudes. In this study, knowledge was positively related to appreciation of pythons but negatively related to acceptance of pythons. These relationships are not necessarily causal. Providing people with information may influence some types of attitudes (e.g. ecologistic, scientistic) but not others (e.g. negativistic).

- The conflict-to-coexistence continuum is useful as a general framework to initiate research in a range of socio-ecological contexts. However, in practice, we recommend following Treves (2012) who conceptualized attitudes (intolerance to tolerance) and behaviours (conflict to coexistence) as existing on two separate continua which do not necessarily intersect. Assuming that attitudes and behaviours towards wildlife will correspond can lead to erroneous and misleading conclusions (Harvey et al. 2017). Empirical examinations of the attitude–behaviour relationship must ensure that the two variables are measured at corresponding levels of specificity.

13.5 References

Ajzen, I. & Fishbein, M. (1977). Attitude–behaviour relations: A theoretical analysis and review of empirical research. *Psychological Bulletin*, 84, 888–918.

Baker, R., Brick, J. M., Bates, N. A., Battaglia, M., Couper, M. P., Dever, J. A., Gile, K. J. & Tourangeau, R. (2013). *Report of the AAPOR Task Force on Non-Probability Sampling*. American Association for Public Opinion Research.

Bremner, A. & Park, K. (2007). Public attitudes to the management of invasive non-native species in Scotland. *Biological Conservation*, 139, 306–14.

Bruskotter, J. T. & Fulton, D. C. (2012). Will hunters steward wolves? A comment on Treves and Martin. *Society & Natural Resources*, 25, 97–102.

Bruskotter, J. T., Singh, A., Fulton, D. C. & Slagle, K. (2015). Assessing tolerance for wildlife: Clarifying relations between concepts and measures. *Human Dimensions of Wildlife*, 20, 255–70.

Convention on Biological Diversity (CBD) Secretariat. (2002). Decision VI/23: Alien species that threaten ecosystems, habitats and species. Document UNEP/CBD/COP/6/23. CBD Secretariat, Montreal, Canada.

Davis, M. A., Chew, M. K., Hobbs, R. J., Lugo, A. E., Ewel, J. J., Vermeij, G. J., Brown, J. H., Rosenzweig, M. L., Gardener, M. R., Carroll, S. P., Thompson, K., Pickett, S. T. A., Stromberg, J. C., Del Tredici, P., Suding, K. N., Ehrenfeld, J. G., Grime, J. P., Mascaro, J. & Briggs, J. C. (2011). Don't judge species on their origins. *Nature*, 474, 153–4.

Dickman, A. J. (2010). Complexities of conflict: The importance of considering social factors for effectively resolving human–wildlife conflict. *Animal Conservation*, 13, 458–66.

Eagly, A. H. & Chaiken, S. (1993). *The Psychology of Attitudes*. Fort Worth, TX: Harcourt Brace.

Ericsson, G. & Heberlein, T. A. (2003). Attitudes of hunters, locals, and the general public in Sweden now that the wolves are back. *Biological Conservation*, 111 (2), 149–60.

Estévez, R. A., Anderson, C. B., Pizarro, J. C. & Burgman, M. A. (2014). Clarifying values, risk perceptions, and attitudes to resolve or avoid social conflicts in invasive species management. *Conservation Biology*, 29, 19–30.

Fischer, A., Selge, S., van der Wal, R. & Larson, B. M. H. (2014). The public and professionals reason similarly about the management of non-native invasive species: A quantitative investigation of the relationship between beliefs and attitudes. *PLoS ONE*, 9(8), e105495.

Fischer, A. & van der Wal, R. (2007). Invasive plant suppresses charismatic seabird: The construction of attitudes towards biodiversity management options. *Biological Conservation*, 135, 256–67.

Fisher, M. (2016). Whose conflict is it anyway? Mobilizing research to save lives. *Oryx*, 50(3), 377–8.

Frank, B. (2016). Human–wildlife conflicts, the need to include tolerance and coexistence: An introductory comment. *Society & Natural Resources*, 29(6), 738–43.

García-Llorente, M., Martín-López, B., González, J. A., Alcorlo, P. & Montes, C. (2008). Social perceptions of the impacts and benefits of invasive alien species: Implications for management. *Biological Conservation*, 141, 2969–83.

Hart, K. M., Schofield, P. J. & Gregoire, D. R. (2012). Experimentally derived salinity tolerance of hatchling Burmese pythons (*Python molurus bivittatus*) from the Everglades, Florida (USA). *Journal of Experimental Marine Biology and Ecology*, 413, 56–9.

Harvey, R. G., Briggs-Gonzalez, V. S. & Mazzotti, F. J. (2017). Conservation payments in a social context: Determinants of tolerance and behavioural intentions towards wild cats in northern Belize. *Oryx*, 51(4), 730–41.

Harvey, R. G. & Mazzotti, F. J. (2016). *Public Knowledge, Attitudes, and Behaviours toward Invasive Lionfish: Pre- and Post-Campaign Surveys*. Final report to the Florida Fish and Wildlife Conservation Commission. 42 pp.

Harvey, R. G., Perez, L. & Mazzotti, F. J. (2015). *Not* seeing is *not* believing: Volunteer beliefs about Burmese pythons in Florida and implications for public participation in invasive species removal. *Journal of Environmental Planning and Management*, 59, 789–807.

Hayman, R. B., Harvey, R. G., Mazzotti, F. J., Israel, G. D. & Woodward, A. R. (2014). Who complains about alligators? Cognitive and situational factors influence behaviour toward wildlife. *Human Dimensions of Wildlife*, 19, 481–97.

Heberlein, T. A. (2012). *Navigating Environmental Attitudes*. New York: Oxford University Press.

Huang, P. & Lamm, A. J. (2016). Identifying invasive species educational needs in Florida: Opportunities for extension. *Journal of Extension*, 54(5), 5RIB7.

Hunter, M. E., Oyler-McCance, S. J., Dorazio, R. M., Fike, J. A., Smith, B. J., Hunter, C. T., Reed, R. N. & Hart, K. M. (2015). Environmental DNA (eDNA) sampling improves occurrence and detection estimates of invasive Burmese pythons. *PLoS ONE*, 10(4), e0121655.

Kaltenborn, B. P., Bjerke, T., Nyahongo, J. W. & Williams, D. R. (2006). Animal preferences and acceptability of wildlife management actions around Serengeti National Park, Tanzania. *Biodiversity and Conservation*, 15(14), 4633–49.

Kellert, S. R. (1984). American attitudes toward and knowledge of animals: An update. In M. W. Fox & L. D. Mickley, eds., *Advances in Animal Welfare Science 1984/85*. Washington, DC: The Humane Society of the United States, pp. 177–213.

Kellert, S. R. (1996). *The Value of Life: Biological Diversity and Human Society*. Washington, DC: Island Press.

Kellert, S. R. & Berry, J. (1981). *Knowledge, Affection, and Basic Attitudes toward Animals in American Society*. Report no. 024–010- 00–625-1. Washington, DC: U.S. Government Printing Office.

Knight, A. J. (2008). Bats, snakes and spiders, oh my! How aesthetic and negativistic attitudes, and other concepts predict support for species protection. *Journal of Environmental Psychology*, 28, 94–103.

Mazzotti, F. J., Rochford, M. R., Vinci, J. J., Jeffery, B. M., Ketterlin Eckles, J., Dove, C. & Sommers, K. P. (2016). Implications of the 2013 Python Challenge(^{TM}) for ecology and management of the Burmese Python (*Python molurus bivittatus*) in Florida. *Southeastern Naturalist*, 15(sp8), 63–74.

McCleery, R. A, Sovie, A., Reed, R. N., Cunningham, M. W., Hunter, M. E., Hart, K. M. (2015). Marsh rabbit mortalities tie pythons to the precipitous decline of mammals in the Everglades. *Proceedings of the Royal Society B*, 282, 20150120.

Morgan, J. M. & Gramann, J. H. (1989). Predicting effectiveness of wildlife education programs: A study of students' attitudes and knowledge toward snakes. *Wildlife Society Bulletin*, 17(4), 501–9.

Naughton-Treves, L., Grossberg, R. & Treves, A. (2003). Paying for tolerance: Rural citizens' attitudes toward wolf depredation and compensation. *Conservation Biology*, 17, 1500–11.

Öhman, A. (2009). Of snakes and faces: An evolutionary perspective on the psychology of fear. *Scandinavian Journal of Psychology*, 50(6), 543–52.

Pagani, C., Robustelli, F. & Ascione, F. R. (2007). Italian youths' attitudes toward, and concern for, animals. *Anthrozoös*, 20(3), 275–93.

Peterson, M. N., Birckhead, J. L., Leong, K., Peterson, M. L. & Peterson, T. R. (2010). Rearticulating the myth of human–wildlife conflict. *Conservation Letters*, 3, 74–82.

Pittman, S. E., Hart, K. M., Cherkiss, M. S., Snow, R. W., Fujisaki, I., Smith, B. J., Mazzotti, F. J. & Dorcas, M. E. (2014). Homing of invasive Burmese pythons in south Florida: Evidence for map and compass senses in snakes. *Biology Letters*, 10, 20140040.

Prokop, P., Ozel, M. & Usak, M. (2009). Cross-cultural comparison of student attitudes toward snakes. *Society and Animals*, 17, 224–40.

Redpath, S. M., Bhatia, S. & Young, J. (2015). Tilting at wildlife: Reconsidering human–wildlife conflict. *Oryx*, 49, 222–5.

Reed, R. N. & Rodda, G. H. (2011). Burmese python and other giant constrictors. In D. S. Simberloff & M. Rejmánek, eds., *Encyclopedia of Biological Invasions*. Berkeley: University of California Press, pp. 85–91.

Russell, J. C. & Blackburn, T. M. (2017). The rise of invasive species denialism. *Trends in Ecology & Evolution*, 32(1), 3–6.

Schüttler, E., Rozzi, R. & Jax, K. (2011). Towards a societal discourse on invasive species management: A case study of public perceptions of mink and beavers in Cape Horn. *Journal for Nature Conservation*, 19, 175–84.

Sharp, R. L., Larson, L. R. & Green, G. T. (2011). Factors influencing public preferences for invasive alien species management. *Biological Conservation*, 144, 2097–2104.

Sharp, R. L., Larson, L. R., Green, G. T. & Tomek, S. (2013). Comparing interpretive methods targeting invasive species management at Cumberland Island National Seashore. *Journal of Interpretation Research*, 17(2), 23–43.

Shine, R. & Doody, S. (2011). Invasive species control: Understanding conflicts between researchers and the general community. *Frontiers in Ecology & the Environment*, 9, 400–6.

Simberloff, D., Alexander, J., Allendorf, F. et al. (2011). Non-natives: 141 scientists object. *Nature*, 475, 36.

Simberloff, D., Martin, J.-L., Genovesi, P., Maris, V., Wardle, D. A., Aronson, J., Courchamp, F., Galil, B., García-Berthou, E., Pascal, M., Pyšek, P. Sousa, R., Tabacchi, E. & Vilà, M. (2012). Impacts of biological invasions: What's what and the way forward. *Trends in Ecology & Evolution*, 28(1), 58–66.

Snow, R. W., Johnson, V. M., Brien, M. L., Cherkiss, M. S. & Mazzotti, F. J. (2007). *Python molurus bivittatus* (Burmese python): Nesting. *Herpetological Review*, 38, 93.

Thompson, K. (2014). *Where Do Camels Belong? The Story and Science of Invasive Species*. London: Profile Books.

Thorpe, S. J. & Salkovskis, P. M. (1997). Animal phobias. In G. C. L. Davey, ed., *Phobias: A Handbook of Theory, Research and Treatment*. Chichester: Wiley, pp. 81–106.

Treves, A. (2012). Tolerant attitudes reflect an intent to steward: A reply to Bruskotter and Fulton. *Society & Natural Resources*, 25, 103–4.

Vaske, J. J. (2008). *Survey Research and Analysis: Applications in Parks, Recreation and Human Dimensions*. State College, PA: Venture Publishing, Inc.

Vavreck, L. & Rivers, D. (2008). The 2006 cooperative congressional election study. *Journal of Elections, Public Opinion and Parties*, 18(4), 355–66.

Wald, D. M., Jacobson, S. K. & Levy, J. K. (2013). Outdoor cats: Identifying differences between stakeholder beliefs, perceived impacts, risk and management. *Biological Conservation*, 167, 414–24.

Warren, C. R. (2007). Perspectives on the 'alien' versus 'native' species debate: A critique of concepts, language and practice. *Progress in Human Geography*, 31(4), 427–46.

White, P. C. L., Ford, A. E. S., Clout, M. N., Engeman, R. M., Roy, S. & Saunders, G. (2008). Alien invasive vertebrates in ecosystems: Pattern, process and the social dimension. *Wildlife Research*, 35, 171–9.

White, P. C. L. & Ward, A. I. (2010). Interdisciplinary approaches for management of existing and emerging human–wildlife conflicts. *Wildlife Research*, 37, 623–9.

Wicker, A. W. (1969). Attitudes versus actions: The relationship of verbal and overt behavioural responses to attitude objects. *Journal of Social Issues*, 4, 41–79.

Young, A. M. & Larson, B. M. H. (2011). Clarifying debates in invasion biology: A survey of invasion biologists. *Environmental Research*, 111, 893–8.

Young, J. C., Marzano, M., White, R. M., McCracken, D. I., Redpath, S. M., Carss, D. N., Quine, C. P. & Watt, A. D. (2010). The emergence of biodiversity conflicts from biodiversity impacts: Characteristics and management strategies. *Biodiversity and Conservation*, 19, 3973–90.

Institutions for Achieving Human–Wildlife Coexistence

The Case of Large Herbivores and Large Carnivores in Europe

JOHN D. C. LINNELL AND BJØRN P. KALTENBORN

14.1 COEXISTENCE AS A CONCEPT

This chapter is based on the premise that the future of wildlife conservation in Europe, and across much of the world, is based on the integration of humans and wildlife into shared spaces (Linnell et al. 2015). This may involve varying degrees of land-sparing (in protected and wilderness areas) where such opportunities exist, but will inevitably depend on a high degree of land-sharing (sensu Phalan et al. 2011). This integration is often referred to as *coexistence*. In this chapter, we view coexistence as being a dynamic state that involves a high degree of co-adaptation by both wildlife and humans to each other (sensu Carter & Linnell 2016), and recognize that it represents a continuum (Frank 2016) of states that involve varying degrees of tensions and conflicts, but where wildlife are managed in a way that contains the conflicts within levels that are regarded as being socially, culturally and economically tolerable in the respective societies. Managing these tensions and deciding on tolerance levels requires effective governance and institutions. The conflict-to-coexistence continuum underlines that interactions are complex and diverse and that there is a need to focus on tolerance, yet it heavily focuses on individual people's attitudes. This chapter deals with institutions and about how to manage groups of humans – who may well have very contrasting views on the same interaction. As such it is a complementary view to the individual focus.

14.2 INSTITUTIONS ARE NEEDED TO UPSCALE ACTIONS

Human–wildlife coexistence implies integrating wildlife into human-dominated landscapes. Although many species of wildlife (but far from all) are surprisingly tolerant of human-modified ecosystems, human tolerance towards wildlife is highly variable and complex (see Chapter 5). For example, in many cases wildlife is managed as a valued resource through both consumptive and non-consumptive means. In other cases, the presence of wildlife can often be the source of a wide range of conflicts between humans and wildlife through livestock depredation, damage to crops and forests, and through vehicle collisions. In many cases, different groups of people may well view the presence of wildlife as both a resource and a conflict in the same location, resulting in social conflicts between different groups of people about how wildlife should be managed (Redpath et al. 2013). The literature contains a massive number of case studies of conservation projects that have tested multiple interventions, including technical and procedural types, which have been used to try and reduce conflict, ensure that exploitation is sustainable and facilitate coexistence. However, the vast majority of these examples are rather localized.

To achieve large-scale coexistence there is a need to up-scale these interventions to cover the entire areas of regions, countries or continents where species occurs. Up-scaling is also necessary to mainstream wildlife interests into all administrative and political sectors (such as agriculture, environment, forestry, transport, energy, etc.) and to ensure legitimacy for conservation mandates. This up-scaling requires the institutionalization of interventions (Carter & Linnell 2016) and the effective mainstreaming of wildlife interests across multiple policy sectors (horizontal integration) and across administrative scales (vertical integration) (Carter & Linnell 2016; Karlsson-Vinkhuyzen et al. 2017). Institutions are diverse, and can broadly be categorized as formal (e.g. agencies, ministries, laws) or informal (e.g. established social norms and traditional practices). An important trend in the development of natural resource institutions in recent years is the movement towards more participatory forms of governance (Redpath et al. 2017). This means that informal networks of stakeholders also play an increasing role in management and policy development. Moreover, government-based wildlife management agencies are constantly challenged to deliver results that are acceptable and satisfactory from a public and not purely scientific viewpoint. Formal wildlife institutions often struggle to comprehend that wildlife values are essentially social values; hence, they often do not deliver the results that the public expects.

Coexistence-relevant research has long had an interdisciplinary nature. However, while there has been a great deal of focus on the relationships between humans and wildlife (ecology), the attitudes of individuals (human-dimensions) and the relationships between different groups of humans as individuals or groups (sociology, anthropology), the study of the institutions that have been created to manage these relationships lags far behind (political science or public administration studies). In this chapter, we present a broad description of the structures of formal institutions that are in place to govern wildlife in Europe, and explore how they may influence the ability to achieve coexistence. We focus on the comparison between large herbivore and large carnivore management institutions to explore how species characteristics might influence the outcome. Although these two species groups are intricately linked ecologically (as predators and prey), and despite the obvious potential for comparative analysis, there have been very few such studies (Kaltenborn et al. 2013). While the formal articulation of *coexistence* as a field of study within conservation biology is relatively recent (Carter & Linnell 2016; Frank 2016) the underlying principles have been present in wildlife management institutions for more than a century.

14.3 LARGE CARNIVORES AND LARGE HERBIVORES IN EUROPE

In contrast to many of the tales of decline and extinction associated with wildlife in many parts of the world, the recent history of large herbivores and large carnivores (Table 14.1 in Text Box 14.1) in Europe is mainly one of recovery (Linnell & Zachos 2011; Chapron et al. 2014). By the late nineteenth and early twentieth centuries the populations of both had been decimated across the continent by over-exploitation, deliberate extermination campaigns, competition for forage/prey and massive scale conversion of forests to farmland or pasture land. A lot has changed in the 100–150 years since this low point (e.g. Breitenmoser 1998). Broadly speaking, the first changes occurred in forests. Multiple factors, including the need for strategic timber reserves, an increasing awareness of the need for forests to protect against erosion and avalanches, and a reduction in human grazing pressure (due to emigration and rural–urban migration associated with industrialization) in remoter areas led to an increase in forest cover. This return of wildlife habitat made it possible to begin the recovery of wild herbivores. The early recovery was strongly promoted by hunters through protection of remnant populations and the large-scale reintroduction of new ones (e.g. red deer, roe deer, ibex and chamois), as well as the spreading of several

Text Box 14.1 Pan-European Institutions for Large Mammal Conservation

Europe is a very diverse continent in terms of history, culture, economics, language, climate and topography. However, during recent decades there has been a dramatic attempt to integrate this diversity into unifying political structures like the Council of Europe and the European Union. There are two major international legal instruments that govern wildlife conservation in Europe. The Council of Europe (with forty-seven member countries) manages the Bern Convention (www.coe.int/en/web/bern-convention) and the European Commission (with twenty-eight member countries) manages the Habitats Directive (http://ec.europa.eu/environment/nature/legislation/habitatsdirective/index_en.htm) (Trouwborst 2015). The Bern Convention (1979) predates the Habitats Directive (1992), although different countries have entered them at different times since their development. All members of the European Union are also signatories to the Bern Convention such that they are spatially, and legally, nested (i.e. the Habitats Directive must be compatible with the Bern Convention, but not necessarily vice versa). One key feature is that the legislation applies to the entire European landscape, including both private and public lands, and protected and non-protected areas. Both instruments contain a set of appendices or annexes that list species under different protection regimes. Under the Bern Convention appendix II refers to *strictly protected fauna species* where deliberate killing is prohibited, while appendix III refers to *protected fauna species* where regulated killing is allowed. Under the Habitats Directive annex II refers to an obligation to include core areas of habitat within the Natura 2000 protected area network, annex IV implies a strict protection regime where killing can only be done under certain conditions and annex V where exploitation is possible as long as it is regulated so as to not endanger its *favourable conservation status*. Table 14.1 lists Europe's large herbivores and large carnivores, indicating whether they are native, naturalized or exotic, and their status under Pan-European wildlife legislation. Table 14.1 clearly shows the differences between how large carnivores and large herbivores are represented within European-level legislation.

Table 14.1 *Europe's large herbivores and large carnivores' origin and protection status*

Species	Origin	Bern Con.	Habitats Dir.
Large carnivores			
Wolf (*Canis lupus*)	Native	II & III[1]	II & IV (V)[6, 7]
Brown bear (*Ursus arctos*)	Native	II & III[2]	II & IV[6]
Eurasian lynx (*Lynx lynx*)	Native	III	II & IV (V) [6, 8]
Wolverine (*Gulo gulo*)	Native	II	II

(*continued*)

Table 14.1 (*continued*)

Species	Origin	Bern Con.	Habitats Dir.
Large herbivores			
Red deer (*Cervus elaphus*)	Native	II & III$_3$	II & IV$_9$
Sika deer (*Cervus nippon*)	Exotic		
Roe deer (*Capreolus capreolus*)	Native	III	
Moose (*Alces alces*)	Native	III	
Wild reindeer (*Rangifer tarandus*)	Native	III	II$_{15}$
Fallow deer (*Dama dama*)	Naturalized		
Muntjac (*Muntiacus reevesi*)	Exotic		
Chinese water deer (*Hydropotes inermis*)	Exotic		
White-tailed deer (*Odocoileus virgianus*)	Exotic		
Alpine chamois (*Rupicapra rupicapra*)	Native	III	II & IV (V)$_{10}$
Isard (*Rupicapra pyrenaica*)	Native	II & III$_4$	II & IV (V)$_{11}$
Alpine ibex (*Capra ibex*)	Native	III	V
Spanish ibex (*Capra pyrenaica*)	Native	III	II & IV (V)$_{12}$
Wild goat (*Capra aegagrus*)	Native	II	II & IV$_{13}$
Mouflon (*Ovis ammon*)	Naturalized/ Exotic	III	II & IV$_{14}$
Barbary sheep (*Ammotragus lervia*)	Exotic		
Wild boar (*Sus scrofa*)	Native	III$_5$	
European bison (*Bos bonasus*)	Native	III	II & IV
Muskox (*Ovibos moschatus*)	Exotic	II	

Notes:

1. Several countries have made reservations from the default position of the legislation for wolves (Bulgaria, Finland, Czech Republic, Latvia, Poland, Slovakia, Slovenia, Macedonia), or have chosen to treat them as appendix III in all or part of their national distributions (Lithuania, Spain).
2. Several countries have either made an exception for bears (Czech Republic, Finland, Slovakia, Slovenia), or have chosen to treat them as appendix III in all or part of their national distributions (Croatia).
3. The Corsican subspecies of red deer (*Cervus elaphus corsicanus*) is on appendix II; all others are on appendix III.
4. The Apennine subspecies (*Ruipcapra pyrenaica ornata*) is on appendix II; all others on appendix III.
5. The subspecies *Sus scrofa meridionalis* on Corsica and Sardinia is on appendix III; all others are not listed.

6. Some countries have made reservations for annex II for lynx (Estonia, Finland, Latvia), wolves (Estonia, Finland, Latvia) and for bears (Estonia, Finland, Sweden).
7. Some countries (Bulgaria, Estonia, Finland, Greece, Latvia, Poland, Slovakia, Spain) manage wolves as annex V instead of IV in all or part of their national distribution.
8. Estonia manages lynx as annex V instead of annex IV.
9. *Cervus elaphus corsicanus* only.
10. *Rupicapra rupicapra balcanica* and *R. r. tatrica* only on annex II and IV, all others only on annex V.
11. *Rupicapra pyrenaica oranata* only on annex II and IV, all others only on annex V.
12. *Capra pyrenaica pyrenaica* on annex II and IV, all others on annex V only.
13. Only so-called 'natural populations'.
14. Only so-called 'natural populations – Corsica and Sardinia' of *Ovis gmelini musimon* and *Ovis gmelini ophion* on Cyprus.
15. *Rangifer tarandus fennicus* (Finland) only.

In addition to these legal structures there are a number of other institutions that operate at a European level. For, example, the Large Carnivore Initiative for Europe (LCIE) was started as an advisory group of experts by the World Wide Fund for Nature (WWF) in 1995 (www.lcie.org). Since then the group has grown and developed into a Specialist Group within the International Union for Conservation of Nature (IUCN)'s Species Survival Commission. During its lifetime the LCIE has worked to network large carnivore experts (researchers, non-governmental organizations (NGOs), wildlife managers) so that experience and best practices can be developed and communicated, helping to bridge the science–policy interface. The LCIE has worked closely with both the Bern Convention secretariat and the European Commission as well as multiple national and regional agencies, helping to develop action plans and policy guidelines and facilitate stakeholder processes. As well as providing technical advice, the LCIE advocates a vision of large carnivore conservation that is firmly based around coexistence as a model, where the focus is to be strong on ambition concerning the extent to which large carnivores can be reintegrated into the European landscape, but very pragmatic about the means adopted to achieve this goal. At present, the LCIE is clearly the most visible competence hub for large carnivores in Europe. There is no comparable network for large herbivore management or conservation.

exotic species (e.g. mouflon and sika deer). By the mid-twentieth century these processes had led to dramatic increases in both forest and potential prey for large carnivores. In some areas, large carnivores (especially brown bears) benefited from active conservation efforts in the early twentieth century when hunting regulations were introduced and from a relaxation in predator control during the social chaos of the World Wars. However, it was not until the late 1960s and 1970s that large carnivores received widespread protection from over-exploitation and extermination campaigns. The first reintroductions of Eurasian lynx and brown bears began in the 1970s, and have continued up until the present day (Clark et al. 2002; Linnell et al. 2009). Although some populations are still highly threatened (e.g. small bear populations in southern Europe), the present distributions and numbers of both large herbivores and large carnivores in Europe are probably wider and larger than at any other point in many centuries (Linnell & Zachos 2011; Chapron et al. 2014).

Europe is currently a very crowded continent, with a high human density, and high proportions of heavily modified habitats. Wilderness only remains in small and isolated patches, fragmented by a massive network of transport infrastructure, urban development and intensive agriculture. Although Europe has the world's highest number and density of protected areas, most of these are small, and are also heavily used by humans for multiple purposes (European Environmental Agency (EEA) 2012; Tsiafouli et al. 2013; Orlikowska et al. 2016). The result is that Europe's wildlife currently exists predominantly in highly modified human-dominated landscapes. When viewed solely in terms of wildlife presence in anthropogenic landscapes this would appear to be a dramatic demonstration of ecological coexistence (i.e. wildlife has shown an ability to adapt to humans; sensu Oriol-Cotterill et al. 2015). However, one does not need to dig deeply to find controversy and conflict (i.e. the limits of human tolerance for wildlife, or of other stakeholder groups with different views on wildlife management).

14.4 CONFLICTS ASSOCIATED WITH LARGE CARNIVORES AND LARGE HERBIVORES

Throughout their range, both species groups are associated with a diversity of conflicts with human interests. Those associated with large herbivores tend to be mainly of a material and economic nature. These include damage to agricultural crops and commercially important forestry, their role as vectors of various wildlife and zoonotic diseases, and collisions with vehicles (road and rail) (Langbein et al. 2011; Czanyi et al. 2014). Furthermore, high densities of large herbivores can have negative impacts on the conservation of some plant and insect species and habitats (Reimoser & Putman 2011). Large carnivores are also associated with various economic conflicts through their depredation on domestic livestock and pets, the potential competition with hunters for shared quarry and the occasional attacks on humans. Historically, large carnivores are associated with a deep hostility in many cultures (Boitani 1995). In addition, they are associated with a wide range of psychological (such as fear), social, cultural (identity-based) and political conflicts linked to disagreements between different publics over their management (Redpath et al. 2013). The intensity and manner in which conflicts are expressed varies dramatically between species and between regions, from very low to major issues that can dominate national-level politics (Skogen et al. 2017).

Despite this variation, as a general pattern, the economic impact of large herbivores is far higher than that of large carnivores and is distributed across a wider proportion of the European landscape. Large herbivores are associated with more human mortalities (mainly through vehicle collisions) and morbidity (as disease vectors) than large carnivores. Despite this discrepancy, there is no doubt that large carnivores are far more controversial than large herbivores, with controversies over carnivore management being conducted very openly in public and political arenas at local, national and international levels. These controversies are often very polarized (various groups calling for either total eradication versus full protection). For example, in Scandinavia and Germany there are several NGOs that are campaigning to have wolves exterminated following their recent recolonization. There is no doubt that herbivore-associated conflicts can be very intense too, but they seem to remain at local levels and without the same degree of polarization. Typical herbivore conflicts will concern discussions about the desired size of local populations as landowners with different interests seek to balance hunting opportunities with damage to forest and crops, and vehicle collisions. The question that we will now explore is: to what extent are the differences in management controversies between the species groups explained by the differences in institutional arrangements?

14.5 SOME BASIC ECOLOGICAL DIFFERENCES BETWEEN LARGE HERBIVORES AND LARGE CARNIVORES

The basic ecological difference between these two species groups is in the trophic level that they occupy. All the species in Table 14.1 can be unambiguously assigned to a trophic level with the exception of brown bears that are omnivorous. Surprisingly, the fact that the different species groups occupy different trophic levels does not influence the extent of economic and material conflicts that they cause. Herbivore consumption of crops and trees can be just as economically serious for humans as carnivore depredation on livestock, and as previously stated, herbivores kill and injure far more people through vehicle collisions and as disease vectors than carnivores through direct attacks.

However, the different trophic levels they occupy does influence one fundamental property of great relevance for their management: the issue of scale. Individual large herbivores typically occupy home ranges that are measured in hectares or a few square kilometres, although in higher latitudes and altitudes their seasonal ranges can be separated by migration routes of many tens of kilometres (Cagnacci et al. 2016). The exception concerns wild reindeer that can roam over ranges of hundreds, or even thousands, of square kilometres in more or less seasonally predictable fashions (Cagnacci et al. 2016). On the other hand, all large carnivores roam year-round over massive individual home ranges, varying between $100 \, km^2$ and $3{,}000 \, km^2$ (Duncan et al. 2015). The relevance of these spatial differences is further confounded by differences in social organization. Most large carnivores are either solitary, territorial or both, whereas most herbivores demonstrate some degree of sociality, or at least tolerance for the presence of conspecifics in shared ranges. The combination of these two factors (home range size and social system) leads to dramatic differences in population densities. Herbivores typically occur at densities ranging from 1 to 40 per km^2, whereas common large carnivore densities are in the range of 0.2 to 2 per $100 \, km^2$, almost three orders of magnitude lower. The implication is that herbivore populations typically operate on scales best measured in tens or hundreds of km^2, whereas carnivore populations operate on scales of thousands, tens of thousands or even hundreds of thousands of square kilometres (Linnell & Boitani 2012). These ecological differences have profound effects on the definition of appropriate management scales (Linnell 2015).

14.6 INSTITUTIONAL ARRANGEMENTS FOR MANAGING HERBIVORES AND CARNIVORES

Europe's wildlife management is exceptionally complex from an institutional perspective. On a continental scale, there are two relevant Pan-

European legislative instruments, the Bern Convention and the Habitats Directive (see Text Box 14.1). The national laws and institutions of the respective signatory/member states must fall within these frames. In countries with federal structures or high degrees of regional devolution/ autonomy (e.g. Spain, Switzerland, Germany, Austria, Italy, Bosnia and Herzegovina) most authority for wildlife management issues is further delegated to each of the federal states, regions, provinces or cantons. Within each of these national (or subnational) authorities are a wide range of more local-level institutions from counties to municipalities and down to individual landowners. Legally speaking, these institutions have to consist of nested hierarchies of authority, with the lowest levels limited in their discretionary freedom by the higher-level constraints (Linnell 2015).

The extent to which the Pan-European bodies are involved in conservation varies widely between species groups. Large herbivores are generally subject to less stringent European-level regulation (Table 14.1). Large carnivores are generally listed as species that require *strict protection* across Europe (although some countries have taken out total or partial reservations for wolves). In contrast, most wild herbivores are either not specifically listed, or are included on lists with lower degrees of protection that permit regulated exploitation. The exceptions are endangered species like European bison, single subspecies of red deer and populations of ibex or chamois with limited distributions, which are listed and afforded greater protection. In sum, this implies that most herbivore populations are only covered by the most general restrictions at the European level. In contrast, large carnivores are generally covered by the strictest levels of protection offered by both instruments. These designations impose very tight restrictions on how national- and subnational- level authorities manage large carnivores. Accordingly, most authority for large carnivore management is made at a national level in member states. Various degrees of decision-making power are then delegated to lower levels in some countries, but their discretionary powers are greatly limited by the frames imposed by international and national authorities.

In recognition of the massive scales at which large carnivores operate, recent years have seen considerable effort invested in trying to increase the degree of transboundary cooperation in large carnivore management (Boitani & Ciucci 2009; Linnell & Boitani 2012). The main progress has come in the form of technical cooperation between wildlife management agencies, NGOs, some local protected areas that span international borders, and expert networks (see Text Box 14.1 for an example). Examples include the adoption of standardized monitoring activities, joint research projects and development of best-practice guidelines. Although the European Commission has developed guidelines for transboundary management planning (Linnell et al. 2008) there has been very slow uptake on the political level. Perhaps the greatest advances have occurred with the frames of a more limited international instrument (covering eight countries), the Alpine Convention (www .alpconv.org) which is making concrete advances in coordinating wolf management at a political level.

The situation for large herbivores is very different. The general lack of major international constraints on their management has allowed the continual development of institutions where a great deal of authority is delegated to local levels. National (or federal state) legislation and authorities provide some oversight and technical guidance, but a great deal of authority lies at county or municipal levels, and even at the level of individual landowners or hunting clubs. There is a bewildering array of different structures and systems in place across Europe for large herbivore management. Putman (2011) describes five main management systems that operate, but each of these exists within many forms to the extent that many countries can have many different management forms in place within different regions, or for different species, or both (see the chapters in Apollonio et al. (2010) for more details). Despite these variations on the theme, they all consist of well-established structures for regulation, planning, decision-making, dialogue, action, reporting and enforcement; almost always within the frames of game (hunting) management structures that are typically placed at local scales and involve a high degree of local stakeholder representation.

Large herbivores and large carnivores differ dramatically in the way they are managed. Large herbivores are generally managed as game species across Europe. The few exceptions are some specific populations with poor conservation status and a very small number of administrations that do not permit any recreational hunting (e.g. the Netherlands, the Swiss canton of Genève and a few protected areas). Although hunting is a multifunctional activity (Fischer et al. 2013) the key point is that it provides multiple tangible and immediate benefits to some of the people being asked to coexist with the species in shared landscapes. These benefits are diverse and include economic benefits (i.e. sales of meat, hunting licences, hunting-based tourism products such as accommodation and spin-offs) as well as recreational, cultural and social benefits associated with hunting (i.e. maintaining tradition, promoting social interactions, physical activity). In rural communities, these non-economic benefits can be crucial for social cohesion and quality of life (Kaltenborn et al. 2017), and represent the continuation of a traditional relationship with the land and wildlife that has been developing for centuries.

Although large herbivores are associated with many costs and conflicts, most of these tend to fall at the same scales as the benefits they provide. In some systems where the farmers and foresters (i.e. the owners of the habitat where conservation is being done) also have the hunting rights, the costs and benefits both fall on the same individuals. In other systems, they tend to fall on neighbours or members of the same community. Across the continent there is a great diversity in locally developed systems to redistribute costs and benefits at this local level: for example, in many countries the hunters pay *ex post facto* compensation (Schwerdtner & Gruberb 2007) to farmers who experience crop damage. In some other countries, the state pays compensation for herbivore damage, whereas in others nobody pays. There are also many institutional systems in place to negotiate between local stakeholders who may have conflicting objectives, for example between hunters, farmers and foresters about desired population sizes. Unfortunately, there has been almost no formal comparative scholarship of these diverse systems, but it is clear that they all have their pros and cons. No management system can totally prevent or avoid all conflict. However, in these herbivore systems the conflicts are generally retained at local levels where it is easier to ensure participation and negotiate workable compromises. The fact that these systems have fostered the recovery and sustainable use of a diverse community of large herbivores across a crowded continent is an indication that there are elements of these systems that work. The greatest tensions appear in systems where the herbivores move over large areas, creating mismatches between social and ecological scales (Sandström et al. 2013), and especially when populations move across administrative borders (Fonseca et al. 2014). Norwegian wild reindeer are a clear example of this situation as there is a clear mismatch between the large scale at which the populations are managed (i.e. harvest quotas) and the more local scales at which land-use planning is decided (Kaltenborn et al. 2014).

Large carnivores were included in similar game management systems in most of Eastern Europe during the second half of the twentieth century under their communist regimes. By the time of the collapse of communism (1989–91) many of these former East Bloc countries were home to most of the large populations of wolves, bears and Eurasian lynx remaining in Europe. Most of the other countries in central and northern Europe either had regimes with strict protection of small remnant populations, or very centrally controlled and limited hunting. Since the end of communism there has been a large degree of harmonization of legislation across Europe due to the expansion of the European Union and the resulting application of the Habitats Directives (Text Box 14.1). This has led to multiple challenges to the management of large carnivores as game species, although the practice persists in some countries (through a system of exceptions from protection called derogations that can be granted under certain conditions). The extent to which this is permitted remains highly contentious (e.g. Linnell et al. 2017), and has been subject to court cases (at the Court of Justice of the European Union) and many national-level proceedings. Compensation for large carnivore damage is now almost always paid for by the state. Most states use an *ex post facto* system, although a few have begun experimenting with incentives, or risk-based payments (Zabel et al. 2014). The consequences of these changes are threefold. First, large carnivores are being essentially removed from the more local wildlife management institutions that have either traditionally managed them or are still used to manage the large herbivores in the same landscapes. Second, the involvement of European-level authorities in national-level proceedings has proven to be intensely controversial with rural stakeholders (Hiedanpää 2013) because they are perceived as being external actors. As such there are currently many active conflicts about both the way large carnivores are managed, and who should make the decisions about their management (Skogen et al. 2017). Third, rural stakeholders have lost, or risk losing, a way to regain some benefits (economic and social) from the presence of large carnivores (Knott et al. 2014). Although there is some potential in large carnivore-based tourism, this will tend to economically favour only some actors, and probably not the ones who lose out from a decline in hunting, and will also probably only be relevant for the more scenic areas, not the whole landscape, making it an incomplete substitute. In places where species populations are viable, and where society supports hunting of those species as traditional practice, a carefully regulated carnivore hunt could be considered. While 'open hunting' is not the answer to conflict resolution for any species, using regulated and institutionalized hunting as a wildlife management tool may provide enough benefit for local residents to be willing to coexist with controversial species populations. (See Linnell et al. 2017 for an exploration of the diverse potential benefits.)

14.7 SUMMARY OF DIFFERENCES BETWEEN HERBIVORE AND CARNIVORE INSTITUTIONS

The benefits arising from large herbivores (viewing experiences, meat, hunting experiences, economic benefits from sales of hunting licences) are far more immediate, tangible and evenly spread than those arising from large carnivores (issues of intrinsic value, diffuse ecosystem services, limited and localized tourism). The benefits also tend to fall on a more local level, such that the balance between costs and benefits is much more internalized within the context of local communities. The costs from large herbivores are also less *bloody* or *visceral* than those caused by large carnivores (e.g. a dead sheep triggers a greater reaction than a browsed tree).

Decision-making for large herbivores is subject to fewer external restrictions and is conducted at a more local level, which permits both a greater degree of involvement of local stakeholders who are sharing their landscape with the species, and allows freedom to adopt locally preferred, or traditionally established, management models. Although there can be many complexities involved in managing conflicts at this level (Sandström et al. 2013; Storie & Bell 2017) the social scale is on a level where social networking theory predicts greater chances for success (Sandström and Lundmark 2016) and minimizes the risks of an organized, large-scale anti-conservation rural backlash (Skogen & Krange 2003; Manfredo et al. 2017).

Large carnivore management is heavily (and increasingly) constrained by external authority, and decision-making is usually taken at large-scale levels that are further away from the stakeholders on the ground who live in proximity to the species. However, completely moving carnivore management to a local scale like that used for herbivores is not an option as it is incompatible with large carnivore ecology. This is perhaps the central dilemma of modern-day large carnivore management (Linnell 2015). How can we design institutions that simultaneously give a voice to local communities who carry the costs associated with these species presence (Redpath et al. 2017) while taking into account the massive spatial scales at which large carnivore populations live (Linnell & Boitani 2012) and the need for technical support and value pluralism that comes from larger spatial scales? Clearly, no single institution can manage this task, so the answer by necessity will involve some form of multi-level, nested, hierarchy of institutions with two main challenges. One is to establish cross-sectoral legislation that enables large-scale cooperation between agencies and policy coordination. The other is to ensure effective communication between key stakeholders at a local level and the larger policy levels at regional, national and international scales (Linnell 2015).

There has been a great deal of investment in the Nordic countries to test out various forms of collaborative or participatory management for both large herbivores (Sandström et al. 2013; Kaltenborn et al. 2014, 2015) and large carnivores (Sandström et al. 2009). Although the models have varied in the different countries the essential principle has been to create arenas where diverse stakeholders and/or decision-makers can interact, exchange experience and either produce formal advice or make decisions within frames provided by higher authorities. The various models applied to wolves, bears, Eurasian lynx and wolverines in Fennoscandia have received a good deal of research attention (e.g. Sandström et al. 2009, 2016; Pellikka & Sandström 2011; Lundmark et al. 2014; Lundmark & Matti 2015; Krange et al. 2016). These evaluations have all identified some benefits and some shortcomings of the different systems, but scholarship has not yet come far enough to conclude about how to best design a multilevel, nested system. However, all the studies underline the complexity of designing institutions to manage species groups as challenging as large carnivores (and highly mobile herbivores). It is very unlikely that institutions that work for species like roe deer or red deer will work for wild reindeer or large carnivores. This implies that there is a clear need to manage expectations concerning the form of *coexistence*, which can be achieved with the different species (Carter & Linnell 2016). It is inevitable that some social conflicts will continue. The goal is to contain them within acceptable levels using procedures that the public can accept as being fair and just (Jacobsen & Linnell 2016).

The various institutional frameworks that exist across Europe have generally succeeded in fostering the recovery and sustainable use of both large herbivores and large carnivores during the twentieth century, each in their own way. There are many remaining questions about the details of how many species are managed, for example through concerns about the long-term impacts of selective hunting (Mysterud 2010; Bischof et al. 2018). However, the bottom line is visible in the wide distributions of these species in the European landscape which contrasts dramatically with the fate of large mammals in Africa, Asia and South America. Typically, these European institutions have operated on one species at a time, worked at relatively local scales and dealt mainly with a limited number of stakeholders (hunters, farmers, foresters). The question is, will these institutions operate well in a changing world that requires more integration, an ecosystem approach and consideration of a wider selection of stakeholder interests?

14.8 ADAPTING TO THE FUTURE

At the heart of social conflicts and political controversies are groups of people with different sets of norms and values, which will essentially lead to differences in opinions about how species and landscapes should be managed, i.e. what are considered appropriate uses of nature and wildlife (Chapter 1). Public values associated with wildlife have been changing dramatically in recent decades (Boitani 1995; Sykes & Putman 2014; Manfredo et al. 2017). These changes are most apparent for the case of large carnivores, where policies swung from extermination to conservation literally overnight in many countries during the latter half of the twentieth century. The late twentieth century was dominated by a conservation discourse that was concerned with issues such as population viability and sustainability. At present the discourses have fragmented into multiple threads (Mace 2014). One is very much focused on conserving or restoring the ecosystem functions of carnivores and herbivores which is heavily anchored in potentially unrealistic expectations of applying hands-off wilderness management ideals to human-dominated multi-use landscapes (Kuijper et al. 2016; Allen et al. 2017). Another seems to be far more concerned with the principle of if it is morally correct to manage large carnivores as game rather than concerns over whether current harvests are sustainable and populations viable or not (Fischer et al. 2013). There is also an emerging rural backlash against carnivore conservation fuelled at least in part by conflicts (Skogen & Krange 2003; Manfredo et al. 2017; Skogen et al. 2017). In many ways, this reflects the changing relationship that society has with both wildlife and traditional rural activities like hunting.

The controversies that we currently see with carnivores may be a taste of what is to come for herbivores. Herbivores are not immune to controversy about hunting, and many areas are seeing conflicts emerge over certain hunting methods. Furthermore, intensive management of herbivore populations by hunters, such as using supplementary feeding

to increase population density, is fuelling both perceptional and economic conflicts associated with herbivores. The increasing need to adopt
a more ecosystem-centric approach (rather than the previous species-
centric approaches) is exposing the challenges associated with
broadening the deliberative base and trying to open for greater value
plurality (Sandström et al. 2013), even for species like moose that are
essentially uncontroversial per se.

There is a wide range of other issues that are emerging which will
also challenge wildlife management (Apollonio et al. 2017). Climate
change, changes to agricultural and forestry policies and the steady
fragmentation of habitats by transport, energy, residential and recreational infrastructure will alter the ecology of the species. The expansion
of human infrastructure is especially challenging for highly mobile
species (large carnivores and wild reindeer) or migratory species of
wildlife (like red deer and reindeer) as it involves both integrating
multiple sectorial interests in considering land-use planning and areas
that exceed the conventional population management units (Kaltenborn
et al. 2014). However, societal change brings some opportunities as well
as challenges. There is currently a rapidly expanding market for non-
consumptive uses of wildlife as a tourist and recreational product,
although it is uncertain how these benefits compare in magnitude and
distribution (between areas and actors) as compared to more traditional
consumptive use.

Overall, we are in a time with many contradictions. While legislation
like the Convention on Biological Diversity and the Ecosystem Services
approach gain traction in policy and promote utilitarian approaches to
nature and biodiversity in general, there is clearly a rise in protectionist
ideologies that question the moral basis of hunting charismatic wildlife.
The concept of stakeholders is also changing, with distant publics
(global and national) demanding a say in how local wildlife populations
are managed. Social change, for example through urbanization, is dramatically changing the behaviour and values of all our societies, often in
diverging ways. We are entering a period when an increasingly large,
distant and diverse set of publics are making a greater diversity of
demands on limited resources. There are also increasing demands for
public engagement and participation in democratic processes. Balancing
all these competing views in a manner that secures the future of wildlife
populations is going to challenge all existing wildlife management institutions. There is a dramatic need for wildlife agencies to recognize that
their fundamental task is to manage social values, and deliver results

that fulfil public expectations. Paradoxically we are in a period that requires the application of high precision and interdisciplinary knowledge, yet this is now being referred to as the post-fact era, when the role of expert knowledge is being increasingly challenged. Similarly, in a period when we really need a well-functioning political system to address conflicts between different publics, the whole structure of the body politics is coming under threat from the rise of populism.

The future of European wildlife will depend on the continual evolution of effective institutions. At present, we do not have all the answers. Luckily, when looking at how herbivores and carnivores are managed across Europe we can find a huge variation in institutional structures that have been operating for decades. This variation provides an incredible opportunity to conduct comparative studies from which we can hope to learn enough about what works and what does not under different conditions so that we can design effective, legitimate and adaptive institutions to ensure that Europe's wildlife can continue to thrive through the twenty-first century.

14.9 RECOMMENDATIONS AND FUTURE DIRECTIONS

- There have been few systematic and comparative studies of the pros and cons of the different management structures that are currently being used across Europe. Their diversity represents a unique starting point for comparative research about the functionality of different institutional arrangements.
- We are in an era of dramatic social, political and ecological change. There is a need to better understand which institutional structures are best adapted to respond to change. If coexistence is to be achieved at large scales it is essential that the appropriate measures are institutionalized into formal structures. Nothing else can achieve the needed legitimacy, cross-sectorial coordination and widespread adoption. In other words, coexistence has to be mainstreamed because the conservation of these species requires utilizing the entire landscape.
- There is no optimal scale for such institutions. Different decision-making responsibilities and operational activities will have to lie at different scales, ranging from the continental to the local. The challenges lie in (1) getting the balance right between freedom and constraints at each level, and (2) ensuring effective communication between scales. Broadly speaking, as much freedom as possible needs

to be given to lower levels to find their own approaches as long as this does not harm larger-scale goals.

- This need to balance freedom and frames is a universal need in wildlife conservation, covering both subnational and international jurisdictions.
- Different institutions may well work under different social and political circumstances, so that there needs to be an acceptance of a diversity of approaches.

14.10 ACKNOWLEDGEMENTS

This work was funded by the Research Council of Norway (grant 251112).

14.11 References

Allen, B. L., Allen, L. R., Andrén, H., Ballard, G., Boitani, L., Engem, R. M., Flemming, P. J. S., Ford, A. T., Haswell, P. M., Kowalczyk, R., Linnell, J. D. C., Mech, L. D. & Parker, D. M. (2017). Can we save large carnivores without losing large carnivore science? *Food Webs*, 12, 64–75.

Apollonio, M., Andersen, R. & Putman, R. (2010). *European Ungulates and Their Management in the 21st Century*. Cambridge: Cambridge University Press.

Apollonio, M., Belkin, V. V., Borkowski, J., Borodin, O. I., Borowik, T., Cagnacci, F., Danilkin, A. A., Danilov, P., Faybich, A., Ferretti, F., Gaillard, J. M., Hayward, M. W., Heshtaut, P., Heurich, M., Hurynovich, A., Kashtalyan, A., Kerley, G. I. H., Kjellander, P., Kowalczyk, R., Kozorez, A., Matveytchuk, S., Milner, J. M., Mysterud, A., Ozolins, J., Panchenko, D. V., Peters, W., Podgorski, T., Pokorny, B., Rolandsen, C., Ruusila, V., Schmidt, K., Sipko, T. P., Veeroja, R., Velihurau, P. & Yanuta, G. (2017). Challenges and science-based implications for modern management and conservation of European ungulate populations. *Mammal Research*, 62(3), 209–17.

Bischof, R., Bonenfant, C., Rivrud, I. M., Zedrosser, A., Friebe, A., Coulson, T., Mysterud, A. & Swenson, J. E. (2018). Regulated hunting re-shapes the life-history of brown bears. *Nature Ecology & Evolution*, 2, 116–23.

Boitani, L. (1995). Ecological and cultural diversities in the evolution of wolf human relationships. In L. N. Carbyn, S. H. Fritts & D. R. Seip, eds., *Ecology and Conservation of Wolves in a Changing World*. Alberta: Canadian Circumpolar Institute, pp. 3–12.

Boitani, L. & Ciucci, P. (2009). Wolf management across Europe: Species conservation without boundaries. In M. Musiani, L. Boitani & P. C. Paquet, eds., *A New Era for Wolves and People: Wolf Recovery, Human Attitudes, and Policy*. Calgary: University of Calgary Press, pp. 15–40.

Breitenmoser, U. (1998). Large predators in the Alps: The fall and rise of man's competitors. *Biological Conservation*, 83, 279–89.

Cagnacci, F., Focardi, S., Ghisla, A., van Moorter, B., Merrill, E. H., Gurarie, E., Heurich, M., Mysterud, A., Linnell, J., Panzacchi, M., May, R., Nygard, T., Rolandsen, C. & Hebblewhite, M. (2016). How many routes lead to migration? Comparison of methods to assess and characterize migratory movements. *Journal of Animal Ecology*, 85, 54–68.

Carter, N. H. & Linnell, J. D. C. (2016). Co-adaptation is key to coexisting with large carnivores. *Trends in Ecology & Evolution*, 31, 575–8.

Chapron, G., Kaczensky, P., Linnell, J. D. C., von Arx, M., Huber, D., Andrén, H., López-Bao, J. V., Adamec, M., Álvares, F., Anders, O., Balčiauskas, L., Balys, V., Bedő, P., Bego, F., Blanco, J. C., Breitenmoser, U., Brøseth, H., Bufka, L., Bunikyte, R., Ciucci, P., Dutsov, A., Engleder, T., Fuxjäger, C., Groff, C., Holmala, K., Hoxha, B., Iliopoulos, Y., Ionescu, O., Jeremić, J., Jerina, K., Kluth, G., Knauer, F., Kojola, I., Kos, I., Krofel, M., Kubala, J., Kunovac, S., Kusak, J., Kutal, M., Liberg, O., Majić, A., Männil, P., Manz, R., Marboutin, E., Marucco, F., Melovski, D., Mersini, K., Mertzanis, Y., Mysłajek, R. W., Nowak, S., Odden, J., Ozolins, J., Palomero, G., Paunović, M., Persson, J., Potočnik, H., Quenette, P.-Y., Rauer, G., Reinhardt, I., Rigg, R., Ryser, A., Salvatori, V., Skrbinšek, T., Stojanov, A., Swenson, J. E., Szemethy, L., Trajçe, A., Tsingarska-Sedefcheva, E., Váňa, M., Veeroja, R., Wabakken, P., Wölfl, M., Wölfl, S., Zimmermann, F., Zlatanova, D. & Boitani, L. (2014). Recovery of large carnivores in Europe's modern human-dominated landscapes. *Science*, 346, 1517–19.

Clark, J. D., Huber, D. & Servheen, C. (2002). Bear reintroductions: Lessons and challenges. *Ursus*, 13, 335–46.

Czanyi, S., Carranza, J., Pokorny, B., Putman, R. & Ryan, M. (2014). Valuing ungulates in Europe. In R. Putman & M. Apollonio, eds., *Behaviour and Management of European Ungulates*. Dunbeath: Whittles Publishing, pp. 13–45.

Duncan, C., Nilsen, E. B., Linnell, J. D. C. & Pettorelli, N. (2015). Life history attributes and resource dynamics determine intraspecific home range sizes in Carnivora. *Remote Sensing in Ecology & Conservation*, 1, 39–50.

European Environmental Agency (EEA). (2012). *Protected Areas in Europe – An Overview*. European Environmental Agency Report 5/2012, Copenhagen.

Fischer, A., Kerezi, V., Arroyo, B., Mateos-Delibes, M., Tadie, D., Lowassa, A., Krange, O. & Skogen, K. (2013). (De)legitimising hunting: Discourses over the morality of hunting in Europe and eastern Africa. *Land Use Policy*, 32, 261–70.

Fischer, A., Sandström, C., Delibes-Mateos, M., Arroyo, B., Tadie, D., Randall, D., Hailu, F., Lowassa, A., Msuha, M., Kerezi, V., Reljic, S., Linnell, J. & Majić, A. (2013). On the multifunctionality of hunting: An institutional analysis of eight cases from Europe and Africa. *Journal of Environmental Planning & Management*, 56, 531–52.

Fonseca, C., Torres, R. T., Santos, J. P. V., Vingada, J. & Apollonio, M. (2014). Challenges in the management of cross-border populations of ungulates. In R. Putman & M. Apollonio, eds., *Behaviour and Management of European Ungulates*. Dunbeath: Whittles Publishing, pp. 192–207.

Frank, B. (2016). Human–wildlife conflicts and the need to include tolerance and coexistence: An introductory comment. *Society & Natural Resources*, 29, 738–43.

Hiedanpää, J. (2013). Institutional misfits: Law and habits in Finnish wolf policy. *Ecology & Society*, 18(1), art. 24.

Jacobsen, K. S. & Linnell, J. D. C. (2016). Perceptions of environmental justice and the conflict surrounding large carnivore management in Norway: Implications for conflict management. *Biological Conservation*, 203, 197–206.

Kaltenborn, B. P., Andersen, O. & Gundersen, V. (2014). The role of wild reindeer as a flagship species in new management models in Norway. *Norsk Geografisk Tidsskrift–Norwegian Journal of Geography*, 68, 168–77.

Kaltenborn, B. P., Andersen, O. & Linnell, J. D. C. (2013). Is hunting large carnivores different from hunting ungulates? Some judgments made by Norwegian hunters. *Journal for Nature Conservation*, 21, 326–33.

Kaltenborn, B. P., Hongslo, E., Gundersen, V. & Andersen, O. (2015). Public perceptions of planning objectives for regional level management of wild reindeer in Norway. *Journal of Environmental Planning & Management*, 58, 819–36.

Kaltenborn, B., Mehmetoglu, M. & Gundersen, V. (2017). Linking social values of wild reindeer to planning and management options in southern Norway. *Arctic*, 70, 129–40.

Karlsson-Vinkhuyzen, S., Kok, M. T. J., Visseren-Hamakers, I. J. & Termeer, C. (2017). Mainstreaming biodiversity in economic sectors: An analytical framework. *Biological Conservation*, 210, 145–56.

Knott, E. J., Bunnefeld, N., Huber, D., Reljic, S., Kerezi, V. & Milner-Gulland, E. J. (2014). The potential impacts of changes in bear hunting policy for hunting organisations in Croatia. *European Journal of Wildlife Research*, 60, 85–97.

Krange, O., Odden, J., Skogen, K., Linnell, J. D. C., Stokland, H. B., Vang, S. & Mattisson, J. (2016). Evalueringav regional rovviltforvaltning. *NINA Rapport*, 1268, 1.

Kuijper, D. P. J., Sahlen, E., Elmhagen, B., Chamaille-Jammes, S., Sand, H., Lone, K. & Cromsigt, J. (2016). Paws without claws? Ecological effects of large carnivores in anthropogenic landscapes. *Proceedings of the Royal Society B*, 283(1841), 20161625.

Langbein, J., Putman, R. & Pokorny, B. (2011). Traffic collisions involving deer and other ungulates in Europe and available measures for mitigation. In R. Putman, M. Apollonio & R. Andersen, R., eds., *Ungulate Management in Europe: Problems and Practices*. Cambridge: Cambridge University Press, pp. 215–59.

Linnell, J. D. C. (2015). Defining scales for managing biodiversity and natural resources in the face of conflicts. In S. M. Redpath, R. J. Guitiérrez, K. A. Wood & J. C. Young, eds., *Conflicts in Conservation: Navigating towards Solutions*. Cambridge: Cambridge University Press, pp. 208–18.

Linnell, J. D. C. & Boitani, L. (2012). Building biological realism into wolf management policy: The development of the population approach in Europe. *Hystrix*, 23, 80–91.

Linnell, J. D. C., Breitenmoser, U., Breitenmoser-Würsten, C., Odden, J. & von Arx, M. (2009). Recovery of Eurasian lynx in Europe: What part has reintroduction played? In M. W. Hayward & M. J. Somers, eds., *Reintroduction of Top-Order Predators*. Oxford: Wiley-Blackwell, pp. 72–91.

Linnell, J. D. C., Kaczensky, P., Wotschikowsky, U., Lescureux, N. & Boitani, L. (2015). Framing the relationship between people and nature in the context of European nature conservation. *Conservation Biology*, 29, 978–85.

Linnell, J. D. C., Salvatori, V. & Boitani, L. (2008). *Guidelines for Population Level Management Plans for Large Carnivores in Europe.* A Large Carnivore Initiative for Europe report prepared for the European Commission (contract 070501/2005/424162/MAR/B2).

Linnell, J. D. C., Trouwborst, A. & Fleurke, F. M. (2017). When is it acceptable to kill a strictly protected carnivore? Exploring the legal constraints on wildlife management within Europe's Bern Convention. *Nature Conservation*, 21, 129–57.

Linnell, J. D. C. & Zachos, F. E. (2011). Status and distribution patterns of European ungulates: Genetics, population history and conservation. In R. Putman, M. Apollonio & R. Andersen, eds., *Ungulate Management in Europe: Problems and Practices.* Cambridge: Cambridge University Press, pp. 12–53.

Lundmark, C. & Matti, S. (2015). Exploring the prospects for deliberative practices as a conflict-reducing and legitimacy-enhancing tool: The case of Swedish carnivore management. *Wildlife Biology*, 21, 147–56.

Lundmark, C., Matti, S. & Sandström, A. (2014). Adaptive co-management: How social networks, deliberation and learning affect legitimacy in carnivore management. *European Journal of Wildlife Research*, 60, 637–44.

Mace, G. M. (2014). Whose conservation? *Science*, 345, 1558–60.

Manfredo, M. J., Teel, T. L., Sullivan, L. & Dietsch, A. M. (2017). Values, trust, and cultural backlash in conservation governance: The case of wildlife management in the United States. *Biological Conservation*, 214, 303–11.

Mysterud, A. (2010). Still walking on the wild side? Management actions as steps towards 'semi-domestication' of hunted ungulates. *Journal of Animal Ecology*, 47, 920–5.

Oriol-Cotterill, A., Valeix, M., Frank, L. G., Riginos, C. & Macdonald, D. W. (2015). Landscapes of coexistence for terrestrial carnivores: The ecological consequences of being downgraded from ultimate to penultimate predator by humans. *Oikos*, 124, 1263–73.

Orlikowska, E. H., Roberge, J. M., Blicharska, M. & Mikusinski, G. (2016). Gaps in ecological research on the world's largest internationally coordinated network of protected areas: A review of Natura 2000. *Biological Conservation*, 200, 216–27.

Pellikka, J. & Sandström, C. (2011). The role of large carnivore committees in legitimising large carnivore management in Finland and Sweden. *Environmental Management*, 48, 212–28.

Phalan, B., Onial, M., Balmford, A. & Green, R. E. (2011). Reconciling food production and biodiversity conservation: Land sharing and land sparing compared. *Science*, 333, 1289–91.

Putman, R. (2011). A review of the various legal and administrative systems governing management of large herbivores in Europe. In R. Putman, M Apollonio & R. Andersen, eds., *Ungulate Management in Europe: Problems and Practices.* Cambridge: Cambridge University Press, pp. 54–79.

Redpath, S. M., Linnell, J. D. C., Festa-Bianchet, M., Boitani, L., Bunnefeld, N., Dickman, A., Gutierrez, R. J., Irvine, R. J., Johansson, M., Majic, A.,

McMahon, B. J., Pooley, S., Sandstrom, C., Sjolander-Lindqvist, A., Skogen, K., Swenson, J. E., Trouwborst, A., Young, J. & Milner-Gulland, E. J. (2017). Don't forget to look down: Collaborative approaches to predator conservation. *Biological Reviews*, 92, 2157–63.

Redpath, S. M., Young, J., Evely, A., Adams, W. M., Sutherland, W. J., Whitehouse, A., Amar, A., Lambert, R. A., Linnell, J. D. C., Watt, A. & Gutierrez, R. J. (2013). Understanding and managing conservation conflicts. *Trends in Ecology & Evolution*, 28, 100–9.

Reimoser, F. & Putman, R. (2011). Impacts of wild ungulates on vegetation: Costs and benefits. In R. Putman, M. Apollonio & R. Andersen, eds., *Ungulate Management in Europe: Problems and Practices*. Cambridge: Cambridge University Press, pp. 144–91.

Sandström, A. & Lundmark, C. (2016). Network structure and perceived legitimacy in collaborative wildlife management. *Review of Policy Research*, 33, 442–62.

Sandström, C., DiGasper, S. W. & Ohman, K. (2013). Conflict resolution through ecosystem-based management: The case of Swedish moose management. *International Journal of the Commons*, 7, 549–70.

Sandström, C., Pellikka, J., Ratamäki, O. & Sande, A. (2009). Management of large carnivores in Fennoscandia: New patterns of regional participation. *Human Dimensions of Wildlife*, 14, 37–50.

Schwerdtner, K. & Gruberb, B. (2007). A conceptual framework for damage compensation schemes. *Biological Conservation*, 134, 354–60.

Skogen, K. & Krange, O. (2003). A wolf at the gate: The anti-carnivore alliance and the symbolic construction of community. *Sociologia Ruralis*, 43, 309–25.

Skogen, K., Krange, O. & Figari, H. (2017). *Wolf Conflicts: A Sociological Study*. Oxford: Berghahn Books.

Storie, J. T. & Bell, S. (2017). Wildlife management conflicts in rural communities: A case-study of wild boar (*Sus scrofa*) management in ErgluNovads, Latvia. *Sociologia Ruralis*, 57, 64–86.

Sykes, N. & Putman, R. (2014). Management of ungulates in the 21st century: How far have we come? In R. Putman & M. Apollonio, eds., *Behaviour and Management of European Ungulates*. Dunbeath: Whittles Publishing, pp. 267–89.

Trouwborst, A. (2015). Law and conservation conflicts. In S. M Redpath, R. J. Gutiérrez, K. A. Wood & J. C. Young, eds., *Conflicts in Conservation: Navigating towards Solutions*. Cambridge: Cambridge University Press, pp. 108–18.

Tsiafouli, M. A., Apostolopoulou, E., Mazaris, A. D., Kallimanis, A. S., Drakou, E. G. & Pantis, J. D. (2013). Human activities in Natura 2000 sites: A highly diversified conservation network. *Environmental Management*, 51, 1025–33.

Zabel, A., Bostedt, G. & Engel, S. (2014). Performance payments for groups: The case of carnivore conservation in northern Sweden. *Environmental & Resource Economics*, 59, 613–31.

15

Worldviews and Coexistence with Coyotes

SHELLEY M. ALEXANDER AND DIANNE L. DRAPER

It has been acknowledged that traditional approaches in resolving human–wildlife conflict or achieving coexistence require a new focus (Lorimer 2015; Pooley et al. 2017). Redirecting our inquiry from animals, and especially carnivores, to human behaviour has been recognized for years to be vital for improving our collective understanding of coexistence with wildlife (Linnell et al. 2001; Treves & Bruskotter 2014). Contemporary scientists are now calling for novel intersectional approaches, as well as entire restructuring of our concepts of nature (Lorimer 2015) and methods of knowledge construction about our relationships with animals and conservation (Havorka 2016). More recently, Pooley et al. (2017, p. 514) remind us that 'studies that explore the roles of culture and values in human–wildlife coexistence remain rare, and the humanities are almost entirely absent from the field'. They call for research that evaluates the drivers of coexistence (Pooley et al. 2017). It was in this spirit that we began the Foothills Coyote Initiative (FCI) in 2015 (see Text Box 15.1). In this chapter, we present our first conceptualization of worldviews (Koltko-Rivera 2004; Joyce 2010) and how they relate to coyotes (Canis latrans). The definition of worldview varies by discipline: our use here aligns with the philosophical worldview, which is broadly understood to reflect the underlying beliefs held by individuals (or groups of individuals) about fundamental aspects of reality that ground or influence all one's perceiving, thinking, knowing and doing (Cresswell 1996; Koltko-Rivera 2004). This is sometimes compared to a philosophy of life. To our knowledge the worldview framework has not been applied to understanding human–coyote (or wildlife) coexistence before.

The term *coexistence* originated in a human context and has a range of definitions; it can be described as a spectrum, ranging from passive to

Text Box 15.1 The Foothills Coyote Initiative Overview

The FCI (www.ucalgary.ca/canid-lab) was launched in 2015. We employ mixed methods (ecological and human dimensions, qualitative and quantitative measures) to identify the differences in people's beliefs, actions, ethical frameworks, how they vary geographically (space, demographics, culture, time) and how these differences affect the outcomes for coyotes. One novel component is an integration of concepts of worldviews as they pertain to understanding of and reactions to coyotes.

Study Area

Research is ongoing in the Foothills Parkland Natural Region (FPR) and adjacent lands, located near Calgary, in southern Alberta (Canada) (Figure 15.1). The FPR is one of the smallest natural regions in Alberta, historically populated by agricultural landowners, including settlers and ranchers, cattlemen, sheep producers and crop producers, working small- and large-scale operations. While this agricultural mosaic still predominates, the region is changing rapidly. In fact, this is one of the fastest-urbanizing areas in Canada, a process driven largely by wealthier *urbanites* transplanting themselves into the agricultural setting and occupying small *ranchettes* (i.e. parcels smaller than 20 acres with comparatively expansive homes). The focal sites include rural residential (peri-urban) and agricultural lands within this region. As a geographic phenomenon, agricultural, peri-urban and urban land-uses may be distinguished by the amount of human infrastructure relative to natural or cultivated *habitat*.

Background

While regional urban sites are not our focus, their growth helps illustrate the magnitude of change. For instance, the City of Calgary (Figure 15.1), which is expanding into the FPR, experienced 2.9 per cent growth by 2013 when the population topped 1 million people (CBC News, Calgary 2013). Cochrane (west of Calgary and central to the FPR) experienced 10.4 per cent growth alone between 2013 and 2014 (Town of Cochrane 2014). This expansion of predominately urban origin landowners into a space that was historically populated by agricultural landowners offers a unique opportunity to explore geographical differences in human behaviour towards coyotes and feedback loops amongst residents. Urbanization such as this also may have implications for wildlife behaviour and human–coyote interactions. Coyotes can respond to human stimuli (i.e. food, built shelters) and manifest behaviour changes that lead to conflict with people (Lukasik & Alexander 2012). Previously, Alexander's research (Calgary Coyote Project 2005–12) found that human behaviours, such as leaving out garbage in unprotected bins, feeding coyotes unintentionally (e.g. leaving fallen tree fruit, bird seed or domestic animal food outdoors), or deliberately feeding coyotes and other wildlife were usually the root cause of coyote conflict with people and pets. In fact, a study showed that 100 per cent of coyotes that attacked people (bites or scratches) were found to have been regularly feeding on human food (Alexander & Quinn 2012). That same body of research showed that

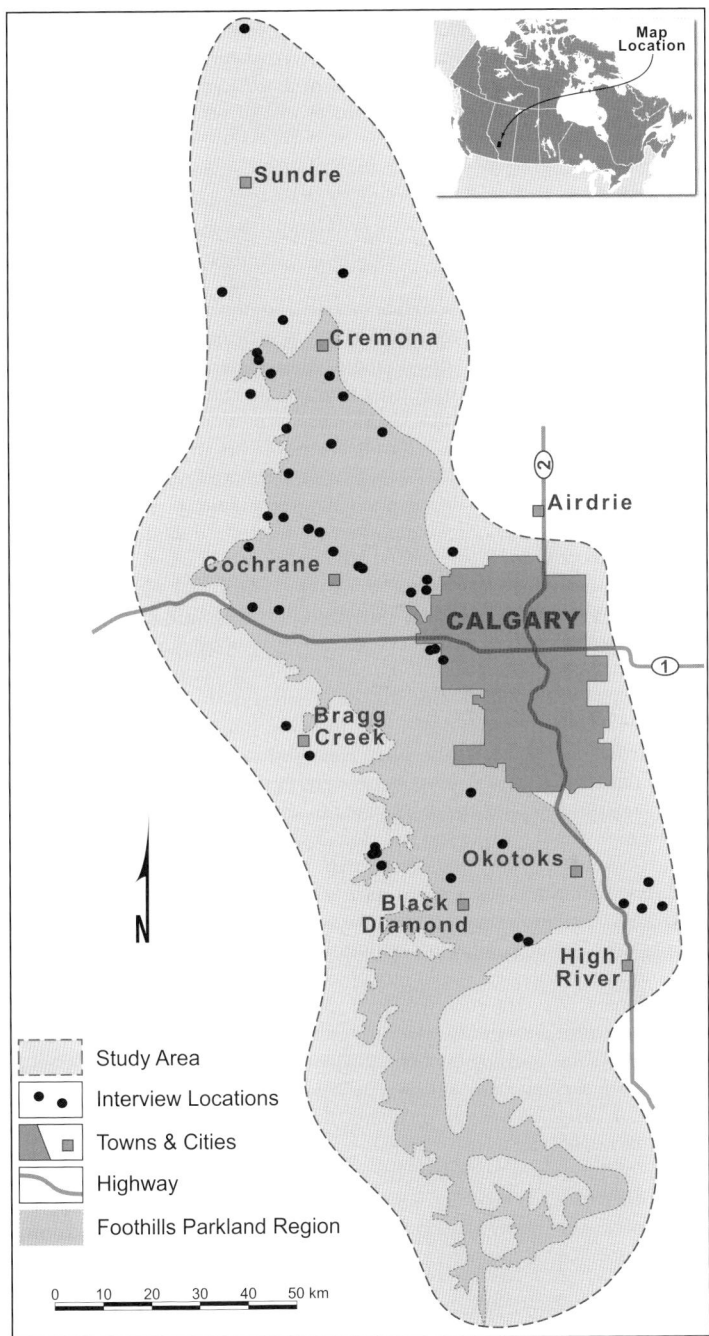

Figure 15.1 Foothills Coyote Initiative study area. *A black and white version of this figure will appear in some formats. For the colour version, please refer to the plate section.*

people's fear of damage to self or property (i.e. pets, livestock) by coyotes resulted in unprovoked killing of coyotes. Based on our FCI observations, killing coyotes before and after they become a problem remains the standard management tool deployed by landowners and government agencies in the region.

Methods

We conducted in-depth interviews with respondents, who were recruited using purposive and snowball sampling techniques (Morgan 2008). Our target audience was rural residential and agricultural landowners. Due to our long-standing history of research and one co-author owning livestock in the region, we identified two key individuals – one from each target group. As well, following a media release on radio, television and newspaper, we were approached by two individuals living in the region and who were unknown to us. All four of the previous individuals were interviewed and we asked each to provide us with any referrals to neighbours, friends, colleagues or others living in the focal study area that they believed would help us learn more about experiences with coyotes. Forty-one additional respondents were then contacted through a *cold call* (albeit via referral) process.

We posed open- and closed-ended questions during a one to one and a half hour, pre-scheduled, face-to-face interview, usually taking place in the respondents' homes. We framed our questions to capture the types of data explored by Joyce (2010) and others, and we customized our questions using prior knowledge gained from our long-standing ecological and human dimensions research (including that on coyotes) in the region and globally (see www.ucalgary.ca/canid-lab). Examples of questions included: How many coyotes do you kill each year? Have you experienced depredation or loss of domestic animals? What words describe how you feel about coyotes? What behaviours would signal you need to kill a coyote? If you had to kill a coyote, how would it make you feel? Is killing coyotes an effective strategy to reduce conflict in the long term? Our interviews gathered a range of attitudes, values and perceptions data, such as concepts about risk posed by coyotes, the coyotes' and humans' role in the ecosystem and whether to fear coyotes, among many other questions designed to explore cultural, economic and ecological implications informing respondents' relationships with coyotes. We also collected basic demographic data such as time of respondent residency *in situ*, place of birth, age and so on.

Interviews are ongoing, and this chapter is based on those completed at the time of writing (n = 45). We believe our snowballing approach was successful based on sample outcomes that follow. We interviewed thirty-three individuals and twelve married or parent–child pairs. This resulted in a sample of thirty-three males and twenty-seven females. When pairs engaged in the interviews, we recorded answers to all questions for both individuals. We interviewed twenty-two rural residential and twenty-three agricultural landowners. These properties ranged in size from 2 to 3,900 acres. Respondents were engaged in all land-use practices, from rural residential to small hobby farms to small- and large-scale cattle operations. Originally, we were going to use land parcel size to identify agricultural landowners, but this was not always a good indicator of agricultural practice. For example, one rural residential landowner owned over 600 acres. Hence, we distinguished agricultural sites as those with existing or recently (<10 years) terminated livestock operations.

We used the mental models approach (see theoretical context in this chapter) to conceptualize the plurality of behavioural typologies, ethical frameworks and worldviews of participants. To derive our worldviews, we aggregated like responses to identify two components: behavioural typologies and ethical framework. We outline this process in more detail below. Our data set is quite vast, and we only include data relevant to worldviews in this chapter. That is, we examined a set of responses to questions that directly related to actions towards coyotes (e.g. How many do you or another person kill each year on your property? If you had to kill a coyote, what method would you use, and why?), sentiments about or reasons for engaging in this killing (e.g. If you had to kill a coyote, how would that make you feel? Have you experienced depredation in the last 10 years?), and we looked for key words and phrases in responses that encapsulated ethical framework and how the respondent viewed themselves or their role in the ecosystem (e.g. 'we expect three calves to be killed by coyotes each year and so we thin them [coyotes] regularly' ... 'it's important for us to keep the balance of nature'). From these forty-five respondents, we identified similar perceptions of, behaviour towards and relationships with coyotes, which we clustered into units we called typologies. We related typologies to the ethical frameworks: anthropocentrism, biocentrism and ecocentrism (see details below). Combined, these two tangents helped us identify elements of the worldviews of our respondents worldviews (see Figure 15.2).

active tolerance, as described in the conflict-to-coexistence continuum framework (see Chapter 1). We use the term coexistence to mean peaceful negotiations, or humans tolerating the presence of carnivores and refraining from indiscriminate killing in response to emergent conflict. The diversity of challenges to coexistence that humans living with wildlife face globally, make the use of a focal species helpful for this inquiry. The focal species approach (e.g. indicator, umbrella, keystone) is well established in conservation science (Miller et al. 1998; Lindenmayer et al. 2014). Our use of the term *focal species* varies a little from tradition. We use the term focal species to reflect a synergy of species adaptation and human encounters that are reflective of the broader challenges of coexistence. To explain, coyotes have a higher resilience and behavioural flexibility, and tend to encounter people, pets and livestock in more diverse and complicated ways than sensitive carnivores (e.g. wolverines, lynx, etc.). In addition, we because in Alberta, as across Canada, coyotes

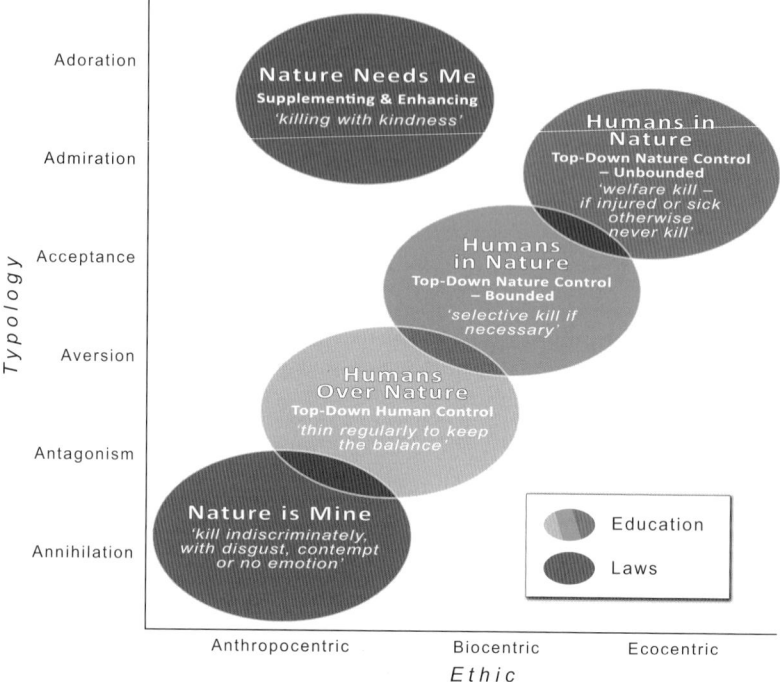

Figure 15.2 Worldviews spectrum showing intersection of behavioural typology, y-axis, and ethic, x-axis, with brief thematic description. Grey represents worldviews open to education, with darker colours more likely to use education. Black with white shows worldviews unlikely to accept science knowledge mobilization. *A black and white version of this figure will appear in some formats. For the colour version, please refer to the plate section.*

are a designated *pest species* (Alberta Environment and Sustainable Resource Development 2012; Alberta Wildlife Act 2014; Canadian Wildlife Act 2014) that label promotes killing, and condones raw human responses to the species. Hence, coyotes encounter people frequently and in varied situations, and actions by humans towards them are unfettered by laws or other mechanisms that might alter people's motivations to kill. It is this confluence that we believe makes coyotes a good focal species – one that may help us better understand foundational motivations to coexistence with carnivores in general.

We define negative encounters as those where humans, pets, livestock or wildlife are injured during or because of an interaction. As with other species, the rate of negative encounters between coyotes and humans is low: on average three people bitten or scratched annually in Canada (Alexander & Quinn 2011). However, attacks on people or the loss of domestic animals have tangible economic and emotional effects (Treves & Bruskotter 2014). In fact, even encounters with coyotes that are not lethal to pets or livestock have been found to evoke grief, anger, fear and sometimes Post-Traumatic Stress Disorder (PTSD) (Alexander & Quinn 2011). Although the numbers of attacks suggest human fears are unfounded, certain individuals and various government agencies deem killing of coyotes (and other carnivores) to be an accepted method to pre-emptively and retroactively manage negative encounters. To exemplify, for decades, hundreds of thousands of carnivores have been killed annually in the United States (USA) (Fox & Papouchis 2005) and in Canada (Alexander & Quinn 2012).

Sometimes the killing of coyotes by individuals or government agencies is selective, but often killing in mass numbers is done in the name of conservation, sport and pre-emptive mitigation of livestock loss – despite scientific evidence that indicates such an approach does little to solve conflict and causes damage to ecological systems (Fox & Papouchis 2005; McManus et al. 2015; Proulx & Rodtka 2015). Other scientific evidence and experience indicate there are important ethical implications to widespread killing of carnivores, particularly if inhumane methods are deployed (e.g. aerial gunning, leg hold traps, poison, snares) (Brook et al. 2015; Dubois et al. 2017). It is now widely accepted that all mammals experience a broad range of emotions, such as: grief, joy, love, anger, disgust, surprise and fear (Bekoff 2013). Despite our uptake on the concept of animal emotions, this is not a new idea. Darwin (1872) already noted that animals differ from humans not in *type* but in *degree*. He argued that humans and animals experience emotions but their expressions vary. Knowing this, it seems that coexistence should be as important amongst people, pets, livestock and coyotes as it is amongst humans themselves.

The question of whether, and how, to use science to redirect people towards coexistence with carnivores, and other wildlife, remains a challenge, because we still do not understand the motivation (or drivers) of coexistence (Pooley et al. 2017). We launched our Foothills Coyote Initiative (FCI) to explore human constructs that motivate encounters and practice towards coyotes – i.e. to explore worldviews. Our FCI research specifically focused on: (1) recording varying experiences with coyotes, (2) exploring the circumstances and impacts of the coyote killing, (3) documenting the coexistence approaches people used, characterizing human behavioural/relationship typologies and (4) combining these and other data to construct a spectrum of worldviews (see

Figure 15.2 in results). While attitudes and values inform worldviews, they are not the focus of this chapter and will be analyzed in subsequent publications. Moreover, attitudes and values have been researched with respect to coyote encounters and coexistence (e.g. Sponarski et al. 2015).

15.1 THEORETICAL CONTEXT FOR THE WORLDVIEWS APPROACH

Before starting our discussion of the theoretical grounds for our research, we discuss worldviews, followed by other two key terms used throughout our chapter: geo-ethics and typologies.

15.1.1 Why Worldviews?

The multiple truths about, or ways of knowing and understanding the world have been referred to as worldviews (Proctor 1998; Teddlie & Tashakkori 2009). These ways of understanding the world are accompanied with enduring beliefs about and behaviours towards various phenomena (Koltko-Rivera 2004) – in this case towards coyotes. Because worldviews can entrain human behaviour and tend to be fixed, understanding the plurality of these worldviews may allow for more informed and efficient use of science related to coexistence. The inflexibility of worldviews has been witnessed in research on climate change, where people were observed to seek information that supports their worldview and disregard other information, regardless of its credibility (Hulme 2008; Joyce 2010). Joyce (2010) infers that this is because challenging information threatens the individual's construct of their world.

Hence, insight about worldviews may facilitate knowing where, how and if we can communicate effectively to realize coexistence. As noted previously, Pooley et al. (2017) emphasized that a better understanding of the *drivers* of human–wildlife issues (and we consider worldviews to be a possible driver) is a critical step towards coexistence. Such understanding may not only increase efficacy of coexistence strategies based on science mobilization, it also may prevent unintended consequences. For instance, when scientists in Alberta spoke out in the media in 2015 against a coyote killing contest based on scientific grounds, one group of rural residents purposely killed even more coyotes and made a point of letting scientists know, by way of media reporting, that this was a retaliatory act (Edmonton Journal 2015a, 2015b). As Havorka (2016) echoes, whether we are scientists or citizens of a certain community, our understanding of nature, science, truth and environment is affected by values, and does not exist outside social and political influences.

Our worldview analysis employed the mental models approach. Broadly speaking, mental models are conceptual maps that explain how people conceive of topics or subjects, their value and their relationships to other components in their world. An individual's mental model is thought to inform how that person assesses risks and benefits and, more critically, how he or she makes decisions (Wilson et al. 2008). The mental models approach has been applied in a variety of settings from climate change (Chowdhury et al. 2012) to fire management (Zaksek & Arvai 2004), and has been integral to education and outreach efforts, such as management of invasive agricultural weed species (Jabbour et al. 2014). We constructed and honed questions that related to coyotes, based on factors identified in the previously noted research. We iteratively developed a conceptual map of worldviews, based on similar responses to questions that identified behavioural typologies and ethical frameworks (see Text Box 15.1).

15.1.2 Nature As a Human Construct and Geo-Ethics

In our research here, we view constructs of nature, and the associated ethical frameworks, as related to worldviews and outcomes for coyotes and coexistence (i.e. whether they are killed or not). This section is not comprehensive, but designed to encourage readers to consider these theories more broadly as they pertain to coexistence.

The proposition that *nature* is a human construct (Proctor 1998) has raised the ire of some conservation scientists, but as with Lorimer (2015) we still believe the idea is a helpful starting point in the discussion of coexistence. Proctor's supposition can be interpreted generally to mean that nature is a concept developed by the human mind, and therefore does not truly exist as something physical or fixed that we can protect, and it is not separate from humans. This is disturbing to conservationists who desire to protect a nature separate from people. The idea that nature is only a human construct was challenged for ignoring the material being of animals, biodiversity and that if we give in to the notion that humans are part of nature then actual animals (not constructs) are going to keep dying and biodiversity will continue down a negative spiral of loss. The construction of nature, as we have employed it, begins with how people view the species *coyote* (e.g. is it part of nature, has an ecosystem importance, where does it belong) and how the role of this species relates to the respondent's broader concept of nature. For example, coyotes are often killed when they cross some imaginary boundary between 'nature out there' and *human property over here* (Alexander & Lukasik 2016). The latter conveys notions of *place*: how we conceptualize where that thing (e.g. animal) or event (e.g. hunting) should occur can shape our behaviour in response (Creswell 1996). In Western cultures at least, our constructs tend to inform a belief that wild animals belong in the hinterland, livestock in the fringes of cities and domestic/companion animals belong in our homes (Philo & Wilbert 2000). Yet when a species adapts to human landscapes, exploits habitat in cities or kills domestic animals, from a biological construct the animal simply lives. The notion of *place* is not expanded upon further in this chapter, as it is the topic of recent and forthcoming publications (Alexander & Lukasik 2016; Alexander & Draper unpublished data). However, place is used herein when exploring worldviews, as it pertains to our respondents' ideas about how close to home (however that is defined spatially) or what type of behaviour the coyote exhibits (in relation to property or self) that affect how coyotes are treated by the respondents.

This idea of transient or shifting ethics is captured by Urbanik (2012), who describes the concept as geo-ethics. Geo-ethics refers to the notion that ethical frameworks deployed by individuals are influenced or vary by place, location and culture (Urbanik 2012). This term emphasizes that where (geo) things happen is fundamental to what is moral within that context, and recognizes that there are 'multiple moral centres' (Urbanik 2012, p. 39). Urbanik (2012) cautions that applying one single analytical ethic (such as our personal views as scientists towards coexistence) to multiple human–animal interactions is too rigid. This appears theoretically consistent for our study area. Geography has dictated settlement patterns, communities and subcultures, and each of these influences human responses to the environment in general, and to coyotes specifically (Alexander & Draper 2017). Arguably, if we do not pay attention to geo-ethics and are single-minded in what we prescribe to be ethical or scientifically defensible ways to engage with carnivores, the thing (e.g. animal) may go awry – as with the example of the coyote-killing backlash following the media outreach described earlier.

In our analysis, we used one ethical framework, as described by Urbanik (2012), which pertains to animals and the environment. Ethical orientation ranges from anthropocentric (e.g. only humans matter and humans control the ecosystem) to ecocentric (e.g. individual beings and ecosystems matter – humans and livestock have entered the ecosystem and are part of it, and the ecosystem has moral consideration) to biocentric (e.g. all or most beings have moral consideration). Our assertion here is that the observed ethic, as articulated by each respondent, helps us situate their behavioural typology in a worldview.

15.1.3 Typologies

We use the term typology to refer to the clusters of similar behaviours/ relationships to coyotes. The behavioural typologies (typologies hereafter) that we chose reflect our understanding of the ways our respondents engaged with or understood coyotes in relation to themselves, other animals and the ecosystem. Because they are based on direct reports of behaviour towards coyotes, the following typologies reflect decreasing tolerance for coyotes:

- Adoration – characterized by a love or reverence of coyotes (and all animals), full integration with coyotes on their property (i.e. living and engaging with them at close range almost daily) and would never kill nor ask anyone to kill a coyote.
- Admiration – the relationship is positive, they are respectful of coyotes as a species and would not kill unless an animal is sick or injured. This differs from adoration, as this group would kill a coyote, and they respect safe boundaries of encounter (e.g. these people do not encourage coyotes to live with them).
- Acceptance – characterized by a tolerance of coyotes and understanding that they are part of the ecosystem, but coyotes that behave unnaturally by attacking people, pets and/or livestock are believed to pose a threat and should be selectively killed; the preferred method is what is believed to be the most humane (i.e. described to be a clean shot, a bullet).
- Aversion – distrust of coyotes, comprehension of them as part of the ecosystem but a necessary evil, generally associated with targeted (*in situ*) thinning of coyotes (from 3–5 individuals to 100 depending on context/method) with the rationale it is necessary to prevent injury or loss of domestic animals and livestock. Methods of killing vary from poison to bullet.
- Antagonistic – characterized by a dislike and active hostility towards coyotes, killing individuals or packs *in* and *ex situ*, and regardless of perceived or actual threat, killing is done with traps, poison and

bullet, and with the general idea that killing is necessary to control populations.

- Annihilation – this represents a typology of person who engages in killing indiscriminately without ethic. The annihilation category captures individuals who kill for no reason but sport, pleasure, to inflict pain and with no expression of remorse. This is distinct from someone who is antagonistic towards coyotes.

15.2 FOOTHILLS COYOTE INITIATIVE WORLDVIEW SPECTRUM

Combining ethical frameworks with the previously listed typologies, we conceptualized the spectrum of worldviews described below (see Figure 15.2). Specifically, the FCI worldviews are a function of: (1) typology on the y-axis and (2) ethics on the x-axis. Worldviews are shown within Figure 15.2 as shaded circles with (mostly) overlapping boundaries, which we did to highlight that there is some fluidity across the spectrum. We use the terms bounded and unbounded in reference to whether the worldview is generalized to all 'places' and species (explained below). In Figure 15.2, we include a brief phase that captures the essence of the typology/ethic that informs the worldview. To visualize the interconnections of worldviews with social frameworks (policy or action) commonly used to help realize coexistence, we colour coded worldviews to reflect which worldviews were most likely influenced by *education* (orange) or *law* (purple). The worldviews are described in decreasing order of tolerance for coyotes.

15.2.1 Nature Needs Me (Supplementing and Enhancing), Killing with Kindness

In our sample, this was singularly a rural residential phenomenon. There are variants of the extreme case we present here. Our example is of a respondent who allowed coyotes repeatedly to maintain a den in close range of the house (<10 m). The coyotes integrated household items (e.g. tarps, ropes, downspouts, pavement blocks) into daily play, interacted regularly with the household pets and subsidized diet with planted fruits, gopher colony and vegetables on the property. There was no attempt by the landowner to discourage denning or remove potential attractants or play items. The concept of place was best conceptualized as: 'We all share nature, we all belong – except most wild animals do not belong in the house – and all animals are good, accepted or managed by changing human behaviour to enhance coyotes' experience.' The level of integration in the case we describe was high and without incident; it raised questions about whether our understanding of the capacity to coexist is too limited. This typology adores coyotes and loves that they can cross human–animal thresholds. The ethic stems from a desire to support the ecosystems and species (ecocentric–biocentric), but we label it anthropocentric. This is because the desire to care for nature appeared to be tightly linked to how the respondents feel about themselves and their self-identity, more than it appeared to reflect the reality of animal or ecosystem health. Some examples of words used by respondents included the description of their property as *the ark*, *refuge* and a *safe place for coyotes*. This view reflects a minority of the rural residential respondents, and no agricultural respondents, even if the latter appreciated that coyotes could den on their property without posing a safety issue to pets or livestock. In our prior experience, denning in such proximity does pose risks for coyotes, pets, livestock and humans in the wider area. (Others are unlikely to know how to be that integrated without attack.) It seems likely that some coyotes may be able to coexist with humans to this very close extent, but it raises concerns. This may have more to do with the relationship between individual coyotes and individual people who are very tolerant of coyotes. It is context specific. We expect there is a high likelihood that by changing the human context the relationship likely would become confrontational. The same coyote approaching a different human's back window may be seen as a threat and killed.

15.2.2 Humans in Nature (Top-Down Nature Control – Unbounded) – Kill Only If Injured or Sick/Never Kill

This group describes themselves as borrowing the land, and needing to keep the balance. They recognize that nature keeps the balance, not people, and some describe that if they had to kill a coyote they would feel that they had failed: 'failed the coyote and my livestock'. A common sentiment in describing their use of the land was, 'it is the coyote's home first'. They exhibit the behavioural typologies of admiration and are predominately biocentric, often ecocentric, in ethical approach. This cohort identifies as part of the ecosystem and describes the ecosystem as something needing protection. This worldview will kill out of compassion: they will kill a coyote 'if it is sick or injured' and will tend not to kill coyotes even if they depredate a dog or livestock. There is also a general belief that coyotes that kill may do so because they are not healthy. They view their house and very near surrounds (e.g. within 100 m of the house) to be human places, and that the rest belongs to nature. These respondents articulate that all wild animals belong in those natural spaces, and do not kill wildlife, except apparently on welfare grounds. While level of education is not a precursor to the worldview, what is evident is that science and education material is readily adopted. Most in this worldview made statements about 'research [they] had read' on the topic. Loss of domestic animals to coyotes is believed to be the responsibility of the individual (respondent) and is not blamed on the coyote, and the use of culls or killing as a method of coexistence is rejected based on science. One respondent summarized the relationship to coyotes aptly, stating 'what do I feel about coyotes? When I see a coyote it makes me feel like everything is right in the ecosystem.' This worldview was well represented by respondents in both agriculture and rural residential cohorts.

15.2.3 Humans in Nature (Top-Down Nature Control – Bounded) – Selective Kill If Necessary

This worldview has some attributes of antagonism and acceptance, but their typology is most frequently acceptance. This group also describes themselves as borrowing the land, state that nature keeps the balance, but we describe their views as bounded. That is, it can change by location and behaviour. The presence of coyotes (location) does not factor into the decision to kill as much as what behaviour occurs in which spaces. Coyotes are killed if their behaviour is deemed unnatural or poses too high of a risk to livestock, pets or people. Killing of coyotes is selective and happens only after some threshold of location or behaviour has been crossed. Often the behaviour of domestics/livestock in response to coyotes is used to decide if the behaviour is unnatural. Ethic tends to align with biocentric, but again that is bounded based on species or locational thresholds – for some species (listed at the end of this paragraph) the ethic is anthropocentric. The human places where coyotes do not belong include the house and specific distances from that site (e.g. '400 m from my home', wherever livestock exist on the property) or behaviours of coyotes in each of these sites – such as stalking or attacking livestock in a cow pasture, or 'coming up onto the front porch'. However, we observed that for respondents in this worldview, while coyotes are perceived in this manner, other species are 'out of place' in all the locations regardless of their behaviour. Species such as badger, moles, voles, ground squirrels, ravens, magpies or crows are all killed, but not everyone kills all these species. Most describe feeling either badly or not feeling at all (suspension of emotion due to the task at hand). This is different to what we refer to as killing without remorse. There is agreement that killing is never done for fun or celebrated, but that killing is necessary in some instances. The viewpoint is very practical and utilitarian.

15.2.4 Humans over Nature (Top-Down Human Control) – Thin Regularly to Keep the Balance

Killing coyotes regularly or expecting a government agency to kill coyotes when needed was the hallmark of this group. Some trust and some distrust the government to properly manage coyotes. Direct killing was exclusively reported by agricultural respondents, who tended not to trust the government in control. Also included here are rural residential landowners who articulated support for killing coyote 'if the numbers justified it', but the decision to kill was left to government agents. All owned or leased land (either where the person lived or far from their homesite) was conceptualized as human places, and even if ecological views were expressed, certain species did not belong and must be controlled. Presence of the coyotes, and not behaviour, was used to determine if a coyote was *out of place*, posed a risk to livestock and guided the choice to kill. The ethic for this group is anthropocentric, and agricultural landowners in this category tend to self-describe as not engaging in indiscriminate killing, but regular targeted thinning. Thinning was almost uniformly described to mean killing 3–5 coyotes, annually. While the respondents in this group were themselves not likely to kill in mass numbers, some allowed Waltzing Matildas (a term given by one respondent to a group – typically men – who go door to door asking to kill coyotes) to cull coyotes on their land. Agricultural respondents who indicated they allowed such hunting did not typically consider that part of their individual killing, and reported that Matildas took between 17 and 100 coyotes per month when active. Respondents in this worldview used language like 'it is important to keep the balance of nature' (and humans must kill coyotes or predators to keep the balance), or 'it's life', 'it's our land', 'we live and let live, as long as the coyotes stay where they should'. Killing coyotes is done because it is viewed as critical to economic livelihood and reflects a cultural practice and belief that coyote populations do not self-regulate and this is the only way to protect livestock, pets and people.

15.2.5 Nature Is Mine (aka Nature Does Not Matter) – Kill Indiscriminately, with Disgust, Contempt or No Expressed Remorse

This worldview encapsulates those who expressed opinions, beliefs and ethics that nature is solely for human consumption and exploitation. The concept of place appears to manifest as the idea that all locations and species belong to humans first. The typology is annihilation; they kill coyotes indiscriminately, for sport and not necessarily for defence of property. Some delight in mutilating coyotes. Distinct from any other group who kill coyotes, these respondents articulated a disdain for coyotes, a satisfaction with killing coyotes and other carnivores, participated in excessive killing and most critically some bragged about inhumane kills or did not recognize their approach to be inhumane. Respondents from other worldviews described this group as engaging in weekend family bonding trips to kill coyotes, unsanctioned competitions and bounties, maiming of coyotes (e.g. one *Humans in Nature* respondent described the pack of four coyotes on his property, three of which were missing different paws) and using coyotes for target practice. While the ethic is shown as anthropocentric (Figure 15.2), at times there appeared to be no ethic. For example, one respondent poisoned coyote pups in the den after seeing the mother coyote leave, and the respondent noted, while grinning, that she 'got 'em'. This type of verbal reference is germane – coyotes are objectified, like tin cans. Another respondent said that killing things is something one 'has to do to feel right' – 'coyotes are one thing that you can kill whenever you want'. He went further to note that killing with poison rather than a neck snare 'is women's work' – i.e. soft and 'cowardly'. Another respondent (female) expressed that coyotes 'deserve' to suffer in leg hold traps. These last examples highlight the importance of not making assumptions about gender in relation to wildlife coexistence and animal welfare.

15.3 WORLDVIEWS – TO WHAT END?

Our worldviews will continue to be nuanced, but we believe we have reached saturation and are unlikely to find additional categories. There were contradictions and surprises that we will continue to explore and use to refine our worldviews. For example, during our FCI interviews we met people who, despite their expressed comprehension of coyotes as critical to ecosystem function and a desire to maintain them (apparent ecocentric ethic), simultaneously believed it was vital to kill coyotes regularly to protect their pets or livestock (anthropocentric ethic). Our future research will tap all our additional interview data to better grasp the drivers of this and other types of *disconnect*.

In addition, there was no clear division between rural residential and agricultural sectors within this framework. There is often a first assumption that all agricultural people kill predators. This is a typical bias in researcher/advocate/manager expectations that must be overcome in coexistence planning. Certainly, while we expected some polarity amongst agricultural landowners (Sponarski et al. 2013), we were surprised by the extreme valences: some killed regularly, some never killed and believed killing represented *failure*, while others were inhumane and sadistic towards coyotes and other carnivores. Likewise, there was a much less uniform valence amongst rural residential respondents: some adored and named coyotes, others would behave inhumanely towards coyotes if they did not feel constrained by law and still others disregarded the laws and killed them using illegal methods. We observed that several agricultural respondents held a negative view towards urban encroachment. There was a belief expressed by agricultural respondents in general that anyone of urban origin was of the Nature Needs Me worldview and that such individuals increased coyote problems for livestock producers by providing them with supplemental food and inflating the carrying capacity of the land. A detailed evaluation of the former is a topic of ongoing FCI research, but our data yields evidence that this urban supplementation effect is at times correct but is not universally true. Many rural residents had parallel views to agricultural landowners. While there were a few exceptions, rural residential landowners did not consider the expansion of urban infrastructure into agricultural lands as a *problem* for anyone or anything.

One of our aims in developing the spectrum of worldviews was to facilitate coexistence strategies. Typically, there are two dominant ways to increase coexistence: education or law. Indeed, Figure 15.2 shows colour coding that indicates where we believe education (orange) or law (purple) will be most effective. We believe that education must be targeted to the needs identified within each context and group, especially for those worldviews that will be unwilling to accept novel information (Joyce 2010). We identified that those on the extremes of the spectrum will likely be resistant to education – their actions appear very tightly coupled with their identity rather than their use of landscape (e.g. whether they raise livestock or farm). The specifics of such education are beyond what we can contribute at this juncture and will be the topic of future analysis of our data.

Annihilation and adoration typologies appear strongly indoctrinated by culture or personal belief about humans' role in the environment. While it may seem counter-intuitive to link these dichotomous world-views – one who kills with disdain and one who adores coyotes – arguably both worldviews can easily lead to coyote death. It is possible to 'love coyotes to death' by setting them up for failure when the habituated (possibly conditioned) animals change contexts and enter relationships with different humans. Importantly, given the propensity of the Nature Needs Me worldview to care deeply for animals, there is a higher likelihood that with the right delivery these individuals may incorporate new knowledge and change their behaviour, because doing so will help animals more.

Those in the central categories are likely to accept new knowledge (some regularly do integrate contemporary science), but we believe information sharing must be catered to specific needs and delivered via different avenues. We show this worldview in a lighter shade of orange, because cultural practice was identified to be a strong driver of behaviour and, therefore, the worldview may be resistant to uptake of new knowledge. It is possible they may be influenced by more research that indicates killing results in more problems than it solves (e.g. McManus et al. 2015). We also do not believe this group is completely rigid, because of responses to other questions like 'do you use poison, why or why not'. In the latter, it emerged in this group's responses that consistent messaging by government and educators had a tangible effect on stopping the indiscriminate and targeted use of poisons – or, that education messaging overcame cultural norms in the past. Understanding how messaging about poison resulted in behavioural change might be insightful to craft messaging about killing. A caveat of course is that one reason poison use was halted is because of humans, particularly children, and pets dying. In contrast, those who align with Humans in Nature (unbounded) attributed their behaviour to personal experience, self-education and integration of new science. These individuals often described themselves as continually refining/enhancing their relationship with nature: they talked about specific scientific documents they read, and appeared to seek out information and integrated it. We represented this group as the darkest orange in

Figure 15.2, to reflect they should be most receptive to education, even though they may not need it. Importantly, we believe this group should continually be supported with new research, because many tend to be from old families, respected community members, and thus may mobilize science to the broader community. Our FCI agricultural respondents commented there are strong community connections and norms that can entrain behaviour.

15.3.1 The Importance of *In Situ* Interviews

During our interviews, we entered into the homes and *lives* of respondents for approximately one to three hours. Doing so resulted in a momentary re-positioning of ourselves within the context of the *other* individual and insight into their perspective and lived experience. We believe this provided a unique (and *better*) vantage than could be offered by any remote interview method (e.g. online), and it highlights the critical importance of qualitative research (Rust et al. 2017). While it obviously results in fewer interviews, we found the depth and richness of experience was critical to more rapidly interpreting responses and developing a concept of how and why people interact with coyotes. We are not suggesting that remote interview or quantitative techniques are unhelpful, but that many critical nuances were gleaned *in situ*. For example, some respondents who did not regularly kill coyotes, nor state that they killed other animals, had the pelts or mounts of rare species in their homes. This helped us conceive of the idea of *bounded* and *unbounded* engagement with wildlife species. This was an emergent property of the interviews, not an a priori concept. *In situ* interviews also helped us to understand how scientists are viewed as 'outsiders' who simply cannot grasp the experience of a landowner. Moreover, because free discussion results at junctures, we learned about non-target issues that helped explain relationships to coyotes, coyote ecology and ecological trends that we could not have conceived of asking about prior to the interviews. As an example, the role of culture was described by one respondent as tribal (i.e. you must adhere to the rules of the tribe or not be a member), which provided us a new language and understanding of the social processes that may govern people's responses to coyotes.

Tackling this research *in situ* also caused us to interrogate and reposition some of our own firmly held personal *truths* about what behaviours are *correct* or *moral* in a coexistence paradigm. Specifically, we have challenged our own notion that coexistence means no killing of coyotes, ever. It was also clear that context matters and contradictions abound for us, as for the community of respondents. Urbanik (2012) described these inherent contradictions in people's everyday practice with animals, giving examples like the contradiction of pet ownership where an animal is simultaneously a family member and captive species, or perhaps killed if it is too aggressive. This presents some *food for thought* related to delivering coexistence messages. If scientists, practitioners or advocates who have never faced the challenge of watching a pet or domestic animal die after a coyote attack, have never had to make a living off the land or have a different construct of *home* insist that coyotes never be killed, then how might this affect our end goal – could our approach backfire?

We began with the notion of coyote as a focal species. Looking at the worldviews, typologies of engagements and outcomes for coyotes that we documented, we believe our research demonstrated that coyote is an excellent species by which to view the breadth and extreme nature of engagements with carnivores in general. Coyote, perhaps, sheds light on the role of law in constraining more widespread heinous behaviour towards carnivores. We believe our work supports and can refine the notion of a three-pronged approach, such as that used by Project Coyote, USA: it deploys science, education and advocacy (i.e. for legal reform) (www .projectcoyote.org/). While our findings are specific to FCI, we believe they reflect a broader understanding of coyote–human (and carnivore–human) interactions. We hope that our worldview spectrum may streamline coexistence efforts and increase effectiveness of science knowledge mobilization by encouraging practitioners to get the right information to the right people, and provoking reflection on how this approach might be improved to benefit coexistence.

In this final section, we identify some action items that we believe are important to changing the status quo for coyotes and realizing coexistence.

15.4 RECOMMENDATIONS AND FUTURE DIRECTIONS

- Research human–coyote interactions, worldviews and drivers of behaviour and worldviews to foster coexistence between humans and this species.
- Identify the moral bounds of encounter and killing. Such an understanding can be beneficial for fostering coexistence also with other species, because it explains the overall reasons why people choose to kill or not, in general.

- Consider the role of *place* and *transgression* in human relationships and conflicts with carnivores. Transgression describes people's understanding of when a species has crossed a person's idea of the right 'place' for it to be. For example, some respondents felt a coyote was out of place and needed to be killed when it came closer than 400 m to a house. In that case the person acts as if the coyote has transgressed a boundary that the animal should not have crossed.
- Examine the role of law in relation to behaviour towards carnivores. For instance, what criteria are used to decide if a species is a *pest*, and are those criteria appropriate? What would occur if we changed those laws?
- Explore the social and political drivers of carnivore killing. For example, we mentioned that one group of landowners retaliated against scientists in 2015, and killed more coyotes. That event could be understood as an attempt of rural residents to regain power over urban residents. Coyotes in this case are killed not because they are coyotes, but because they provide a means to message or reassert control.
- Education programmes might be enhanced by integrating the results of this research. Our hope is that the worldviews framework could improve the delivery of science – get the right information to the right people. At the same time, perhaps our results could improve effectiveness by identifying where the dissemination of education may be useless (based on resistant worldviews).
- Advocate to government agencies to incentivize the use of or to implement non-lethal approaches. This can also empower people more generally to explore alternatives. Some respondents were either unaware they had any other option than to kill coyotes, or felt it was not feasible to implement alternatives.
- Work with groups like Project Coyote and Coyote Watch Canada that have a scientifically grounded, well-tested practice in all the above areas.
- Shift researcher and management approach to lead by example: we need to embrace compassionate conservation (do no harm) in research and management, and reject de facto killing paradigms.

15.5 References

Alberta Environment and Sustainable Resource Development. (2012). *Alberta Guide to Hunting Regulations.* Available from www.albertaregulations.ca/huntingregs/gameregs.html (accessed September 2014).

Alberta Wildlife Act. (2014). Available from www.qp.alberta.ca/documents/Regs/1997_143.pdf (accessed September 2014).

Alexander, S. M. & Draper, D. L. (2017). Field notes. Foothills Coyote Initiative. Available from www.ucaglary.ca/canid-lab (accessed November 2018).

Alexander, S. M. & Lukasik, V. M. (2016). Re-placing coyote. *Lo Squaderno*, 42, 19–22.

Alexander, S. M. & Quinn, M. S. (2011). Coyote (*Canis latrans*) interactions with humans and pets reported in the Canadian print media (1995–2010). *Human Dimensions of Wildlife*, 16(5), 345–59.

Alexander, S. M. & Quinn, M. S. (2012). Portrayal of interactions between humans and coyotes (*Canis latrans*): Content analysis of Canadian print media (1998–2010). *Cities and the Environment (CATE)*, 4(1), art. 9.

Bekoff, M. (2013). *Ignoring Nature No More: The Case for Compassionate Conservation*. Chicago: University of Chicago Press.

Bostrom, A., Morgan, M. G., Fischoff, B. & Read, D. (1994). What do people know about global climate change? *Risk Analysis*, 14(6), 959–70.

Brook, R., Cattet, M., Darimont, C., Paquet, P. & Proulx, G. (2015). Maintaining ethical standards during a conservation crisis. *Journal of Canadian Wildlife Biology & Management*, 4(1), 72–9.

Canadian Wildlife Act. (2014). Available from http://laws-lois.justice.gc.ca/eng/acts/W-9/ (accessed September 2014).

CBC News, Calgary. (2013). Calgary's population hits 1.15M people. *CBC News, Calgary*, 25 July. Available from www.cbc.ca/news/canada/calgary/calgary-s-population-hits-1-15m-people-1.1385386 (accessed September 2014).

Chowdhury, P. D., Haque, C. E. & Driedger, S. M. (2012). Public versus expert knowledge and perceptions of climate change-induced heat wave risk: A modified mental model approach. *Journal of Risk Research*, 15(2), 149–68.

Cresswell, T. (1996). *In Place, Out of Place: Geography, Ideology and Transgression*. Minneapolis: University of Minnesota Press.

Darwin, C. (1872). *The Expression of the Emotions in Man and Animal*, 1st edn. London: John Murray.

Dubois, A., Fenwik, N., Ryan, E., Baker, L., Baker, S., Beausoleil, N., Carter, S., Cartwright, B., Costa, F., Draper, C., Griffin, J., Grogan, A., Howald, G., Jones, B., Littin, K., Lombard, A., Mellor, D. Ramp, D., Schuppli, A. & Fraser, D. (2017). International consensus principles for ethical wildlife control. *Conservation Biology*, 31(4), 753–60.

Edmonton Journal. (2015a). Hunters, conservationists square off over coyote hunt. *Edmonton Journal*, 8 January. Available from http://edmontonjournal.com/news/local-news/hunters-conservationists-square-off-over-coyote-hunt (accessed January 2017).

Edmonton Journal. (2015b). Debate over coyote hunts runs hot. *Edmonton Journal*, 30 January. Available from www.edmontonjournal.com/Debate+over+coyote+hunts+runs/10775509/story.html (accessed January 2017).

Fox, C. H. & Papouchis, C. M. (2005). *Coyotes in Our Midst: Coexisting with an Adaptable and Resilient Carnivore*. Sacramento, CA: Animal Protection Institute.

Havorka, A. (2016). Animal geographies 1: Globalizing and decolonizing. *Progress in Geography*, 41(3), 1–13.

Hulme, M. (2008). Geographical work at the boundaries of climate change. *Transactions of the Institute of British Geographers*, 33, 5–11.

Jabbour, R., Zwickle, S., Gallandt, E. R., McPhee, K. E., Wilson, R. S. & Doohan, D. (2014). Mental models of organic weed management: Comparisons of New England US farmer and expert models. *Renewable Agriculture & Food Systems*, 29(4), 319–33.

Joyce, C. (2010). Belief in climate change hinges on worldview. *NPR*, 23 February. Available from www.npr.org/templates/story/story.php?storyId=124008307 (accessed January 2017).

Koltko-Rivera, M. (2004). The psychology of worldviews. *Review of General Psychology*, 8(1), 3–58.

Lindenmayer, D. B., Lane, P. W., Westgate, M. J., Crane, M., Michael, D., Okada, S. & Barton, P. S. (2014). An empirical assessment of the focal species hypothesis. *Conservation Biology*, 28(6), 1594–1603.

Linnell, J. D. C., Swenson, J. E. & Andersen, R. (2001). Predators and people: Conservation of large carnivores is possible at high human densities if management policy is favourable. *Animal Conservation*, 4, 345–9.

Lorimer, J. (2015). *Wildlife in the Anthropocene: Conservation after Nature*. Minneapolis: University of Minnesota Press.

Lukasik, V. M. & Alexander, S. M. (2012). Spatial and temporal variation of coyote (*Canis latrans*) diet in Calgary, Alberta. *Cities and the Environment (CATE)*, 4(1), art. 8.

McManus, J. S., Dickman, A. J., Gaynor, D., Smutts, B. H. & McDonald, D. W. (2015). Dead or alive? Comparing costs and benefits of lethal and non-lethal human–wildlife conflict mitigation on livestock farms. *Oryx*, 49(4), 687–95.

Miller, B., Reading, R., Strittholt, J., Carroll, C., Noss, R., Soulé, M., Sanchez, O., Terborgh, J., Brightsmith, D., Cheeseman, T. & Foreman, D. (1998). Using focal species in the design of nature reserve networks. *Wild Earth Special Issues*, Winter 1998/9, 81–92.

Morgan, D. L. (2008). *The SAGE Encyclopedia of Qualitative Research Methods*. Thousand Oaks, CA: SAGE Publications, Inc.

Philo, C. & Wilbert, C. (2000). *Animal Spaces, Beastly Places: New Geographies of Human–Animal Relations*. London: Routledge.

Pooley, S., Barua, M., Beinart, W., Dickman, A., Holmes, G., Lorimer, J., Loveridge, A. J., Macdonald, D. W., Marvin, G., Redpath, S., Sillero-Zubiri, C., Zimmerman, A. & Milner-Gulland, E. J. (2017). An interdisciplinary review of current and future approaches to improving human–predator relations. *Conservation Biology*, 31(3), 513–23.

Proctor, J. D. (1998). The social construction of nature: Relativist accusations, pragmatist and critical realist responses. *Annals of the Association of American Geographers*, 88(3), 352–76.

Prouxl, G. & Rodtka, D. (2015). Predator bounties in western Canada cause animal suffering and compromise wildlife conservation efforts. *Animals*, 5 (4), 1034–46.

Rust, N., Abrams, A., Challender, D., Chapron, G., Ghoddousi, A., Glikman, J. A., Gowan, C., Hughes, C., Rastogi, A., Said, A., Sutton, A., Taylor, N., Thomas, S., Unnikrishnan, H., Webber, A., Wordingham, G. & Hill, C. (2017). Quantity does not always mean quality: The importance of qualitative social science in conservation research. *Society & Natural Resources*, 30(10), 1304–10.

Sponarski, C., Semeniuk, C., Glikman, J. A., Bath, A. & Musiani, M. (2013). Heterogeneity among rural resident attitudes toward wolves. *Human Dimensions of Wildlife*, 18, 239–48.

Sponarski, C., Vaske, J. & Bath, A. (2015). The role of cognitions and emotions in human–coyote interactions. *Human Dimensions of Wildlife*, 20(3), 238–54.

Teddlie, C. & Tashakkori, A. (2009). *Foundations of Mixed Methods Research: Integrating Quantitative and Qualitative Approaches in the Social and Behavioral Sciences*. Los Angeles: Sage Publications.

Town of Cochrane. (2014). Cochrane growth rate up. Available from www.cochrane.ca/ (accessed September 2014).

Treves, A. & Bruskotter, J. T. (2014). Tolerance for predatory wildlife. *Science*, 344(6183), 476–7.

Urbanik, J. (2012). *Placing Animals: An Introduction to the Geography of Human–Animal Relations*. Lanham, MD: Rowman & Littlefield.

Wilson, R. S., Tucker, M., Hooker, N. H., LeJeune, J. T. & Doonan, D. (2008). Perceptions and beliefs about weed management: Perspectives of Ohio grain and produce farmers. *Weed Technology*, 22, 339–50.

Zaksek, M. & Arvai, J. L. (2004). Toward improved communication about wildland fire: Mental models research to identify information needs for natural resource management. *Risk Analysis*, 24(6), 1503–14.

16

Conservation Marketing As a Tool to Promote Human–Wildlife Coexistence

DIOGO VERÍSSIMO, BROOKE TULLY AND
LEO R. DOUGLAS

Ensuring the coexistence between humans and wildlife is a common concern within conservation science. This is however an inherently complex goal because it frequently involves both disputes between groups of people about wildlife, and undesirable interactions between people and wildlife. As the human footprint expands and the extent and quality of natural habitats decreases, new ways of facilitating the coexistence between wildlife and humans (sensu Carter & Linnell 2016) become increasingly important, requiring non-traditional methods of understanding and managing human–wildlife interactions. Conservationists have gone about this in a myriad of ways, with most emphasis historically being placed on interventions focused on the behaviour of the animals (e.g. erecting fences to limit movement). These early interventions were largely top-down and overly technical in focus, commonly paying little attention to the role of human perceptions, culture and behaviour (Dickman 2010). However, there is a growing body of work that has focused on the human side of human–wildlife coexistence (Gandiwa et al. 2013; Hazzah et al. 2014). This shift in scope has led to the implementation of interventions to change human behaviour: from legislation to control the effects of people on wildlife, which has perhaps been the most common measure (Redpath et al. 2013), to other less coercive activities such as education programmes (Espinosa & Jacobson 2012; Heberlein 2012; Sakurai et al. 2015) and conservation marketing initiatives (Saypanya et al. 2013; Booker & Maycock 2015).

Education-based interventions emphasize goals around learning and knowledge (Jacobson et al. 2015) and are more likely to be successful where the target audience is motivated (i.e. believes that change is in their best interest) and able to change (i.e. has the capacity to change in

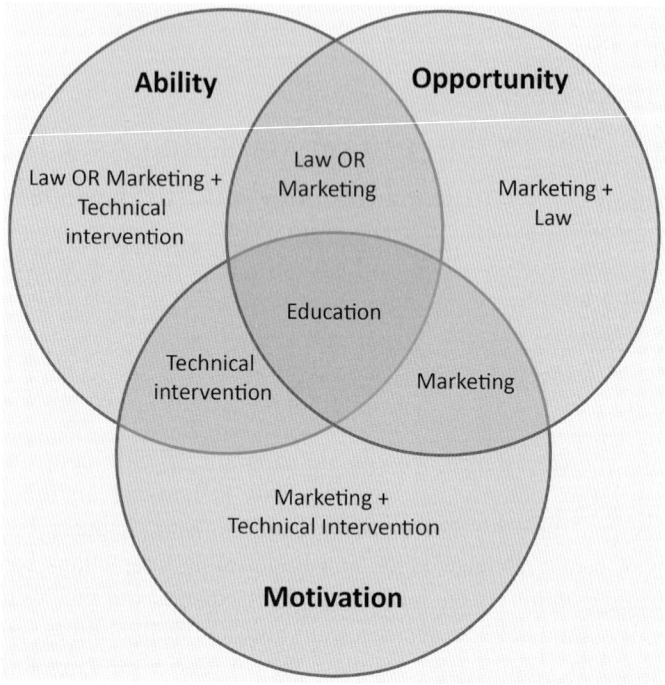

Figure 16.1 Types of behaviour change interventions suited to different contexts, defined by the ability, opportunity and motivation of the target audience to change.
Adapted from Rothschild (1999), Santos et al. (2011) and Smith et al. (2019).

the face of factors such as inertia, social norms, peer pressure, etc.). On the other hand, law-based interventions could be implemented where the target audience has no motivation to change (Figure 16.1). Yet these paradigms in isolation can only cover a portion of the different contexts where behaviour change may be needed (Rothschild 1999). This is why technical interventions[1] can be a vital part of behaviour change interventions (Figure 16.1), especially when there are physical or environmental barriers to change that may deny the target audience the opportunity to change (e.g. logistical challenge or lack of access to technology). Marketing can thus fill an important gap, both by itself and by working with other approaches (Figure 16.1).

---

[1] Technical interventions are defined as those aspects such as technology, infrastructure or equipment, that while material in nature can be critical to enable behaviour change to take place.

16.1 CONSERVATION MARKETING AND HUMAN–WILDLIFE COEXISTENCE

Conservation marketing, a subfield of social marketing,[2] is the ethical use of marketing concepts and principles to influence a target audience towards the adoption of more environmentally sustainable behaviours that benefit the individual as well as society (Wright et al. 2015; Veríssimo & McKinley 2016). Social marketing approaches are heavily based on commercial marketing, which employs a customer-oriented approach guided by a core set of factors called the *marketing mix* or *4Ps* (Product, Price, Place and Promotion). These factors are used to design and describe new behaviour change interventions.

Conservation marketing interventions are set out explicitly to influence behaviour as the ultimate objective of transforming relationships between people and nature. This is a critical difference as changes in knowledge or attitudes tend to be poor proxies for changes in behaviour (Heberlein 2012; Kollmuss & Agyeman 2002). The use of marketing concepts and principles towards social good had its beginning in the late 1960s (Kotler & Levy 1969) with marketing professionals starting to develop strategies to deal with public health concerns around sexually transmitted diseases, smoking and road safety. This led to the creation of the field of social marketing, which eventually expanded to consider environmental issues. For example, DeWan et al. (2013) used conservation marketing coupled with a technical intervention in the form of fuel-efficient stoves to limit competition between human and snub-nosed monkey populations in China, while Saypanya et al. (2013) used marketing and enforcement to reduce bushmeat hunting in Laos. In view of this, and the growing awareness of the potential of marketing to contribute towards conservation science (Wright et al. 2015; Veríssimo & McKinley 2016; Bennett et al. 2017), conservation marketing is increasingly considered an important tool to promote human wildlife coexistence (Carter & Linnell 2016). By emphasizing common interests and benefits for both humans and wild species, conservation marketing can help shift mental dispositions and behaviours from the negative to the neutral/positive side of the conflict-to-coexistence continuum.

---

[2] According to the International Social Marketing Association, social marketing seeks to develop and integrate marketing concepts with other approaches to influence behaviours that benefit individuals and communities for the greater social good.

Although marketing-based approaches have been used in a diversity of sensitive or contentious social contexts, such as sexually transmitted diseases (Pfeiffer 2004), teenage pregnancy (Marsiglio 1985) and even armed conflicts (Fattal 2014), there are surprisingly few documented interventions that have explicitly addressed conservation conflicts (notable exceptions include Saypanya et al. 2013; Booker & Maycock 2015). This is perhaps a sign of how recently marketing concepts and principles have arrived to the biodiversity conservation field. Based on the experience harnessed in other sectors, we expect conservation marketing to be able to play a vital role in moving humans towards coexistence at multiple levels, from addressing disputes over material damages caused by wildlife, to deep-rooted conflict that overwhelmingly focuses on interactions between stakeholders about wildlife management (see Madden & McQuinn 2014). In the first instance, conservation marketing can increase community buy-in and ownership of technical solutions used to address impacts caused by wildlife. In the second instance, conservation marketing can influence social norms that can drive behaviour. Conservation marketing can help reframe how issues and stakeholders are perceived and open the door to dialogue, which can support, for example, bottom-up conflict transformation paradigms (Madden & McQuinn 2014).

16.2 WHAT CONSERVATION MARKETING CAN OFFER

As a subfield of social marketing, conservation marketing interventions follow the same six defining benchmarks described by Andreasen (2002), namely:

1. Use of a Marketing Mix
2. Behavioural Focus
3. Audience Research
4. Audience Segmentation
5. Attractive Exchanges with Target Audience
6. Attention to Competition.

While not every intervention will have these elements in equal measure, they are a requirement for an intervention to have the label of conservation marketing. Below we link each benchmark with its role in addressing human–wildlife coexistence. Additionally, we illustrate these benchmarks by reflecting on the case of the *Share a place to live* behaviour change campaign (Text Box 16.1) in which the Philippine cockatoo was perceived by some stakeholders as an agricultural pest.

Text Box 16.1 Conservation Marketing Inspires Coexistence between Local Farmers and the Philippine Cockatoo

The Coexistence Challenge

The Philippine cockatoo (*Cacatua haematuropygia*), known locally as *Katala*, is endemic to the Philippines. The species is listed as Critically Endangered, largely due to the fast population reduction that has taken place since the 1980s. This decline has been caused by loss of the bird´s lowland habitat, trapping for the cage trade and persecution as a crop pest (BirdLife International 2016). One of the last species strongholds include the municipality of Dumaran, Palawan Island. However, most farmers on Dumaran still use slash-and-burn agricultural practices in upland areas, which affect an increasing amount of forested areas on steeper slopes. This farming practice was identified by local stakeholders as one of the primary threats facing the Philippine cockatoo (Lacerna-Widmann 2005).

Programme Objectives

- Increase perception among Dumaran population that cockatoos and people can live together in harmony to increase support for forest protection and more sustainable behaviours.
- Formally protect two of the remaining cockatoo habitats under municipal laws.
- Shift farming practices away from traditional slash-and-burn methods towards alternative farming techniques that do not cause forest habitat destruction.

Conservation Marketing Strategy

The strategy was to promote the conservation of resources that both people and the cockatoo rely upon (i.e. forests, water, space to live) and create a stronger connection between humans protecting their own home and protecting the cockatoo's home. The following campaign slogan was created for this purpose: *Share a place to live* (Lacerna-Widmann 2005) (see Table 16.1 for targeted audience).

Implementation Activities

The campaign used a total of fifteen marketing materials (i.e. posters, billboards, songs with key messages) and implemented twenty-two activities (i.e. week-long Katala festival, vegetable farming trainings) over an 18-month period (Lacerna-Widmann 2005). Engaging the press was key to reaching a wider audience: the Palawan residents. The press was instrumental in disseminating information, broadcasting the Katala festival, conducting live interviews and playing the song conceived as the campaign anthem on both radio and TV stations. In Palawan, the press is a trustworthy source of information, so its coverage added legitimacy to the campaign.

Methods

To measure changes in knowledge, attitudes and practices among the primary target audience of farmers and the wider community (Table 16.1),

Table 16.1 *Target audiences for the conservation marketing campaign* Share a place to live *in the Philippines*

Audiences	Desired behaviour
Primary target audience: Farmers of Dumaran Municipality	Adopt more sustainable farming practices instead of slash-and-burn.
Secondary audience: Wider Dumaran community (all citizens, including school children)	Support the protection of Katala by applying social pressure to farmers to adopt more sustainable practices and positively support measures that legally protect Katala and its habitat.
Enforcers: Wildlife wardens, local police, local government officials	Effectively enforce and implement laws and regulations without fear or apprehension.
Influencers: Priests, teachers	Include mention of environmental laws in sermons and classroom lectures to both famers and the wider community.
Policy-makers: Municipal government officials	Formally protect remaining cockatoo habitats under municipal law.

Figure 16.2 Poster used in the conservation marketing campaign promoting coexistence with the Philippine cockatoo, in Dumaran, Palawan Island, Philippines.

a pre- (before campaign activities began) and post-campaign (after campaign activities were completed) quantitative survey was conducted. Surveys were conducted in all sixteen villages (known locally as *barangays*) of Dumaran through the support of nineteen trained enumerators. Both the pre- and post-campaign survey included 638 respondents.

Results
- After the campaign, farmers believing that coexistence with the Katala was possible increased from 42% (n = 198) to 75% (n = 199).
- The Municipality of Dumaran secured two forest areas as formal protected areas under municipal legislation. Six volunteer wardens were trained to monitor the protected areas, with special attention given to the cockatoo's breeding season.
- In the two areas where the local protected areas were to be established, the number of community members that have reported illegal activities relative to the environment and wildlife increased from 8.5% (n = 47) and 3.2% (n = 31) pre-campaign, to 17.8% (n = 45) and 45.2% (n = 31) respectively.

Conclusion
This conservation marketing programme leveraged the Philippine cockatoo, a species historically involved in human–wildlife conflict, as a conservation flagship to increase community support in protecting the species through new legislation and shifts in behavioural practices around farming. Although conservation can often be received with scepticism and concern by local communities, as it typically represents obstacles and sacrifices, conservation marketing approaches can communicate positive benefits to the target audience for taking action. As seen in the Palawan case study, deeply understanding audiences and what they perceive to be attractive exchanges can help to accelerate adoption of other conservation interventions (legislation, adoption of alternative farming methods, etc.)

16.2.1 Marketing Mix

The marketing mix is a concept originally developed in commercial marketing that gives structure to the way a marketing intervention is planned. Although many variations exist, the most used format includes place, price, product and promotion (4 Ps). In the context of conservation marketing it is defined as follows:

- Product includes the desired behaviour and the products and services needed to help the target adopt it. This includes also the brand dimension of the product, which acts as an identity that summarizes the key features of the conservation marketing effort (Veríssimo et al. 2014).

- Price is the cost that the target audience associates with adopting the new behaviour. This cost most frequently is not monetary but measured in time, effort or social capital.
- Place is where and when the target audience will perform the desired behaviour and receive any associated services. This is important in terms of identifying the most opportune locations and times to reach your intended audience with conservation marketing messages.
- Promotion comprises the communication activities used to communicate the product benefits, its value in relation to the competitors and the place where it is available.

The marketing mix will aid human–wildlife coexistence programmes by providing an overall framework around which a comprehensive behaviour change intervention can be designed. This is particularly helpful in the context of social marketing where often the product being promoted is more abstract than in commercial marketing, where the focus is most often in tangible products.

16.2.2 Behavioural Focus

Conservation marketing interventions are built on the conviction that only through changes in behaviour can threats to biodiversity be mitigated. This focus on human behaviour means that conservation marketing interventions are built on a few key principles that define how these interventions are designed and evaluated. Regarding design, the first principle is that human behaviour is dynamic and can change repeatedly and in different ways, not only across time and space but also across different social contexts, which means that behaviour change is a continuous process without a clear endpoint. The second principle is that human groups tend to be heterogeneous when it comes to behavioural patterns, which creates the need to tailor an intervention to the specific group it intends to target (and to the social and geographic context the group lives in). The third principle is that many of the behaviours we engage in daily are driven by habit or social norm and not by conscious judgement, which means that breaking with long-established behavioural patterns can be difficult. The fourth and last

Table 16.2 *Example of the marketing mix for the farmer target audience of the conservation marketing campaign* Share a place to live

Product	1. Adoption of more sustainable farming practices, instead of using slash-and-burn, that will keep farming area healthier for longer periods of time (farmers do not have to continually find new space). 2. A free training in alternative farming practices.
Price	Time: Sustainable farming practices require more time and energy to complete, compared to the *quick fix* of slash-and-burn. Attending a training takes time away from work and family. Effort: New and different farming practices require acquiring new skills in farming, which takes longer to learn and develop efficiencies. Efficacy: Feeling confident in new farming skills comes after much practice, and not feeling confident at first is a cost and barrier to changing practices. Money: Loss of farming time and shifting practices may result in lower crop yield, which represents a loss of income for farmers.
Place	Adoption of more sustainable practices to take place at each farmer's land area. Training done at a single farm near to the municipal nursery and market area.
Promotion	Materials and messages targeting farmers will be placed in locations and channels most frequently accessed by this audience segment (based on data gathered during pre-campaign surveys and qualitative research): posters, fact sheets, waterproof stickers handed out to farmers during site visits; songs and programmes on local radio; festival in the centre of the municipality; placing billboards and signposts on major roads travelled by farmers; and conducting site visits at each farm location.

principle is that much of the decision-making that drives our behaviour is driven by our emotions rather than rational decision-making, which is why marketing strategies have to go beyond filling information gaps to connect emotionally with their target audiences (Brenkert 2002). In terms of evaluation, the focus on behavioural outcomes, as opposed to others focused on knowledge or attitudes, is also critical to ensure that lessons learned are meaningful. This means that conservation marketing interventions avoid the common pitfall of using changes in knowledge or attitudes as proxies for changes in behaviour (Kollmuss & Agyeman 2002). This will help to build, over time, an evidence base of which interventions are effective in supporting coexistence and why (Pfeiffer 2004; Redpath et al. 2013).

The above insights mean that conservation programmes being designed for human–wildlife coexistence (including, but not restricted to, conservation marketing) should begin with a clear understanding of the drivers of the relevant behaviours and the behavioural shifts needed to achieve conservation goals. Additionally, desired behaviours (e.g. not killing wildlife) must be clearly identified, along with motivators of why local stakeholders would be interested in adopting alternatives. In the *Share a place to live* campaign (Text Box 16.1), the behaviours targeted for modification were forest destruction for farming, and the killing of the Philippine cockatoo. Alternative behaviours promoted were high-value vegetable farming, and encouraging local stakeholders to report illegal forest or wildlife activities (such as slash-and-burn agriculture or the killing of the Philippine cockatoo).

16.2.3 Audience Research

Audience research is a central tenet of marketing. Extensive qualitative and quantitative research, in addition to the use of secondary data sources such as popular media sources, is important to gain insights on the target audience's existing beliefs, perceived benefits and barriers to change. Yet assigning enough time and resources to this initial stage is often a challenge for conservationists, who often are not trained in social science methods and tend to view conservation science as a crisis discipline demanding immediate action.

In conservation marketing, audience research serves three main purposes: (1) understand the target audience prior to the design of the intervention, (2) pre-test the intervention elements at the end of the design stage and (3) monitor the implementation of the intervention (Andreasen 2002). Audience research surrounding the *Share a place to live* (Text Box 16.1) campaign revealed the socio-cultural reasons for forest-depleting activities and the long-held belief that the existence of the Philippine cockatoo was incompatible with the local agrarian livelihoods. This research illustrated a lack of awareness about the cockatoo's uniqueness, and highlighted how environmental changes such as forest destruction, water unavailability and soil quality reduction were threatening local lives and livelihoods.

The use of research to understand the values, social norms and behaviours of the target audience is critical to ensure that an intervention is adequately adapted to the local social, political and cultural context. Pre-testing is therefore a fundamental process, as it allows the elements of a campaign to be tested and ensure that nothing is either missed or misinterpreted during the formative research states. This is crucial when dealing with sensitive subjects, as subtle changes in the elements of a campaign, such as the wording of a message or the aesthetics of a logo, can trigger wildly different interpretations, possibly conditioning the acceptance and impact of a campaign. One example that has received increasing attention is how the use of particular kinds of metaphors used to frame an issue determine how stakeholders may perceive each other (Campbell & Veríssimo 2015). Using military metaphors to describe the conflict around migratory bird hunting in Malta has largely reframed the issue as an animal welfare concern, thus made hunting in its entirety morally reproachable and further polarizing the issue (Campbell & Veríssimo 2015; Veríssimo & Campbell 2015). It should be noted that both pre-testing and formative research involve the use of qualitative research methods such as focus groups and multi-species ethnographies, which can be perceived as less reliable by conservation scientists (Bennett et al. 2017; Pooley et al. 2017). It is however crucial that this qualitative focus is maintained as only these methods allow for an in-depth exploration of how stakeholders relate to the complex issues surrounding coexistence. One example of this complexity is the symbolic nature that species and other stakeholder groups can assume, which reflects itself in the way stakeholders often redirect deeply rooted historical grievances towards local species and other stakeholders groups, even if those are not the drivers of the conflict (Douglas & Veríssimo 2013).

Lastly, in terms of implementation, audience research can help support the monitoring of the implementation of the intervention and ensure that a campaign is being adequately executed and well received. This is particularly important in interventions that are implemented over the medium or long term, since during conservation marketing efforts social, cultural and political conditions may change abruptly, something that a formal impact evaluation process would take too long to detect.

16.2.4 Audience Segmentation

Human groups tend to be heterogeneous in terms of their knowledge, values, attitudes and behaviours. When it comes to human–wildlife interactions, different groups are likely to have distinct probabilities of being affected by any given conflict and varying abilities to manage risk

Table 16.3 *Audience research conducted during the conservation marketing* Share a place to live *campaign in the Philippines*

Timing and purpose	Type of audience research	Summary of insights
Before campaign began: Formative research to better understand the audiences	Quantitative surveys (details on this included in case study introduction, Text Box 16.1). Focus group discussions with key stakeholders.	Community members and farmers did not believe that coexistence is possible, and are not sure why the Philippine cockatoo or its habitat should be protected. Major threat to the Philippine cockatoo is unsustainable farming practices.
Before launch of campaign materials and messages: Pre-testing of creative concepts	Focus groups and in-depth interviews with target audience members.	More accurate translation of slogan in local dialect. Enthusiasm around *Share a place to live* direction. High comprehension and understanding of messages and call-to-action.
During and after campaign implementation: Monitoring of campaign progress and evaluation of impact	In-depth interviews with all audience segments on perceptions of activities. Estimate of number of people reached by the campaign materials and post-campaign quantitative results on audience changes relative to baseline.	Most effective campaign activities were the Katala festival and poster/billboards, as they were easy to understand, relevant to the local context and surrounded the audiences with key messages. The least effective activity was the drawing contest and corresponding calendar, due to poor timing in implementation and low distribution after production. Percentage of community who had not heard anything about the Philippine cockatoo dropped from 41.2% (n = 263) at start of campaign to 16.1% (n = 90) after campaign.

and cope with potential damage (Pooley et al. 2017). Yet it is common to see conservation outreach efforts targeting either the *general public* or generic *local communities* as monolithic entities (Kanagavel et al. 2014). It is thus critical to realize that targeting everyone is targeting no one, as heterogeneity between groups precludes a campaign from being tailored to everyone's needs.

Conservation marketing interventions avoid the *one size fits all* fallacy by using the audience research described above to identify subgroups (or segments) of a population that share strategically important characteristics. This means audience segmentation should go beyond demographic variables (e.g. gender, age, income, education) or spatial data (e.g. nationality, postcode), since these rarely have a strong link to behavioural patterns. Thus, conservation marketing interventions should focus on behavioural variables, such as past behaviours and willingness to change, and psychographic variables, such as those relating to values, attitudes or belief systems. While it is true that there will always be a large number of ways to segment a group, the key will be to define the variables that are the most important to the conflict being considered, which will have to be done on a case-by-case basis.

Once a population is segmented, the next step is to prioritize these segments according to strategic criteria. These can be of a psychographic and/or behavioural nature, as described above, but can also be more geared towards implementation factors such as segment size, accessibility, openness, etc. One available approach to identifying the priority segments within a population is the TARPARE method (Donovan et al. 1999), which uses criteria such as accessibility, size and exposure to risk to inform prioritization (Table 16.4). The move from a highly heterogeneous population to more homogenous segments allows for a more effective tailoring of the intervention, therefore increasing the chances of success. At the same time, focusing only on the group or groups that are more strategically relevant for a given objective, makes it possible to avoid diluting the existing resources over a much larger group, therefore maximizing the effective reach of the available resources.

The *Share a place to live* campaign targeted mostly two audience segments: (1) farmers with messages about alternative farming practices, and (2) the wider community, including schoolchildren who could reinforce the developing social norm around sustainable farming practices that protect the Philippine cockatoo.

Table 16.4 *The TARPARE method of prioritizing audience segments.*

Adapted from Hopwood & Merritt (2011).

T	Total segment size: Is it large enough? Is it too large?
AR	At-risk: Proportion of those who would benefit
P	Persuasion: Is the segment easily persuaded? Is it likely to influence others?
A	Accessibility: How accessible is the segment?
R	Resources: What is required to influence behaviour?
E	Equity: What are the barriers for disadvantaged segments?

16.2.5 Attractive Exchanges with Target Audience

A critical conceptual aspect of marketing is the belief that a campaign is, in essence, a proposed exchange to the target audience, where the target audience is offered some benefits, tangible or intangible, for carrying out the desired behaviour. Framing outreach efforts in this way ensures that the perspective of the target audience is placed at the centre of the intervention design process.

The ultimate task of the conservation marketing intervention is to ensure that the benefits, both tangible and intangible, associated with the desired behaviour outweigh the costs. This can be achieved by increasing the benefits of adopting the new behaviour but also by reducing the barriers and costs of its adoption. Nevertheless, it is important that both alternatives are considered in the long term, as often the costs of maintaining a new behaviour, for example, can differ substantially from those associated with just its initial adoption. For example, in a behaviour change intervention designed to save natural habitats from deliberately set grass fires in Wales, the *Grass Is Green, Fire Is Mean* campaign offered effective alternative product-activities (graffiti, survival and CD recording classes) to offending youth. In the long term, however, these alternatives offered by the state proved to be too costly to maintain (McKenzie-Mohr et al. 2011). A final consideration in balancing cost and benefits of behavioural adoption, is to also understand the intangible benefits and costs. Those factors may differ substantially, and may be particularly salient, especially when they entail local stakeholder groups' concerns about or desire for identity, power or prestige. This is particularly the case when wildlife becomes a status symbol or iconic for the elite's power in local affairs (Peterson et al. 2002), or when wildlife-related conflicts cause significant psychological stresses or food insecurity (Peterson et al. 2002; Barua et al. 2013).

An underlying assumption is that the benefits of adopting the desired behaviour should accrue to the individual or group, making the change as compelling as possible. It is often an error in conservation interventions to emphasize how a change in behaviour will benefit wildlife, as for many stakeholder groups this is not compelling enough to motivate a change in behaviour. In the *Share a place to live* campaign, for example, programmes emphasized the common threats that both the cockatoos and farmers faced (i.e. unsustainable forest destruction, degradation of watersheds and agricultural soil). Additionally, research prior to intervention design identified that local residents were interested in putting their *island on the map*. Building on their desire for a more recognized identity, the campaign offered an attractive source of pride and rebranded the Philippine cockatoo as a symbol of national uniqueness.

16.2.6 Attention to Competition

Many conservationists often see themselves holding the moral high ground and in a position where, according to their personal values, there simply is no defensible alternative to the behaviour change they are promoting (Redford & Sanderson 2006; Macdonald et al. 2016). Yet the reality is that this is seldom true. Not only because conservationists tend to hold values that are not representative of the wider population, but also because habit or social norms impose heavy social costs on change, even if an individual is motivated to embrace it. This means that anyone attempting to influence human behaviour must have a clear idea of what other alternatives the target audience has to the desired behaviour, including continuing with the current behaviour (Buchanan et al. 1994). An important starting point in this process is to have a clear understanding of what compels the target audience to behave in the way they currently do as this will provide insights as to what factors are valued by the group. This information can then be used to gain competitive advantage over the current behaviour. The next step is to look for other potential behaviours that may compete with the alternative behaviour being promoted. It should be noted that these competitors may be direct or indirect and can be very different from the current behaviour or the proposed desirable behaviour.

To overcome the competition, conservation marketing interventions need to consider the costs and benefits described in the section above, and compare those to the costs and benefits (real and perceived) of the behaviour the target audience currently engages in. In the *Share a place to live* campaign the cost of unsustainable hunting and forest destruction was made more salient by increasing the enforcement of environmental laws and using influential community members to reinforce the social norm around the unacceptability of hunting the cockatoo. The campaign also took into account the economic appeal of the slash-and-burn agricultural practices by actively promoting economically viable alternative behaviours, such as the planting of vegetable plots that did not require forest destruction.

16.3 ON THE LEGITIMACY OF INFLUENCING HUMAN BEHAVIOUR

Some agree that human behaviours that are harmful to wildlife conservation should be respectfully challenged, even if based on culture and tradition (Dickman et al. 2015). Yet conservation and social marketing interventions have been reproached for being paternalistic or even manipulative (Andreasen 2002). This criticism is most commonly centred around the fact that any intervention to influence behaviour assumes that the target group is not behaving according to its own self-interest and that someone external to the target audience knows what is best, not only for that group but also for society as a whole.

These critiques are somewhat ironic in the context of conservation marketing, as historically conservationists have favoured more coercive paths to influencing behaviour, such as fines and legislation (Redpath et al. 2013). Conservationists have a long history of lobbying governments to enact laws that dictate the terms of use of a given species, resource or area (West & Brockington 2006). While in theory it is easy to deflect any accusations using the legitimacy of a government mandate, the reality is that this lobbying has included many governments known for their lack of a democratic mandate (see e.g. Ho 2001).

Conservation marketing does not curtail the freedom of choice of the target audience. Yet it is true that in most instances the goals of a conservation behaviour change intervention (e.g. increase the population of a species, reduce the use of a natural resource) are set prior to consultation with local stakeholders, who are usually consulted only about the best way to achieve those goals (Brenkert 2002). Given that conservation marketers are not democratically elected, this raises issues around the legitimacy of conservation marketing interventions, as goal setting for conservation marketing requires value judgements around what human behaviours should be changed and what coexistence should look like (Pielke Jr 2007). Conservationists should therefore acknowledge their role as stakeholders, not neutral arbiters (Redpath et al. 2015) and be transparent about the underlying values that motivate their goals. The most common option to overcome these obstacles is to work in partnership with democratic institutions (Fox & Kotler 1980), but in many contexts those may either not be available or suffer from governance weaknesses such as elite capture.[3] As such, it is key that conservation marketing interventions include a wider and meaningful consultation of stakeholders and ensure that evidence is available about the needs being addressed by a particular intervention (Brenkert 2002).

Lastly, culture is dynamic and will inevitably change over time through the action of agents such as civil society and the media. Thus, all those involved in any intervention to influence human behaviour would be aware of the possibility that through some unexpected causal pathway, the conflict that was to be mitigated worsens, which is particularly an issue in protracted and complex challenges such as those that commonly surround human–wildlife coexistence. However, it means that if conservationists refrain from engaging as other agents of societal change do, then there is little hope for them to become agents of societal change.

16.4 FIRST DO NO HARM

The implementation of conservation marketing programmes entails several pitfalls that can undermine their impact. In many conservation contexts, the failure of an intervention represents the onerous loss of time and resources. However, in the case of human–wildlife coexistence, there are multiple ways through which interventions could even have unintended negative consequences, further fuelling or simply displacing the conflict conservationists were hoping to mitigate. It is thus worth understanding these pitfalls, and considering them when designing and implementing any behaviour change intervention aimed at promoting human–wildlife coexistence.

A potential challenge emerging from focusing on a specific human–wildlife interactions is hyper-saliency, where events that were previously

---

[3] Elite capture is a process where a minority group of individuals of superior social, economic, political, educational or ethnic status misuse, usually for their own gain, resources designated for the benefit of the wider community.

perceived to be part of a livelihood are reinterpreted as unacceptable problems due to the focus placed on the issue by an outside actor (Pooley et al. 2017). This situation can be further intensified if the conflict becomes associated with those hoping to mitigate it, with for example problem animals becoming the responsibility of conservationists (Macdonald et al. 2010; Douglas & Veríssimo 2013). Conservation marketing can reduce this risk by avoiding top-down communication, and instead use more inclusive approaches that unite communities around a new norm.

Another pitfall that can fuel conflict is oversimplification, by for example reducing to material damage a conflict that may be predominately taking place in the psychosocial realm (Ginges et al. 2007). This fallacy tends to be driven by a belief in Maslow's *hierarchy of needs* (Maslow 1943) which posits that until one's physiological (e.g. food, water, shelter) and security (e.g. physical, employment, health) needs are met, people are less concerned with 'higher level' social and psychological needs (e.g. self-esteem, belonging). Despite seeming intuitive, Maslow's framework has been widely refuted (Madden & McQuinn 2014). Psychological drivers of belonging and self-efficacy are vitally important factors in motivating changes in behaviour, and should thus be explored when designing marketing messages for human–wildlife coexistence campaigns.

Lastly, there is a potential to displace conflict when the development of pro-conservation behaviours towards a species may reduce the importance given to other elements of biodiversity. For example, on the island of Dominica, a conservation marketing programme had the unintended consequence of reframing how local stakeholders perceived a wildlife conflict. While more favourable attitudes and behaviours developed towards the imperial parrot (*Amazona imperialis*), a species marketed as the conservation flagship, a sister species, the red-necked parrot (*Amazona arausiaca*) became socially marginalized, magnifying perceptions of the later species as a pest for crops (Douglas & Winkel 2014). These effects underscore the fact that in the long term effective behaviour change programmes require continual audience research to remain relevant to changing socio-cultural contexts (Andreasean 2002).

16.5 OPPORTUNITIES AND CHALLENGES

The uses of marketing in the environmental realm have so far taken place in contexts where the goals were broadly consensual (e.g.

recycling, transportation, energy conservation), which meant that push-back from stakeholders was less likely and any unintended consequences more modest in magnitude (Langford & Panter-Brick 2013; Pfeiffer 2004). Improving the way conservation marketing is used in the context of human–wildlife coexistence will require better impact evaluation as to improve the quality of the learning and avoid repeating mistakes that could have important social and political costs (Pfeiffer 2004; Veríssimo et al. 2018).

Currently there is widespread reliance on self-reported behavioural indicators, which while easy and cheap to collect, have been proven to be poor proxies for actual behaviour even in situations where the behaviour is not contentious (Kormos & Gifford 2014). Given that when working on topics where conflicts exists, social biases are likely to provide strong incentives for respondents not to answer truthfully, it is clear that more accurate techniques are needed. Happily the use of specialized techniques for sensitive questions is growing in conservation (Nuno & St John 2015), something that conservation marketing interventions can draw from.

An additional key improvement when it comes to impact evaluation would be the wider use of qualitative data collection techniques to better understand the casual pathways followed in an intervention (Pfeiffer 2004). Qualitative research is most often used in marketing in the formative stages but it can also be used to avoid interventions becoming *black boxes* where the impacts are accurately measured but their drivers remain unknown (Pfeiffer 2004; Langford & Panter-Brick 2013). Another fundamental challenge for conservation marketing interventions will be ensuring that they do not deepen social inequality. For example, the process of segmentation and targeting can easily exclude those deemed out of reach due to their social status or geographic location (Laczniak & Murphy 1994). At the same time even when included, there are segments of the population that may not be able to change their behaviour for reasons completely outside of their control (e.g. extreme poverty) (Brenkert 2002; Langford & Panter-Brick 2013). This can thus deepen inequality, with those who could most benefit from an intervention often being those least able to take advantage from it. To counter this, conservation marketing interventions will have to be designed bearing in mind the context-specific barriers that may stop groups from engaging in the desired behaviour. While frameworks such as TARPARE (Table 16.4), highlighted above, will likely help, ensuring conservation marketing interventions do not deepen social inequality will require a deeper integration of conservation marketing interventions with public policy. This will create more demand for combining upstream (i.e. policy-makers) and downstream marketing (i.e. resource users) (Langford & Panter-Brick 2013; Lorenc et al. 2013).

Conflicts involving wildlife are an important concern within global conservation efforts. The behaviour change focus of conservation marketing will be an important tool to support the engagement of audiences and facilitate coexistence-oriented approaches. As the tool of conservation marketing gets applied towards more environmental challenges, the field will continue to learn, grow and expand the learning and possibilities of application, providing a new window of opportunity for human–wildlife coexistence.

16.6 RECOMMENDATIONS AND FUTURE DIRECTIONS

- Conservation marketing has the potential to help mitigate a diverse set of conflicts with and about wildlife. By focusing on clear behavioural objectives, based on thorough audience research, by recognizing the heterogeneity within any human population and by looking at behaviour change programmes as meaningful exchanges, interventions that follow conservation marketing principles can avoid pitfalls that have undermined past outreach programmes.
- The use of marketing around biodiversity conservation has been limited and has mostly focused on relatively consensual issues where opposition from stakeholders was likely to be limited. To tackle the additional challenges around human–wildlife coexistence conservationists will need to improve their ability to both gain a deep understanding of stakeholders and to robustly evaluate the impact of their interventions. This first will allow for a better understanding of actual and perceived benefits and costs of different behaviours. The second will allow for better learning from past practices, avoid the dissemination of ineffective interventions and drive improvement over time.
- This additional social complexity also means that conservationists need to be aware of the potential for unintended consequences stemming from a behaviour change intervention. Where negative, these can be highly costly in social terms and raise ethical questions around the legitimacy of conservationists to define, for example, what human–wildlife coexistence should look like. These challenges will require full transparency towards stakeholders from those implementing conservation marketing interventions.

16.7 References

Andreasen, A. R. (2002). Marketing social marketing in the social change marketplace. *Journal of Public Policy & Marketing*, 21(1), 3–13.

Barua, M., Bhagwat, S. A. & Jadhav, S. (2013). The hidden dimensions of human–wildlife conflict: Health impacts, opportunity and transaction costs. *Biological Conservation*, 157, 309–16.

Bennett, N. J., Roth, R., Klain, S. C., Chan, K., Christie, P., Clark, D. A., Cullman, G., Curran, D., Durbin, T. J. & Epstein, G. (2017). Conservation social science: Understanding and integrating human dimensions to improve conservation. *Biological Conservation*, 205, 93–108.

BirdLife International. (2016). *Cacatua haematuropygia*. The IUCN Red List of Threatened Species 2016: e.T22684795A93047114. Available from http://dx .doi.org/10.2305/IUCN.UK.2016–3.RLTS.T22684795A93047114.en (accessed March 2018).

Booker, C. & Maycock, S. (2015). Conservation on island time: Stakeholder participation and conflict in marine resource management. In F. Madden, E. Parsons, J.-B. McCarthy and C. Parsons, eds., *Human–Wildlife Conflict: Complexity in the Marine Environment*. Oxford: Oxford University Press, pp. 21–38.

Brenkert, G. G. (2002). Ethical challenges of social marketing. *Journal of Public Policy & Marketing*, 21(1), 14–25.

Buchanan, D. R., Reddy, S. & Hossain, Z. (1994). Social marketing: A critical appraisal. *Health Promotion International*, 9(1), 49–57.

Campbell, B. & Veríssimo, D. (2015). Black stork down: Military discourses in bird conservation in Malta. *Human Ecology*, 43(1), 79–92.

Carter, N. H. & Linnell, J. D. (2016). Co-adaptation is key to coexisting with large carnivores. *Trends in Ecology & Evolution*, 31, 575–8.

DeWan, A., Green, K., Xiaohong, L. & Hayden, D. (2013). Using social marketing tools to increase fuel-efficient stove adoption for conservation of the golden snub-nosed monkeys, Gansu Province, China. *Conservation Evidence*, 10, 32–6.

Dickman, A. J. (2010). Complexities of conflict: The importance of considering social factors for effectively resolving human–wildlife conflict. *Animal Conservation*, 13, 458–66.

Dickman, A. J., Johnson, P. J., van Kesteren, F. & Macdonald, D. W. (2015). The moral basis for conservation: How is it affected by culture? *Frontiers in Ecology & the Environment*, 13(6), 325–31.

Donovan, R. J., Egger, G. & Francas, M. (1999). TARPARE: A method for selecting target audiences for public health interventions. *Australian and New Zealand Journal of Public Health*, 23(3), 280–4.

Douglas, L. R. & Veríssimo, D. (2013). Flagships or battleships: Deconstructing the relationship between social conflict and conservation flagship species. *Environment & Society*, 4, 98–116.

Douglas, L. R. & Winkel, G. (2014). The flipside of the flagship. *Biodiversity & Conservation*, 23(4), 979–97.

Espinosa, S. & Jacobson, S. K. (2012). Human–wildlife conflict and environmental education: Evaluating a community program to protect the Andean bear in Ecuador. *The Journal of Environmental Education*, 43(1), 55–65.

Fattal, A. L. (2014). *Guerrilla Marketing: Information War and the Demobilization of FARC Rebels*. PhD Dissertation. Cambridge, MA: Harvard University.

Fox, K. F. & Kotler, P. (1980). The marketing of social causes: The first 10 years. *The Journal of Marketing*, 44(4), 24–33.

Gandiwa, E., Heitkönig, I., Lokhorst, A., Prins, H. & Leeuwis, C. (2013). CAMPFIRE and human–wildlife conflicts in local communities bordering northern Gonarezhou National Park, Zimbabwe. *Ecology & Society*, 18, art. 7.

Ginges, J., Atran, S., Medin, D. & Shikaki, K. (2007). Sacred bounds on rational resolution of violent political conflict. *Proceedings of the National Academy of Sciences*, 104(18), 7357–60.

Hazzah, L., Dolrenry, S. Naughton, L., Edwards, C. T., Mwebi, O., Kearney, F. & Frank, L. (2014). Efficacy of two lion conservation programs in Maasailand, Kenya. *Conservation Biology*, 28(3), 851–60.

Heberlein, T. A. (2012). Navigating environmental attitudes. *Conservation Biology*, 26(4), 583–5.

Ho, P. (2001). Greening without conflict? Environmentalism, NGOs and civil society in China. *Development & Change*, 32(5), 893–921.

Hopwood, T. & Merritt, R. (2011). *Big Pocket Guide to Using Social Marketing for Behaviour Change*. London: NSMC.

Jacobson, S. K., McDuff, M. & Monroe, M. (2015). *Conservation Education and Outreach Techniques*. Oxford: Oxford University Press.

Kanagavel, A., Raghavan, R. & Veríssimo, D. (2014). Beyond the 'general public': Implications of audience characteristics for promoting species conservation in the Western Ghats Hotspot, India. *Ambio*, 43(2), 138–48.

Kollmuss, A. & Agyeman, J. (2002). Mind the gap: Why do people act environmentally and what are the barriers to pro-environmental behavior? *Environmental Education Research*, 8(3), 239–60.

Kormos, C. & Gifford, R. (2014). The validity of self-report measures of proenvironmental behavior: A meta-analytic review. *Journal of Environmental Psychology*, 40, 359–71.

Kotler, P. & Levy, S. J. (1969). Broadening the concept of marketing. *The Journal of Marketing*, 33(1), 10–15.

Lacerna-Widmann, I. D. (2005). Final Report. Rare Pride Campaign Dumaran, Palawan, Philippines. Rare Diploma in Conservation Education, University of Kent at Canterbury, United Kingdom, Kent Cohort III.

Laczniak, G. R. & Murphy, P. E. (1994). *Ethical Marketing Decisions: The Higher Road*. Upper Saddle River, NJ: Prentice Hall.

Langford, R. & Panter-Brick, C. (2013). A health equity critique of social marketing: Where interventions have impact but insufficient reach. *Social Science & Medicine*, 83, 133–41.

Lorenc, T., Petticrew, M., Welch, V. & Tugwell, P. (2013). What types of interventions generate inequalities? Evidence from systematic reviews. *Journal of Epidemiology & Community Health*, 67(2), 190–3.

Macdonald, D. W., Johnson, P. J., Loveridge, A. J., Burnham, D. & Dickman, A. J. (2016). Conservation or the moral high ground: Siding with Bentham or Kant. *Conservation Letters*, 9(4), 307–8.

Macdonald, D. W., Loveridge, A. J. & Rabinowitz, A. (2010). Felid futures: Crossing disciplines, borders and generations. In D. W. Macdonald & A. J. Loveridge, eds., *Biology and Conservation of Wild Felids*. Oxford: Oxford University Press, pp. 599–650.

Madden, F. & McQuinn, B. (2014). Conservation's blind spot: The case for conflict transformation in wildlife conservation. *Biological Conservation*, 178, 97–106.

Marsiglio, W. (1985). Confronting the teenage pregnancy issue: Social marketing as an interdisciplinary approach. *Human Relations*, 38(10), 983–1000.

Maslow, A. H. (1943). A theory of human motivation. *Psychological Review*, 50(4), 370–96.

McKenzie-Mohr, D., Lee, N. R., Schultz, P. W. & Kotler, P. (2011). Protecting fish and wildlife habitats. In D. McKenzie-Mohr, N. R. Lee, P. W. Schultz & P. A. Kotler, eds., *Social Marketing to Protect the Environment: What Works*. Thousand Oaks, CA: Sage Publications Inc., pp. 109–32.

Nuno, A. & St John, F. A. (2015). How to ask sensitive questions in conservation: A review of specialized questioning techniques. *Biological Conservation*, 189, 5–15.

Peterson, M. N., Peterson, T. R., Peterson, M. J., Lopez, R. R. & Silvy, N. J. (2002). Cultural conflict and the endangered Florida Key deer. *Journal of Wildlife Management*, 66(4), 947–68.

Pfeiffer, J. (2004). Condom social marketing, Pentecostalism, and structural adjustment in Mozambique: A clash of AIDS prevention messages. *Medical Anthropology Quarterly*, 18(1), 77–103.

Pielke Jr, R. A. (2007). *The Honest Broker: Making Sense of Science in Policy and Politics*. Cambridge: Cambridge University Press.

Pooley, S., Barua, M., Beinart, W., Dickman, A., Holmes, G., Lorimer, J., Loveridge, A., Macdonald, D., Marvin, G. & Redpath, S. (2017). An interdisciplinary review of current and future approaches to improving human–predator relations. *Conservation Biology*, 31(3), 513–23.

Redford, K. H. & Sanderson, S. E. (2006). No roads, only directions. *Conservation & Society*, 4(3), 379–82.

Redpath, S. M., Bhatia, S. & Young, J. (2015). Tilting at wildlife: Reconsidering human–wildlife conflict. *Oryx*, 49(2), 222–5.

Redpath, S. M., Young, J., Evely, A., Adams, W. M., Sutherland, W. J., Whitehouse, A., Amar, A., Lambert, R. A., Linnell, J. D. & Watt, A. (2013). Understanding and managing conservation conflicts. *Trends in Ecology & Evolution*, 28(2), 100–9.

Rothschild, M. L. (1999). Carrots, sticks, and promises: A conceptual framework for the management of public health and social issue behaviors. *The Journal of Marketing*, 63, 24–37.

Sakurai, R., Jacobson, S. K., Matsuda, N. & Maruyama, T. (2015). Assessing the impact of a wildlife education program on Japanese attitudes and behavioral intentions. *Environmental Education Research*, 21(4), 542–55.

Santos, C. O., Simões, A., Atalaia, J., Lopes, N., Maranga, A., Gil, C. & Ribeiro, S. (2011). *Melhorar a Vida: Um Guia de Marketing Social*. Portugal: Fundação CEBI.

Saypanya, S., Hansel, T., Johnson, A., Bianchessi, A. & Sadowsky, B. (2013). Effectiveness of a social marketing strategy, coupled with law enforcement, to conserve tigers and their prey in Nam Et Phou Louey National Protected Area, Lao People's Democratic Republic. *Conservation Evidence*, 10, 57–66.

Smith, R. J., G. R. Salazar, J. Starinchak, L. A. Thomas-Walters, and D. Veríssimo. 2019. Social marketing and conservation in W. J. Sutherland, P. Brotherton, Z. Davies, N. Pettorelli, B. Vira, and J. Vickery, editors. Conservation Research, Policy and Practice. Cambridge University Press, Cambridge, UK.

Veríssimo, D., Bianchessi, A., Arrivillaga, A., Cadiz, F. C., Mancao, R. & Green, K. (2018). Does it work for biodiversity? Experiences and challenges in the evaluation of social marketing campaigns. *Social Marketing Quarterly*, 24(1), 18–34.

Veríssimo, D. & Campbell, B. (2015). Understanding stakeholder conflict between conservation and hunting in Malta. *Biological Conservation*, 191, 812–18.

Veríssimo, D., Fraser, I., Girão, W., Campos, A. A., Smith, R. J. & MacMillan, D. C. (2014). Evaluating conservation flagships and flagship fleets. *Conservation Letters*, 7(3), 263–70.

Veríssimo, D. & McKinley, E. (2016). Introducing conservation marketing: Why should the devil have all the best tunes? *Oryx*, 50(1), 14.

West, P. & Brockington, D. (2006). An anthropological perspective on some unexpected consequences of protected areas. *Conservation Biology*, 20(3), 609–16.

Wright, A. J., Veríssimo, D., Pilfold, K., Parsons, E., Ventre, K., Cousins, J., Jefferson, R., Koldewey, H., Llewellyn, F. & McKinley, E. (2015). Competitive outreach in the 21st century: Why we need conservation marketing. *Ocean & Coastal Management*, 115, 41–8.

17

Leaping Forward

The Need for Innovation in Wildlife Conservation

LEELA HAZZAH, SALISHA CHANDRA AND
STEPHANIE DOLRENRY

As stated in Chapter 1, human–wildlife conflict is one of the most critical threats to wildlife, in particular the large carnivores; when they injure or kill domestic animals and threaten people, more often than not, they are killed in retaliation. For example, African lions (*Panthera leo*) have declined by at least 43 per cent over the past 21 years (Bauer et al. 2016) and indiscriminate killing by people poses the greatest threat (IUCN 2006). Conservationists have worked for decades to reduce the killing of threatened large carnivore species with the aim to increase their populations. The end goal has been to achieve coexistence and sustainable wildlife populations (Woodroffe et al. 2005).

We suggest that a linear view of coexistence is limiting (i.e. there is no end destination) and can reduce conservationists' ability to successfully understand the conservation context and implement effective long-term successful initiatives. To overcome this limitation, it may be useful to borrow insights from the business world and pedagogy of innovation to elucidate our understanding of how to maintain long-term coexistence. Innovation is commonly regarded as a new way, or at least a perception of a new way, of doing something within a given context (Zaltman et al. 1973; Cantwell 1989). In this chapter, we explore two types of innovation, incremental and radical, through the case study of a field-based lion conservation programme, Lion Guardians (www.lionguardians.org). Incremental innovation, which involves a low degree of new knowledge, moderately improves performance (e.g. minor adjustments to current conditions); whereas the higher risk radical innovation, involving a high degree of new knowledge, can enhance outcomes and performance in a dramatic fashion. Radical innovation, therefore, represents a clear departure from existing practice (Duchesneau et al. 1979; Ettlie et al. 1984;

Dewar & Dutton 1986) such as when Amazon leveraged emerging technologies and introduced a completely new business model within the need for physical retail space, or when rhino conservations introduced GPS and Google Glass technologies for anti-poaching efforts (Ortolani 2016). Radical innovation involves the development of a novel idea, demands out-of-the box thinking, yet the chance of success can be difficult to estimate, often resulting in opposition to such ideas (McKeown 2008; Biggs et al. 2010; Un 2010).

Until now, there have been only small pockets of hope in the conservation arena, with failures and evidence indicating a global decline of wildlife populations (Brashares et al. 2014). Since time is of the essence, we suggest that conservationists should restrict their focus on incremental innovations, as the pace of change fostered by this approach is too slow to actually save many of the declining species. Instead, a focus on radical innovation to shake up the conservation agenda is necessary. Pursuing an interdisciplinary approach, borrowing from other advanced fields and ultimately taking bigger risks will provide the impetus for change that the current conservation landscape needs. Furthermore, without substantial long-term committed funds these 'radical' ideas cannot come to fruition. Unlike the business world, it is inordinately more difficult to raise funds for conservation efforts. Not only do businesses have access to various sources of funding, but also their ability to raise these funds is directly linked to results. The better the results, the more funds will flow towards that business. For the most part, this is not the case in conservation, making it difficult for these radical innovations to take hold and have impact on a larger scale.

17.1 THEORETICAL FRAMEWORK

In the simplest form, innovation refers to a change in the way something is done (Carrillo-Hermosilla et al. 2010). Most innovation takes place in the incremental mode (Hellstrom 2007; Van den Bergh et al. 2011). It is increasingly acknowledged in the literature that focusing on incremental innovation along traditional channels does not suffice for attaining challenging environmental goals, such as reducing human–wildlife conflict (Tukker & Butter 2007; Carrillo-Hermosilla et al. 2010). In addition, long-term incremental innovations cannot be sustained without radical innovation since an incremental effort will face decreasing marginal returns (Hellstrom 2007). Similarly, in the business world, CEOs want their organizations to innovate strategically. In particular,

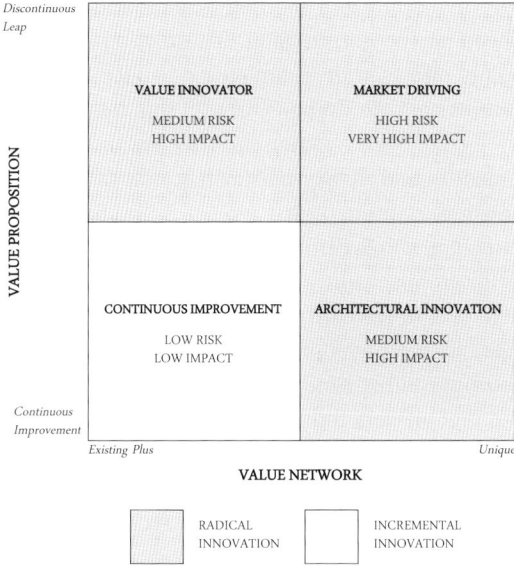

Figure 17.1 Types of strategic innovation.
Adapted from Kumar et al. 2000.

they stress the value of radical innovation because it allows a company to surge past its competitors and deliver sustained growth by creating new markets and/or by changing the rules of the game (Kumar et al. 2000). These strategic innovations are classified into four different quadrants based on how they modify the value proposition (combination of bene-fits and price offered to customers) or how they reconfigure the value network (alignment of activities required to create, produce and deliver the value proposition to the customer) and in special cases how they innovate along both these dimensions (Kumar et al. 2000) (Figure 17.1).

Most firms operate in the lower left-hand quadrant, using careful market research of customer needs to develop a slightly differentiated product or service. Companies such as Nestlé operate rather effectively in this space of *continuous improvement*. Take for example, Coffee Mate™ (creamer) that was introduced by Nestlé over 50 years ago. Since then it has essentially made minor improvements to suit customer changes such as providing non-dairy options (almond and coconut milk-based) and different flavours (caramel, macchiato, etc.). These are essentially minor modifications to the value proposition offered to the customer. Moving along the value proposition dimension, the upper left-hand quadrant is composed of *Value Innovators* or companies that focus on delivering superior value on key factors to their customers, not just on staying marginally ahead of the competition. These companies do not expend their resources to offer certain product and service features just because that is what their rivals are doing. Through this they realize cost savings that are transferred to the customer and they are also able to free up their resources to identify and deliver completely new sources of value (benefits). An interesting example of this type of innovation is Kinepolis – the world's first megaplex that came into existence when the movie theatre industry in Belgium was declining steadily. Many cinema companies shut down and there was fierce competition among the remaining operators. While all other cinema operators turned their cinemas into multiplexes with small viewing rooms, Kinepolis built rooms with 700 seats, and so much legroom that viewers did not have to move when someone passed by. The seats themselves were oversized with individual armrests and configured on a steep slope in the floor to ensure everyone an unobstructed view. The screens measured up to 29 meters by 10 meters and rested on their own foundations so that sound vibrations were not transmitted among screens. Kinepolis was able to offer this radically superior cinema experience without increasing ticket prices because the concept of the megaplex resulted in one of the lowest cost structures in the industry. By going against the grain and providing customers with a discontinuous leap in benefits at the same cost of an inferior product, Kinepolis was able to take over 50 per cent of the market in its first year (Kim & Mauborgne 2004).

Diagonally opposite the *Value Innovator* lies *Architectural Innovation*. Companies in this segment offer a similar product or service to others in the same industry but they differentiate themselves by radically redefining components or the entire method in which the value proposition is delivered to the customer. For instance, easyJet (a low-cost airline launched 20 years ago in the UK) systematically redefined the purchasing, operations, marketing and distribution components within the value network to deliver low prices at a profit. easyJet managed to achieve distribution savings of about 20–25 per cent over other full-service carriers (e.g. British Airways) by encouraging internet sales, not using travel agents, not issuing paper tickets and not participating in industry reservation systems. Furthermore, it used a yield management tool to maximize revenues for each flight based on matching supply and demand. Most importantly, much of the saving in its value network was generated through radically streamlining the operations using fast turnaround times, a single type of airplane and elimination of kitchen and business classes.

Both *Value Innovators* and *Architectural Innovation* are radical innovations on a single dimension. However, on the upper right-hand quadrant lie *Market-Driving* firms such as Amazon, The Body Shop and IKEA which have created new markets and revolutionized existing industries. The success of such *Market-Driving* firms is rooted in radical innovation on two dimensions: a discontinuous leap in the value proposition and the rapid configuration of a unique value network (Kumar et al. 2000). We will delve into the details of this quadrant below by providing a conservation case study, Lion Guardians.

As evidenced by the success of companies such as Amazon, *Market-Driving* innovations bust the status quo leading to higher value for both the business and the customers than any other type of innovation. Nonetheless, the majority of innovation in the business world is incremental because organizations believe that creating a differentiated product or service based on detailed market research of customer leads to success (Kumar et al. 2000). In other words, most companies tend to operate in the bottom left-hand quadrant of Figure 17.1: inching along with *Continuous Improvement*. Similarly, most conservationists are naturally risk averse due to the apprehension of wasting philanthropic donations. In addition, the success or failure of a conservation organization is often directly linked to human life and livelihoods, making it much harder to recover from a financial or reputational loss than in the business world. To put this into the carnivore conservation context, field organizations primarily focus on developing new mitigation techniques to reduce human–wildlife conflict. These incremental innovations often only solve part of the problem or shift the problem. For example, many conservation organizations working on carnivore conflict issues invest resources into building better bomas/livestock corrals. From fortified steel mesh to *lion lights*, many options are being tried and tested across Africa (Okello et al. 2014; Manoa & Mwaura 2016). While there is evidence that this kind of incremental innovation may reduce predators entering homes by fortifying the livestock enclosure (Okemwa 2015; Manoa & Mwaura 2016), it does not solve the issue of human–wildlife conflict outside of the boma, where much of the conflict takes place. Furthermore, it could also shift predatory behaviour to a neighbour's home and/or cause people to become less risk averse to potential predator damage. Carnivores could also adapt to the changes/product improvements, making them obsolete. As a result, the impact of these incremental innovations may not be sustained in the long run.

Conversely, we need to consider how immense the impact of conservation could be if we chose a higher risk, market-driving approach to the issues we are combatting, i.e. operated in the top right-hand quadrant of Figure 17.1. Take for example the Aravind Eye Hospital in India that was founded to *eradicate needless blindness*. It began as a modest eleven-bed hospital in the house of the founder and today is one of the largest eye-care systems in the world with over 4,000 beds, servicing close to 7 million patients, 60 per cent of whom receive the service free of charge. The secret behind this lasting and high impact non-profit organization is a radically innovative approach to funding, marketing and delivery of eye-care to both the wealthy and the impoverished (Rangan & Thulasiraj 2007). Specifically, Aravind developed a self-funding model where the 40 per cent who are able to pay for the service provide enough margin for the hospital to deliver the same level of treatment to the 60 per cent who cannot (Rangan & Thulasiraj 2007). In other words, the hospital provides identical service to all its patients regardless of their ability to pay. In this way, the founder managed to change the rules of the game by providing a high quality and competent service to all patients (a higher value proposition). They also standardized eye-care treatment allowing for mass delivery of services (a unique value network). This spurred a discontinuous leap in the eye-care industry where patients from both extremes of the economic spectrum come to the same hospital for eye-care. In the past, the poor and the rural village dwellers could only go to public hospitals where they may or may not have gotten what they needed. This example illustrates how a non-profit organization, even when the stakes are high, can make lasting and sustainable impact through radical innovation.

It is, however, important to note that both types of innovation are required for sustained impact and that one cannot exist without the other. The association between incremental and radical innovation in conservation can be understood in terms of the adaptive cycle from the social-ecological system (SES) theory, which focuses on interactions between people and ecosystem (Berkes & Folke 2000; Gunderson 2001; Holling 2001; Biggs et al. 2010). This is particularly important when managing complex social and ecological systems, where not enough information is available, since that means projects and organizations have to learn to adapt to uncertainties (Holling 1973; Holling & Meffe 1996). Feedback loops are used to help improve management decisions and adapt to uncertainty (Olsson et al. 2004). The front loop can be seen as incremental innovation that improves and strengthens the current effort and change can be slow and more deliberate; while the back loop provides an opportunity for more radical innovation and unpredictable change is introduced (Plowman et al. 2007; Biggs et al. 2010; Westley & Antadze 2010).

Similar to the continuum of coexistence, there is a continuum of innovation (Hage 1980). In the business world, the history of innovation consists of one-off radical innovations that disrupt industries followed by minor incremental changes. Following the initial burst of radical innovation, similar or slightly superior products enter the market until the next radical innovation redefines the industry and so on. For example, IKEA burst into the furniture retail market in the 1950s with a unique value network and high value proposition, disrupting the way furniture was purchased (Kling & Goteman 2003). It focused on young people and families and developed big stores outside of city centres where customers picked the products themselves in direct contrast with the way traditional furniture stores operated. Not only did IKEA change the way the product was delivered, it provided classic clean Scandinavian designs at affordable prices, thereby also upping the value proposition for the customer. By effectively turning the furniture retail market on its head, IKEA has grown from strength to strength; from a mail order company that used a milk van to deliver its furniture, today IKEA is a multi-channel home furnishing retailer with 183,000 employees, turnover of $37.6 billion and 392 stores worldwide.

Similarly, in the conservation arena when Lion Guardians was initiated in early 2007, it was with a unique approach that also increased the benefits to the communities that live with wildlife. Wildlife conservation traditionally focused on the species in question, not the people. Lion Guardians took the opposite approach and focused on the people. Together, conservationists and communities developed a model that blended traditional knowledge and culture with science that resulted in enhanced and dramatic impact on lion conservation in southern Kenya. The radical innovation that is known as the Lion Guardians model has ultimately transformed people who once killed lions into their guardians and reduced lion killing by more than 90 per cent (Hazzah et al. 2014).

As we delve into the Lion Guardians case study, we will highlight the parallels with the business world, in particular using the IKEA example to further showcase the importance and relevance of precepts such as radical innovation in the conservation arena. Additionally, we will discuss how conservationists may learn and adapt successful innovative approaches from the business world.

17.2 CASE STUDY: LION GUARDIANS

In response to the high level of lion killing (over 160 lions in an eight-year period; Hazzah et al. 2014), we initiated a conservation programme, called Lion Guardians, in which traditional Maasai warriors (henceforth Guardians) were employed. Until this point most conventional conservation organizations were focused on studying lion behaviour, collaring lions and conserving them in protected areas. Just as IKEA set up shop outside of the central business area to reduce cost, Lion Guardians focused on community lands (non-protected areas) and hired individuals with no formal education and with past killing behaviour. In so doing, Lion Guardians essentially changed the rules of the game (the value network; see Table 17.1 for specific details of Lion Guardians' unique value network).

Prior to being appointed as Guardians, many of the warriors were renowned lion killers with vast influence and respect in their communities. Lion killing has traditional significance within the Maasai society. The Maasai have historically valued lions (except when they have attacked livestock) because they provide warriors with a cultural role that reasserts their power and strength as they protect their communities (Hazzah et al. 2017). Protection of community –whether marauding animals or encroaching tribes – is the primary job of a Maasai warrior.

The programme provided incentives through conservation-related employment, training in literacy (the majority of Guardians at time of employment are non-literate) and scientific monitoring, and community assistance, all directly linked to the presence of lions (see Table 17.2). For example, job opportunities as Guardians only become available if and when lion densities are shown to have increased; and if they decreased then jobs would be removed. During employment, the Guardians lived and worked from their home communities and wore traditional clothes as their uniform (see Table 17.2). They took pride in their strong traditional knowledge of their environment, abilities to track lions on foot and to protect their communities (e.g. alerting herders to lion presence to proactively prevent attacks on livestock and assisting in better husbandry practices) (Hazzah et al. 2014). Guardian jobs were in high demand because warriors could live at home and use their specialized tracking skills and their confidence working near large wild animals (Dolrenry et al. 2016). By focusing on the issues affecting the people and their values, Lion Guardians

Table 17.1 *Lion Guardians' unique value network: key differentiators that allowed Lion Guardians to change the way carnivore conservation was effected in the community lands of the Amboseli-Tsavo ecosystem*

	Solution design	Focus area	Delivery	Marketing/ fundraising
Traditional conservation organizations	Fit/customize existing solution External solution imposed (low risk)	Protected areas Species-centric	Trained scientists High cost	Emotional appeal
Lion Guardians	Participatory approach to solution Strong model of behaviour theory change Establishment of trust Problem solving – understanding the root cause of the problem Experimental and adaptive – learn by doing and continue to adapt by using feedback loops (high risk)	Community lands People-centric	Indigenous community members with no formal education Leverage traditional ecological knowledge Cost-effective Targeted employment and volunteer process Marriage of modern science and traditional knowledge	Result-based

was able to provide a discontinuous leap in benefits to the communities – making it attractive for them to conserve lions instead of killing them. The most effective part of this equation is that, like IKEA, Lion Guardians was able to deliver these benefits to the communities while reducing the sacrifices or compromises that community members would have to make. IKEA customers did not have to sacrifice on design or choice even though they were paying significantly less; communities in Lion Guardians areas could continue to live in their traditional and chosen manner while losing less livestock to depredations.

Table 17.2 *Key factors that triggered conservation innovation and the formation of the Lion Guardians conservation model, acted as sources of ideas for alternative approaches (bricolage) and facilitated diffusion of new approaches (contagion) (adapted from Briggs et al. (2010))*

Factors	Lion Guardians example	Illustrative qualitative narratives
Impetus for innovation	1. Eradication of lion population within the ecosystem (minimum of 160 lions killed between 2003 and 2011).	'Killing lions is what I grew up with; there used to be so many lions when I was younger. We almost wiped them out' (Maasai elder 2006)
	2. Disenfranchisement of the warrior age-set and erosion of Maasai cultural practices and traditions.	'If we don't stop killing all the lions then we will end up with none very soon, and this is what our fathers did to us with the rhinos' (Lion Guardians 2007)
	3. Strong cultural drive to kill lions to gain prestige and inadequate legal ramifications imposed by the government for killing.	'We often don't like wildlife because our age-set is left out and we are harassed by the older age-sets who are employed as game scouts. We [warriors] would start to like wildlife if we were given [an] opportunity to work with them' (Maasai warrior 2005)
	4. Historical conflict between pastoralists and government over land and resources restrictions.	'*Olamayio* [traditional lion hunt] was formed only for warriors to show off our strength for women. It brings immense prestige to the warrior who spears the lion first and is very important to Maasai culture' (Maasai warrior 2005)
	5. Strong institutional support via an umbrella organization (Living with Lions) that gave Lion Guardians the breadth to explore and innovate.	'…those foxes [government] just get money from wildlife and they forget about the problems people encounter from wildlife – have taken all our fertile land as protected areas, and those wildlife are killing people, eating our livestock' (Maasai elder 2006)
	6. Funds secured for pilot of the model. However, radical innovation is often difficult to fund. Wildlife conservation society the Great Cats Program (now Panthera) provided the funds to start the pilot in 2006 (initially $25,000).	'Let us warriors help conservationist[s] monitor lions. Our tradition and culture makes us the best and most experienced people to save lions. And most important is I am not removed from my culture and my people' (Maasai warrior 2006)
Bricolage: sources of alternative ideas and approaches	1. Engagement of key stakeholders: spent time living in Maasai communities and studied the drivers/motivations behind lion killing (Hazzah 2006). During this period trust was built with the communities, specifically the warriors. Trust cannot be overemphasized as a critical component for community-based innovation to occur.	

2. Building the model: over fifty brainstorming sessions (informal) with key warrior leaders and lion killers led to discussions about them feeling marginalized in conservation and how they had the traditional knowledge and skills to protect lions if given the opportunity due to their years as herders. Also, how conservation-related employment would allow them to fulfil their traditional role as warriors in the community (protection via mitigation) while also earning a living to provide for their families.

3. Discussions with other conservationists and scientists about how to integrate a high-risk model.

Contagion: adoption and diffusion of new ideas

1. During the pilot we worked with a local long-standing NGO (Maasailand Preservation Trust – now Big Life Foundation) to help the recruitment of warriors/ Guardians and get advice on management issues. After the initial five Guardians were hired we realized that political appointment of Guardians was not providing the highest quality candidates. By January 2007, we shifted to a new hiring model. We informed communities of a vacancy and provided them with a phone number to call in with any lion reports. Warriors would volunteer for months looking for lions and the best candidates – those with exceptional tracking skills, strong work ethic/ commitment and leadership skills – were offered employment.

'We can track lions in the dark, with our eyes closed, and we will never fail at it' (Lion Guardian 2008)

'We live with the community, wear shukas [traditional clothes] and nothing can be hidden from us. We are leaders in our own right and we will use our positions to improve the image of our project and save lions' (Lion Guardian 2013)

'Conservation of wildlife is the mandate of the Kenya Wildlife Service, but this work is too huge for us to achieve alone. We need other stakeholders to support us with this mandate' (Julius Cheptei, Kenya Wildlife Service 2015)

'I never thought as a herder I would ever be given such an opportunity. And what is even better is that this programme is only for traditional warriors who have never been to school – it is our programme' (Lion Guardian 2007)

'Lion Guardians has given us the opportunity to gain formal, gainful employment. It has helped us as individuals and known lion killers, saved us from a life behind bars' (Lion Guardian 2014)

'Before I become a Lion Guardian I killed eight lions, and was known and respected by many far and wide. Over the past eight years I have stopped over 100 lion hunts and risked my life many times to ensure my lions are safe' (Lion Guardians 2015)

(continued)

Table 17.2 (*continued*)

Factors	Lion Guardians example	Illustrative qualitative narratives
2.	For the first two years of the programme we held monthly meetings with all the Guardians to continue to improve and adapt the model, and to ensure it was culturally relevant and impactful. We also met with community leaders/elders each month to hear their suggestions. In addition, every year we conducted a systematic questionnaire throughout the communities to get broad-scale feedback.	'After I heard about Lion Guardians I encouraged people to stop killing lions. I did this because I am happy that my son no longer endangers himself going out to kill lions but is employed instead' (Maasai elder 2007) 'You may find a Lion Guardian who has no livestock is of equal social status to someone who has more than 400 livestock. A person who owns livestock might lose them one day, while the Lion Guardian continue to get benefits from lions' (Lion Guardians 2007)
3.	Ownership – high level of participation (mainly through monthly meetings, weekly field interactions during lion tracking, naming of individual lions) led the warriors to feel a strong sense of ownership over the programme and also their lions. For example, the name Lion Guardians was voted on by the entire team with each member getting one vote.	'The first lion I saw roaring was Sikiria. When we collared Sikiria, I touched him with my own hand and the day after I saw him walking around, which made me very happy. When Sikiria mated with three females they all had many cubs who are now sub-adults lions. I know all these sub-adults by name and I protect them like my own children' (Lion Guardians 2013)
4.	Support and attention – having dedicated people, both Guardians and biologists, on the ground responding daily to reports was critical in driving the programme forward. Particularly, Guardians needed to know they were supported by the main camp and that their reports were being responded to and there were opportunities for ongoing and consistent training.	"Lion Guardians are protectors of lions and livestock – we are part of a programme that develops coexistence' (Lion Guardians 2016) 'I have never been to school and thus never held a pen before. I first held it like a spear. But I am now proud that I can write my name, fill data forms and even use a GPS.

5. Allowing the model to transpire organically. We did not come into the communities with a preconceived solution on how to save lions and help communities. We listened to the communities, most importantly, the warriors, and from their ideas and suggestions the Lion Guardians model began to take form.

6. Scaling Lion Guardians – in 2008 we began getting interest in our model from surrounding Maasai communities; they began to demand Lion Guardians at their sites. In 2009 we expanded the programme to two new areas and in 2010 we did a final expansion covering all the core communities in the Amboseli-Tsavo ecosystem. By 2011, we began sharing the model with other sites in Africa.

This has brought me so much respect in my community' (Lion Guardians 2015)

'Once the community accepts you as a Guardian, they don't hide anything from you. A hunt can be stopped simply because they respect you and don't want to jeopardize your job' (Lion Guardians 2013)

'Lion Guardians is one of the most important answers to lion conservation that I've seen. There's no doubt it has turned the tide for the Amboseli lions, and it has terrific potential to do the same for lions in large parts of savanna Africa' (Luke Hunter 2013)

'The Lion Guardians philosophy is intriguing. In Ngorongoro Tanzania, local communities have not had the chance to participate in wildlife conservation. When I heard about this programme, I saw it as an opportunity for the pastoralists in my area in Tanzania to be fully involved in conservation in a meaningful way' (Maasai elder 2014)

Lion Guardians was piloted in September 2006, and began in earnest in January 2007, covering a 1,229 km² area in the Amboseli-Tsavo ecosystem (Table 17.2). The programme started with five Guardians and over the years has grown to cover over 4,000 km² in Kenya and has been adapted to other sites in Africa with approximately eighty warriors employed in Guardian-like positions. As we saw with the IKEA example, the radical innovation that Lion Guardians brought to carnivore conservation changed the landscape. Since then other organizations in Africa have adapted the Lion Guardians model (e.g. Ruaha Lion Defenders, KopeLion Ngorongoro, Long Shields Zimbabwe). Meanwhile, over the course of the last ten years, Lion Guardians has continued along the path of continuous improvement; through incremental innovation several elements have been added to the model (see lion naming example in Table 17.2). Similarly IKEA has branched out from being purely a furniture store to a one-stop shop for all home furnishings and has started to branch out of its traditional Scandinavian design to appeal to global markets.

To illustrate the key factors that allowed for radical conservation innovation to take place to conserve lions in Southern Kenya, we adapted Table 17.2 from Biggs et al. (2010) and provided qualitative examples. As Biggs et al. (2010) suggested, we grouped the key factors along three specific dimensions that have been highlighted in literature to be important dynamics in social innovation: (1) impetus for innovation – which factors triggered and supported innovation, (2) bricolage – which new ideas were necessary to form a novel approach and (3) contagion – diffusion of the new ideas and how they were adopted/implemented. In addition, the final column is based on interviews with Guardians and other key stakeholders including Maasai warriors (Hazzah 2006, 2011). We pooled quotes that qualitatively supported and facilitated the innovation of Lion Guardians.

Most importantly, radical community-based conservation innovation can only take place when there is trust between the communities and the conservationists (Hahn et al. 2006). Establishing this trust can be difficult, time consuming and nuanced. Every community is different, with varied values, needs and socio-cultural practices. These all need to be understood and respected before productive discussions can ensue. In some places this could take months; in others it could be years; thus, patience and endurance are key. This also includes being respectful about the dress code of a specific culture and engaging in appropriate customary greetings and exchange of news. In the Lion Guardians example, we spent a year listening to stories, participating in community events such as traditional ceremonies and church revivals and helped transport sick community members. We did not talk about lions or conflict; we let those conversations happen organically once the trust was built.

Innovation cannot be directly planned but is stimulated by creating an environment conducive for new innovation (Westley 2002; Biggs et al. 2010). One of the key elements recognized in the business world in creating this conducive environment is *allowing space for serendipity* (Kumar et al. 2000). Successful market-driving firms create an environment where individual creativity can flourish. For example, 3M Company, the American multinational conglomerate corporation, provides a large variety of centres and forums where ideas can be generated, shared and nurtured; employees are enabled to pursue their own research projects. The confluence of one of these research projects and the forums was the birth of the now well-known Post-It Note (Govindarajan & Srinivas 2013).

Much of the formation of Lion Guardians happened organically, and once the warriors generated the idea and took ownership over the model it enabled radical innovation to transpire naturally. Since then we have bolstered it with incremental innovations. For example, once the Guardians started to spend time with the lions, they started to give them Maasai names. In Maasai culture everything important has a name and thus they took pride in naming the lions, as well as in videotaping and photographing them to show to their communities (Dolrenry et al. 2016). They told stories to the elders, women and children using the lions' Maasai names, personalizing the lions to the broader community. No longer were lions simply anonymous enemies; they became individuals even to the community members not directly involved in their monitoring and conservation. A survey conducted in the study communities in late 2012 showed that 55 per cent (n = 85) of randomly sampled respondents across the ecosystem knew the name of at least one lion (Dolrenry et al. 2016). At an important community meeting one Guardian stated: 'There is a very deep connection between the Guardians and the lions we name. This connection can only be compared to the bond between best friends, or the feelings you have for your best cows.' At the time of creating the Lion Guardians model we had no idea that naming lions would have been so important, but because the model was adaptable and the Guardians felt comfortable and were part of the creation process, the environment of serendipity fostered innovation to continue to transpire.

In the Lion Guardians example, we found it important to initially reframe the lion–people relationship with the warriors in particular, but also the broader community, as part of the radical innovation stage. Additionally, we believe we were able to better engage the relevant stakeholders – the warriors that were killing the lions – by fostering a group identity and building a place for them to be recognized for their traditional roles that encourage bravery and courage (Hazzah et al. 2014; Dolrenry et al. 2016). This cannot be understated; part of what made this a successful innovation is the time investment and attention that each Guardian was given from the beginning and the opportunity for them to fully engage both in monitoring/ecological data collection and conservation. We believe that giving consistent respect and attention (i.e. responding to their reports, listening to their suggestions to improve the programme, etc.) was absolutely critical to getting the appropriate buy-in of the programme. The Guardians' perspective was also influenced by exposure to the lions and the informal interactions they had with them (Dolrenry et al. 2016). We observed how the interest in and attachment to the lions grew exponentially through informal experiential activities of tracking and then observing the lions as individuals within their natural settings (Vredenburg & Westley 1997).

At the core of any resilience model are components that focus on facilitating feedback loops and allowing for high levels of participation and monitoring, which often is what allows an initiative to be adaptive (Colfer 2005; Mutimukuru et al. 2006). Despite full participation being customarily emphasized in theory, in practice communities are often marginalized (Cooke & Kothari 2001). If marginalized groups are involved in the decision-making process (about rules and practices) and in monitoring their own resources, then their needs and interests are more likely to be taken into account (Dangol 2005) and a sense of ownership over the resources will emerge.

An example of the strong buy-in was witnessed in 2014, when thousands of warriors armed with their spears intended to kill wildlife in protest over a disagreement of land and human rights issues with the government. They felt the government was not listening to them and the only way they could be heard was by killing the *government's wildlife* (see Table 17.2). At this time, the Guardians were asked to cease work by their leaders to avoid getting injured while carrying out their duties. Even so, the majority of the Guardians found innovative ways to continue to safeguard the lions by using their position and knowledge as a Lion Guardian as they did not want to stand by and see their lions killed

(Table 17.2). Some would send the hunting parties in the wrong direction telling them they saw lions in a specific area the day before, others told hunting parties to avoid a certain pride because the lion's collar would take pictures of the killers and send it immediately to the government, and others more directly protected lions and risked their lives by standing in the path between the lions and the hunters and worked to dissuade their peers through compelling reasons not to kill lions.

In addition, literature has suggested that personality traits affect a person's ability to be entrepreneurial and to execute radical ideas (McCauley & Van Velsor 2004). We targeted lion killers with specific field skills, personality traits and leadership capabilities to carry out the innovation. When hiring Guardians we asked for months of volunteering to fully understand if they held the right traits and skills necessary to track and protect lions. This was a focused selection of a targeted group rather than an opportunistic hiring process. In retrospect, this was a key factor of why the innovation was successful. We believe that the programme will only be as strong as the employees who are on the ground working diligently every day, and therefore, a selective/targeted process is likely the most straightforward way to achieve this.

It is important to note, however, that in spite of being a radical innovation, Lion Guardians has not been able to scale and amplify its impact similar to Aravind Eye Hospital or IKEA. As aforementioned, the reason behind this is the fundamental problem that plagues conservation – a funding model that is not aligned with the types of results achieved in conservation. In the case of IKEA, for instance, the feedback loop between buyer and seller is much clearer and this market advantage is borne out in financial results that make the business sustainable in the long run. With Aravind, the unique approach of streamlined operation costs together with a for-profit arm allowed it to scale without having to rely on altruistic donor funding based on a good deed feeling rather than on a financial return on investment. In conservation, it can take decades to realize and verify benefits. Furthermore, often these benefits are hard to monetize or hard to specifically credit to a particular conservation intervention (Huwyler et al. 2009). Accordingly, Lion Guardians has adjusted its growth model to broaden impact using a knowledge-sharing service delivery concept rather than a franchising of expansions of Lion Guardians' core operations to several sites. This allows the organization to maintain streamlined operations and a smaller cost base as we try to achieve broader impacts. In addition, the knowledge-sharing delivery model could be a fund-generating source as we are providing a tangible service in exchange for appropriate remuneration. However, it is unclear whether the fees generated from knowledge sharing will ever be sizeable enough to support the organization, as our customers themselves are other conservation organizations.

After a decade of operations, given the changing contexts (cultural shifts, human and lion population expansions, and climatic factors) we are now beginning the cycle of gathering new knowledge so as to evaluate whether radical innovation is needed once again to continue to enhance outcomes and performance profoundly as it did during the creation of Lion Guardians. It has been essential to manage human–carnivore conflict through applying incremental innovation since our radical innovation a decade ago. However, another radical innovation may once again be necessary to meet the growing demands and increasing strain that is put on the communities and lions as we move along the coexistence continuum.

17.3 DISCUSSION

Both in the business world and the conservation space, the theory and our examples clearly illustrate that organizations need to be ambidextrous and versatile. They need to be capable of managing both incremental and radical innovation. This is often difficult, given the contrasting needs of these two types of innovation. Maintaining an environment that is conducive to radical innovation is more difficult as organizations become larger and set in their ways. Accordingly, it is incumbent upon the leadership driving the organizations to maintain an environment that nurtures innovation with a focus on outcomes and impact – not on activities and tasks. In addition, the organizational structure and strategy have to be realigned over time to reflect the changing environment. IKEA is known as one of the world leaders in innovation and it continues to maintain its market advantage by innovating, keeping its core strength in mind – its unique value proposition of classy furniture design at affordable prices delivered through a unique value network (Tushman & O'Reilly 1996; Kumar et al. 2000).

Based on our experience at Lion Guardians and examples from the business world, we posit that using incremental innovation and feedback loops is necessary to maintain and sustain coexistence for periods of time. However, at certain times or in areas where species are declining rapidly or there is a dramatic shift in the system (e.g. cultural erosion, subdivision, etc.) radical innovation needs to be prioritized, since time to save species is running out.

For both types of innovation, it is important to have an informal setting where trust has been established and mistakes are accepted and integrated, since an overformal structure may compromise organic innovation and exchange of ideas and trust (Vredenburg & Westley 1997; Gunderson 1999; Hahn et al. 2006; McKeown 2008). A key component of innovation is fitting the programme within existing needs and values of the community; however, many conservation groups focus more on understanding the needs of species or ecosystems (Sorice & Donlan 2015).

Furthermore, we have found it important to remember that innovation and coexistence are both non-linear processes. There are often periods of time when new ideas or approaches have limited adoption or when community tolerance is decreased. Studies show that it takes time to appreciate the value of the ecological attribute that is being lost and to establish the likely cause (Gunderson 1999; Berkes et al. 2008). Additionally, the initial responses to conservation problems are generally done through incremental changes to existing approaches. It is only once it has been established that these responses are insufficient given current contexts, that radical innovative responses are sought (Biggs et al. 2010). We need to promote integrated, collaborative, adaptive environmental management through democratic leadership where parties contribute and take ownership of key decisions (Greenleaf & Spears 2002). And through this collaborative process, we found parties contributed insights which were instrumental in framing and implementing innovation at both levels. At IKEA, for instance, all employees are involved in innovation strategy development from top management to business unit heads including internal innovation experts (ikea.com).

Innovation is also relative to a particular set of contexts and time. What was innovative a decade ago can, in time, generate its own problems or become obsolete. A challenge for all of conservation is to design and redesign models, approaches and institutions that remain innovative and adaptive over time (Gunderson 2001; Berkes et al. 2008), given where local communities and key stakeholders are on the coexistence scale. Therefore, always having a pulse on the local changing contexts and being ready and open for adaptations and new innovations is critical for long-term success of conservation goals. In the Lion Guardians example we presented, prior to implementation, lion killing was at an all-time high and was decimating the lion population (Hazzah et al.

2014). There was a demand from the community, specifically the warriors, to innovate and there was a real need for radical innovation to halt the killing. However, after a decade of innovation (both radical and incremental) we have seen a more than tripling of the lion population in the Amboseli-Tsavo ecosystem that has resulted in a different set of problems – more conflict leading to a lowering of community tolerance. In addition, in the past decade we have seen substantial changes in Maasai cultural values, particularly the warriors; this includes an erosion of traditional ecological knowledge and bush skills with more youth attending school and others finding work in nearby towns. We also faced a challenge of organizational growth with the number of Guardians increasing ten-fold. We struggled with how to manage the changing culture with many more employees – how do we provide the same level of training, attention and time in the field with a growing team across several expansive landscapes while maintaining motivation? Over the years we have engaged several incremental innovations to address these areas of concern, but the change is relentless and culture is dynamic so the need for another radical innovation may be necessary to maintain high levels of performance and desirable outcomes.

Westley et al. (2010) suggest that there has to be a demand for innovation for it to actually happen. This was clear in the Lion Guardians example because the warriors strongly demanded an innovation that was inclusive of their skills and knowledge. Demand for innovation alone is likely not enough given the current conservation space, and thus securing long-term financial invesments is also vital to ensure sustainability. There are think tanks that exist (e.g. Centre for Social Innovation, Skoll Forum, etc.) that have attempted to provide a platform for entrepreneurs to present their work to potential funders. Yet these arenas often necessitate radical innovative ideas rather than incremental ones, which is the norm for conservationists. Radical innovation is often costly and needs sustained financial backing for long periods of time to truly test the validity of an idea (e.g. Van den Bergh et al. 2011).

17.4 RECOMMENDATIONS AND FUTURE DIRECTIONS

- Be brave; since time is of the essence, we suggest that conservationists should restrict their focus on incremental innovations, as the pace of change of this approach is too slow to actually save many of the declining species. Instead, a focus on radical innovation to shake up the conservation agenda is necessary. Do not be afraid to change the rules of the game; where there is risk there is also potential for high reward. Risk can be limited through pilot testing before spreading to wider areas.

- Be patient and build respect; innovation can only take place when there is trust between the communities and the conservationists. Establishing this trust can be difficult, time consuming and nuanced. Every community is different, with varied values, needs and socio-cultural practices. These all need to be understood and respected before productive discussions can ensue. In some places this could take months; in others it could be years, thus patience is key.

- Be supportive; one of the key elements recognized in the business world in creating this conducive environment is *allowing space for serendipity*. Successful market-driving firms create an environment where individual creativity can flourish. We believe that giving consistent respect and attention (i.e. responding to the Guardians' reports, listening to their suggestions to improve the programme, etc.) was absolutely critical to getting the appropriate buy-in of the programme.

- Be inclusive; when designing a community-based conservation model, think about the needs of both the people and species.

- Think big and take risks; it is necessary to inject innovation into long-term conservation funding mechanisms and strategically approach the question of how to secure long-term and substantial support for radical out-of-the-box ideas that conserve species and provide sustained benefits to the communities impacted by wildlife.

- Be adaptive, flexible and quantify your impacts; at the core of any resilience model are components that focus on facilitating feedback loops and allowing for high levels of participation and monitoring, which often is what allows an initiative to be adaptive. Being ambidextrous in your approach with high levels of flexibility will allow for the greatest conservation outcome to develop organically and likely be culturally appropriate. Lastly, it is important to frequently measure your performance on the ground; only then it is possible to understand what areas of the model can be adapted and improved to further broaden conservation impacts.

- One of the major gaps and hindrances to innovation in conservation is the current donor-driven funding model. Both conservation organizations and their donors want to steer clear of the *commoditization* that comes with translating conservation projects into cash flows and products (Huwlyer et al. 2009). However, the current funding models rampant in the conservation world are generally risk averse and hesitant to fund innovations or provide the sustainable financing require to scale innovations from a pilot project/prototype to a fully fledged solution. Financial instruments could possibly bridge this gap between investor interests and conservation needs. However, more research is required in this arena to further establish access to mainstream finance markets for conservation organizations to take innovative risks and grow their impact.

17.5 References

Bauer, H., Becker, M., Begg, C., Bertola, L., Chapron, G., Croes, B., Dricuru, M., Funston, P. F., Groom, R., Henschel, P., Hunter, L., Loveridge, A., Macdonald, D., Packer, C., Petracca, L., Robinson, H., Tende, T., Tumenta, P. F., Venktraman, M., White, P. A. & Winterbach, C. (2016). *Panthera leo*. The IUCN Red List of Threatened Species 2016. e.T15951A79929984.
Berkes, F., Coldin, J. & Folke, C. (2008). *Navigating Social-Ecological Systems: Building Resilience for Complexity and Change*. Cambridge: Cambridge University Press.
Berkes, F. & Folke, C. (2000). *Linking Social and Ecological Systems: Management Practices and Social Mechanisms for Building Resilience*. Cambridge: Cambridge University Press.
Biggs, R., Westley, F. R. & Carpenter, S. R. (2010). Navigating the back loop: Fostering social innovation and transformation in ecosystem management. *Ecology & Society*, 15(2), art. 9.
Brashares, J. S., Abrahms, B., Fiorella, K. J., Golden, C. D., Hojnowski, C. E., Marsh, R. A., McCauley, D. J., Nuñez, T. A., Seto, K. & Withey, L. (2014). Wildlife decline and social conflict. *Science*, 345(6195), 376–8.
Cantwell, J. (1989). *Technological Innovation and Multinational Corporations*. Cambridge, MA: B. Blackwell.
Carrillo-Hermosilla, J., del Río, P. & Könnöläc, T. (2010). Diversity of eco-innovations: Reflections from selected case studies. *Journal of Cleaner Production*, 18(10), 1073–83.
Colfer, C. J. P. (2005). *The Complex Forest: Communities, Uncertainty, and Adaptive Collaborative Management*. Washington, DC: Resources for the Future.
Cooke, B. & Kothari, U. (2001). *Participation: The New Tyranny?*. London: Zed Books.
Dangol, S. (2005). Participation and decision making in Nepal. In C. J. P. Colfer, ed., *The Equitable Forest: Diversity, Community, and Resource Management*. Washington, DC: Resources for the Future and CIFOR, pp. 54–71.
Dewar, R. D. & Dutton, J. E. (1986). The adoption of radical and incremental innovations: An empirical analysis. *Management Science*, 32(11), 1422–33.
Dolrenry, S., Hazzah, L. & Frank, L. G. (2016). Conservation and monitoring of a persecuted African lion population by Maasai warriors. *Conservation Biology*, 30(3), 467–75.

Duchesneau, T. D., Cohn, S. F. & Dutton, J. E. (1979). *A Study of Innovation in Manufacturing: Determinants, Processes, and Methodological Issues.* Orono: University of Maine, Social Science Research Institute.

Ettlie, J. E., Bridges, W. P. & O'Keefe, R. D. (1984). Organization strategy and structural differences for radical versus incremental innovation. *Management Science*, 30(6), 682–95.

Govindarajan, V. & Srinivas, S. (2013). The innovation mindset in action: 3M corporation. *Harvard Business Review*, H00B25-PDF-ENG.

Greenleaf, R. K. & Spears, L. C. (2002). *Servant Leadership: A Journey into the Nature of Legitimate Power and Greatness.* Mahwah, NJ: Paulist Press.

Gunderson, L. (1999). Resilience, flexibility and adaptive management – antidotes for spurious certitude? *Conservation Ecology*, 3(1), art. 7.

Gunderson, L. H. (2001). *Panarchy: Understanding Transformations in Human and Natural Systems.* Washington, DC: Island Press.

Hage, J. (1980). *Theories of Organizations: Form, Process, and Transformation.* New York: John Wiley & Sons.

Hahn, T., Olsson, P., Folke, C. & Johansson, K. (2006). Trust-building, knowledge generation and organizational innovations: The role of a bridging organization for adaptive comanagement of a wetland landscape around Kristianstad, Sweden. *Human Ecology*, 34(4), 573–92.

Hazzah, L. (2006). *Living among Lions (*Panthera Leo*): Coexistence or Killing? Community Attitudes towards Conservation Initiatives and the Motivations behind Lion Killing in Kenyan Maasailand.* MSc Thesis. Madison: University of Wisconsin-Madison.

Hazzah, L. (2011). *Exploring Attitudes, Behaviors, and Potential Solutions to Lion (*Panthera leo*) Killing in Maasailand, Kenya.* PhD Dissertation. Madison: University of Wisconsin-Madison, Nelson Institute of Environmental Studies.

Hazzah, L., Bath, A., Dolrenry, S., Dickman, A. & Frank, L. (2017). From attitudes to actions: Predictors of lion killing by Maasai warriors. *PLoS ONE*, 12(1), e0170796.

Hazzah, L., Dolrenry, S., Naughton-Treves, L., Edwards, C. T., Mwebi, O., Kearney, F. & Frank, L. (2014). Efficacy of two lion conservation programs in Maasailand, Kenya. *Conservation Biology*, 28(3), 851–60.

Hellstrom, T. (2007). Dimensions of environmentally sustainable innovation: The structure of eco-innovation concepts. *Sustainable Development-Bradford*, 15(3), 148.

Holling, C. S. (1973). Resilience and stability of ecological systems. *Annual Reviews in Ecology & Systematics*, 4(1), 1–23.

Holling, C. S. (2001). Understanding the complexity of economic, ecological, and social systems. *Ecosystems*, 4(5), 390–405.

Holling, C. S. & Meffe, G. K. (1996). Command and control and the pathology of natural resource management. *Conservation Biology*, 10(2), 328–37.

Huwyler, F., Käppeli, J. & Tobin, J. (2009). *Conservation Finance from Niche to Mainstream: The Building of an Institutional Asset Class.* Credit Suisse et al. Available from https://assets.rockefellerfoundation.org/app/uploads/20160121144045/conservation-finance-en.pdf (accessed November 2018).

IUCN. (2006). Regional conservation strategy for the lion *Panthera leo* in Eastern and Southern Africa. Gland, Switzerland, IUCN SSC Cat Specialist Group, p. 60.

Kim, W. C. & Mauborgne, R. (2004). Value innovation: The strategic logic of high growth. *Harvard Business Review*, July–August. Available from https://hbr.org/2004/07/value-innovation-the-strategic-logic-of-high-growth (accessed November 2018).

Kling, K. & Goteman, I. (2003). IKEA CEO Anders Dahlvig on international growth and IKEA's unique corporate culture and brand identity. *The Academy of Management Executive*, 17(1), 31–7.

Kumar, N., Scheerb, L. & Kotlerc, P. (2000). From market driven to market driving. *European Management Journal*, 18(2), 129–42.

Manoa, D. O. & Mwaura, F. (2016). Predator-proof bomas as a tool in mitigating human–predator conflict in Loitokitok Sub-County, Amboseli Region of Kenya. *Natural Resources*, 7(1), 28–39.

McCauley, C. D. & Van Velsor, E. (2004). *The Center for Creative Leadership Handbook of Leadership Development*. San Francisco: John Wiley & Sons.

McKeown, M. (2008). *The Truth about Innovation*. India: Pearson Education.

Mutimukuru, T., Kozanayi, W. & Nyirenda, R. (2006). Catalyzing collaborative monitoring processes in joint forest management situations: The Mafungautsi Forest case, Zimbabwe. *Society & Natural Resources*, 19, 209–24.

Okello, M. M., Kiringe, J. W. & Warinwa, F. (2014). Human–carnivore conflicts in private conservancy lands of Elerai and Oltiyiani in Amboseli Area, Kenya. *Natural Resources*, 5(08), 375.

Okemwa, B. O. (2015). *Evaluating Anti-Predator Deterrent against Lions in Group Ranches surrounding Amboseli National Park, Kenya*. MSc Thesis. Kenya: University of Nairobi.

Olsson, P., Folke, C. & Berkes, F. (2004). Adaptive comanagement for building resilience in social–ecological systems. *Environmental Management*, 34(1), 75–90.

Ortolani, G. (2016). Nepal goes high-tech in its fight against rhino poachers. *Mongabay*, 14 November. Available from https://news.mongabay.com/wildtech/2016/11/nepal-goes-high-tech-in-its-fight-against-rhino-poachers/ (accessed November 2018).

Plowman, D. A., Baker, L. T., Beck, T. E., Kulkarni, M., Solansky, S. T. & Villarreal Travis, D. (2007). Radical change accidentally: The emergence and amplification of small change. *Academy of Management Journal*, 50(3), 515–43.

Rangan, V. K. & Thulasiraj, R. (2007). Making sight affordable (innovations case narrative: The Aravind eye care system). *Innovations*, 2(4), 35–49.

Sorice, M. & Donlan, J. (2015). A human-centered framework for innovation in conservation incentive programs. *Ambio*, 44, 788–92.

Tukker, A. & Butter, M. (2007). Governance of sustainable transitions: About the 4(0) ways to change the world. *Journal of Cleaner Production*, 15(1), 94–103.

Tushman, M. L. & O'Reilly, C. A. (1996). The ambidextrous organizations: Managing evolutionary and revolutionary change. *California Management Review*, 38(4), 8–30.

Un, C. A. (2010). An empirical multi-level analysis for achieving balance between incremental and radical innovations. *Journal of Engineering & Technology Management*, 27(1), 1–19.

Van den Bergh, J. C., Truffer, B. & Kallis, G. (2011). Environmental innovation and societal transitions: Introduction and overview. *Environmental Innovation & Societal Transitions*, 1(1), 1–23.

Vredenburg, H. & Westley, F. (1997). Innovation and sustainability in natural resource industries. *Optimum: The Journal of Public Sector Management*, 27 (2), 32–49.

Westley, F. (2002). The devil in the dynamics: Adaptive management on the front lines. In L. H. Gunderson & C. S. Holling, eds., *Panarchy: Understanding Transformations in Human and Natural Systems*. Washington, DC: Island, pp. 333–60.

Westley, F. & Antadze, N. (2010). Making a difference: Strategies for scaling social innovation for greater impact. *Innovation Journal*, 15(2), art. 2.

Woodroffe, R., Thirgood, S. & Rabinowitz, A. (2005). The future of coexistence: Resolving human–wildlife conflicts in a changing world. In R. Woodroffe, S. Thirgood & A. Rabinowitz, eds., *People and Wildlife, Conflict or Coexistence?* Cambridge: Cambridge University Press, pp. 388–405.

Zaltman, G., Duncan, R. & Holbeck, J. (1973). *Innovativeness and Organizations*. New York: John Wiley & Sons.

18

Towards Human–Wildlife Coexistence through the Integration of Human and Natural Systems

The Case of Grey Wolves in the Rocky Mountains, USA

NEIL H. CARTER, JEREMY T. BRUSKOTTER, JOHN
VUCETICH, ROBERT CRABTREE, HANNAH JAICKS,
GABRIEL KARNS, MICHAEL PAUL NELSON,
DOUG SMITH AND JOHN D. C. LINNELL

Increasing evidence indicates we face the sixth mass extinction – an extinction largely the result of human activities (Barnosky et al. 2011). The disappearance of wildlife and their habitats has far-reaching consequences on humans, including the degradation of life-sustaining ecosystem services such as the availability of medicines, control of pests and diseases and provision of clean water and air (De Groot et al. 2002). In addition to services, the biosphere also provides *assets* that benefit both society and the natural resources that sustain humankind (Obst et al. 2016). For example, healthy predator populations provide regulation of prey populations that otherwise can overpopulate. Insomuch as wildlife is significantly attributed with aesthetic, cultural, religious, economic and educational values (Manfredo et al. 2009; Carter et al. 2012), their loss also arguably diminishes humans' quality of life. The well-being of humans and wildlife is thus inextricably linked, necessitating the integration of social and natural sciences to understand human–wildlife interactions, and identify means of promoting coexistence (Liu et al. 2007; Carter et al. 2014).

Efforts to understand the interactions between human systems and natural systems increasingly employ a *social-ecological systems* or *human–environment systems* conceptual framework (Walker & Salt 2006; Folke et al. 2010). A social-ecological system (SES) can be defined as an *integrated system of ecosystems and human society with reciprocal feedback and interdependence* (Folke et al. 2010). Because social and ecological systems are interdependent, changes in one component can affect a variety of other components throughout the system, and human-induced changes directed at a particular outcome (e.g. wildlife management) will almost assuredly have unintended consequences that may ripple throughout the system (Gunderson & Holling 2002). Proponents of systems approaches argue that SESs are characterized by complex, reciprocal – positive and negative – feedback loops spanning spatial, temporal and organizational scales (Liu et al. 2007). An SES approach rejects the idea that systems are characterized by a single state of equilibrium, and uses 'systems thinking … to bridge social and biophysical sciences' (Folke et al. 2010, p. 2). Systems thinking focuses on the linkages and interactions between the system components more so than on attributes of the components themselves.

Given their emphasis on complexity and interdisciplinarity, SES frameworks are ideal for understanding and characterizing the interactions among people, wildlife and the broader human and ecological communities. The general framework in Figure 18.1 illustrates the dynamic interplay between macroscopic properties (e.g. governance, ecosystem patterns) and individual actors (both humans and animals) capable of adapting to fluctuating conditions by changing behaviour, learning from experience or pursuing their own agendas (Levin et al. 2012). Individual humans and animals organize at different levels (e.g. groups, populations) and spatial scales (Figure 18.1), with those multiple dimensions interacting to influence the capacity for human societies

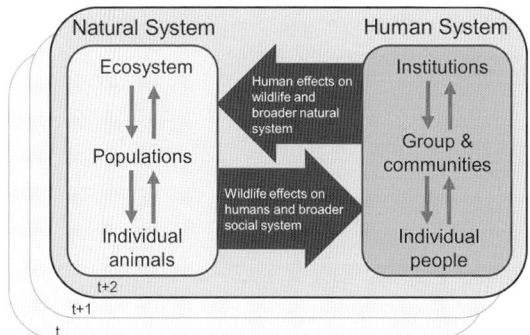

Figure 18.1 A highly simplified, general framework for understanding a social-ecological system (SES). The diagram indicates that individual animals interact with each other and the broader natural system; that individual humans interact with each other and the broader human system; and that the two systems feed back into each other at various spatial, temporal and organizational scales. The SES is at different states at different points in time (e.g. t, t+1, t+2). Based on the overall structure of this diagram, an SES of human–wolf interactions is elaborated in more detail in Figure 18.4.

and wildlife populations to coexist. For example, the behaviours and movements of individual animals can provide individual humans with tangible (e.g. wildlife-viewing opportunities) and other benefits (e.g. spiritual satisfaction), as well as be the source of a diversity of conflicts (Redpath et al. 2013). Those benefits (or costs) influence the shared interests and norms of human groups, which in turn can modify the policies and practices (e.g. through advocacy or litigation) we put in place to manage wildlife and their habitats.

Although a general framework is not well suited for identifying specific policy interventions for different contexts, it is useful for organizing relevant factors identified in theories and empirical research by biophysical and social scientists, and therefore provides a structure for synthesizing data for improving our understanding of human–wildlife interactions in an SES. It also reduces the likelihood that critical human–wildlife interactions are overlooked, which is important when considering that such missed interactions can lead to unanticipated effects, such as increases in illegal killing of wildlife or unexpected increases in livestock losses (Peebles et al. 2013; Chapron & Treves 2016) from large carnivores. Recent work used an SES framework, for example, to pinpoint how and why wildlife anti-poaching interventions differ in their efficacy (Carter et al. 2017).

18.1 SOCIAL-ECOLOGICAL SYSTEMS AND LARGE CARNIVORES

Large-bodied, terrestrial carnivores (hereafter, carnivores) are particularly sensitive to human activities both because of biological characteristics, such as low densities and slow reproductive rates, as well as social factors, such as policies that provide insufficient protection (Linnell et al. 2001) and legal hunting and illegal poaching (Packer et al. 2009; Liberg et al. 2011). Nevertheless, recent attempts to protect and recover carnivores in Europe and North America have resulted in population and range expansions for grey wolves (*Canis lupus*), brown bears (*Ursus arctos*) and other species (Smith & Bangs 2009; Eberhardt & Breiwick 2010; Chapron et al. 2014). These expansions, however, often place carnivores in increased proximity to human populations (Gehrt et al. 2010), prompting more frequent interactions between carnivores and humans, and potentially greater risks for both (Woodroffe 2000; Cardillo et al. 2004). Spatial and temporal overlaps between human and carnivore populations, coupled with continued human population growth, are likely to increase pressure to minimize the negative impacts of carnivores, highlighting the need to uncover ways for people and carnivores to coexist in human-modified landscapes (Carter & Linnell 2016).

Although humans attribute carnivores with a range of beneficial values, including material and spiritual, interactions between humans and carnivores have historically been characterized by competition arising primarily from a common interest to eat wild and domestic ungulates. Human societies have responded to this competition by killing large carnivores, removing their habitat and depleting their food sources, resulting in local and regional extirpations of large carnivore species over the last century (Woodroffe 2000). However, we now know that the eradication of large carnivores from ecosystems can trigger trophic cascades that reduce biodiversity and other life-sustaining ecosystem services otherwise provided by healthy large carnivore populations (Crooks & Soulé 1999; Ripple et al. 2014). These reciprocal interactions indicate that human populations and ecological communities are fundamentally linked. Yet social and ecological studies on carnivores are typically conducted independently, constraining our ability to tackle many problems in carnivore management, such as habitat fragmentation and nuisance complaints (Carter et al. 2014).

In what follows, we draw upon existing evidence from one of the most studied populations of large carnivores – grey wolves in the Rocky Mountains, USA – to build a conceptual model of the interactions among humans, wolves and the broader systems upon which both species depend. We focus on wolves because they are a prime example of how human governance systems – through policy, management and individual action – influence where and at what densities carnivores persist, thereby regulating and limiting the impacts of carnivores on both human and ecological communities (Mech 2012; Muhly et al. 2013). The case of wolves illustrates how broad social and ecological forces can influence how humans live with these animals, and underscores the need for governance systems that can adapt to changing social and ecological conditions. For example, policies that protected grey wolves facilitated their expansion and population growth in eastern Washington, prompting actions by some ranchers and wildlife agencies to remove wolves in order to protect their livestock. Governance systems in these areas of wolf expansion need to adapt to address increasing human–wolf interactions and the controversies they create. Wolves are also the foci of many long-standing controversies due to their limiting effect on game species such as deer, moose (*Alces alces*), elk (*Cervus canadensis*) and caribou (*Rangifer tarandus*). Thus, studying the multi-level, multiscalar interactions between humans and wolves will help provide insights on developing successful coexistence strategies in the Rocky Mountains. Moreover, because co-occurrence between people and wildlife is expected to increase globally, we contend that a social-ecological systems framework can better inform efforts to sustain and recover large carnivore populations while minimizing the negative impacts on human well-being.

18.2 WOLVES IN THE ROCKY MOUNTAINS

In the late nineteenth and early twentieth centuries, government-sponsored eradication programmes designed to protect livestock nearly eliminated wolves from the contiguous USA (Bergstrom et al. 2014). However, in the second half of the twentieth century, wolves became one of the first species listed under the federal Endangered Species Act (ESA) signifying a shift in policy from eradication towards conservation (Mech 1995). Following a reintroduction programme in the mid-1990s, northern Rocky Mountain wolf populations grew by approximately 19 per cent per year (1997–2010; Figure 18.2), and abundance expanded from 100 animals in 1995 to >1,800 in 2011 (Jimenez & Becker 2015).

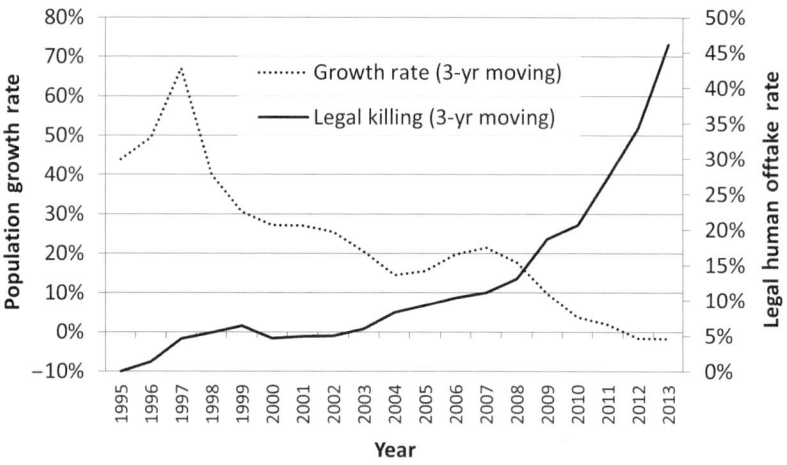

Figure 18.2 Grey wolf population growth rate and legal human offtake rate (legal harvest, plus lethal control) for the northern Rocky Mountain States of Idaho, Montana and Wyoming (1995–2013).
Data from United States Fish and Wildlife Service (UFWS).

Human communities experienced a range of benefits and risks associated with the growing wolf populations. For example, wolf-watching activities in Yellowstone National Park are estimated to bring $35 million annually to Idaho, Montana and Wyoming (Duffield et al. 2006). On the other hand, from 1989 to 2008, nearly 1,000 livestock depredations by wolves occurred in those same three states, and as a consequence 326 partial packs (2.2 wolves on average) and 48 full packs were lethally removed by wildlife agencies (Bradley et al. 2015). Although wolf depredation comprises a small fraction of total livestock mortality each year (Creel & Rotella 2010), in regions where livestock producers and wolf populations overlap, some individual livestock producers experience significant losses (Muhly & Musiani 2009).

The real and perceived risks from wolves can motivate some people to illegally kill them. Poaching is likely the single biggest cause of adult wolf mortality in the USA (Bangs & Shivik 2001; Treves et al. 2017). The heterogeneous distribution of costs and benefits from wolves to human society, and the consequences of these interactions on both natural and social systems, has contributed to a polarizing public discourse about wolf conservation and made it very difficult to develop non-controversial legislation. After a highly contentious process including numerous lawsuits, wolves in the American West were removed from federal protection several years ago (Bruskotter et al. 2014), and now some states allow legal hunting while others do not.

In its recent attempts to remove wolves from federal ESA protections, the USFWS acknowledged the importance of human behaviour towards wolves and the policy mechanisms that govern behaviour:

> ... attitudes toward wolves is the main reason the wolf was listed under the [ESA] ... [p]ublic hostility toward wolves led to the excessive human-caused mortality that extirpated the species from the [northern Rockies] ... Because of the impact that public attitudes can have on wolf recovery, we are requiring adequate regulatory mechanisms to be in place that will balance negative attitudes toward wolves in the places necessary for recovery
>
> (74 FR 15175)

In this brief passage, the USFWS argues that the successful conservation of wolves depends upon adequate *regulatory mechanisms* (i.e. public policy and enforcement) that limit human behaviours (i.e. both legal hunting and illegal poaching of wolves) that negatively impact wolf populations. In other words, the USFWS recognized that human communities, through their collective actions, substantively impact wolf populations. Indeed, wolves were quickly and systematically eliminated in the Western USA (Riley et al. 2004), yet when policies shifted towards protection and restoration, the reintroduction and restoration of wolves to parts of the West proved successful, attesting to their ability to thrive with a sufficient prey base and under sufficiently low human-caused mortality (Smith et al. 2010). In contrast, despite an adequate prey base, legal (mostly control actions) and illegal killing of wolves has essentially prevented the species from occupying suitable habitat in Utah, eastern Wyoming, eastern Montana and western Colorado. Even in the northern Rockies, human killing of wolves from 2003 to 2010 rose sharply, and wolf populations stabilized or declined (Bruskotter et al. 2010, Figure 18.2).

Our history with wolves in the western USA demonstrates that government policies were largely effective both at eradicating and re-establishing wolves. It also serves to illustrate the ability of public policy to directly determine where wolves will be present and where they will remain absent. In this sense, public policy helped achieve legal restoration of wolves, but also prevents wolves from reaching ecologically functional densities in much of the contiguous USA. Moreover, because human and ecological communities are so tightly interconnected, seemingly unrelated policies may have unintended consequences for large carnivores, as well as other species. For example, under current federal policy large portions of federal public lands managed by the USA Forest Service and Bureau of Land Management allow grazing of domestic livestock, which places livestock in proximity to wolves, instigating conflicts (both real and perceived) that often result in the lethal removal of wolves (Thrower 2009). Removal of wolves, in turn, can impact wolf pack structure and behaviour (Brainerd & Andrén 2008; Borg & Brainerd 2015) and, under some conditions, might actually *increase* livestock depredation rates (Wielgus & Peebles 2014; Bradley et al. 2015; Treves et al. 2016). Lethal removal of wolves and other apex predators can also potentially result in increased populations of smaller mesopredators that are suppressed by top carnivores (Prugh et al. 2009). Similarly, regulated human hunting of wild ungulates can influence the abundance of wolves' primary prey (Vucetich et al. 2005), which, in turn, is likely to impact wolf populations and densities. Even water policy may impact the availability and quality of forage for both wild and domestic ungulates, thereby affecting wolves' use of habitat (for evidence, see Muhly et al. 2013). Land-use policies related to forestry, agriculture and the development of transport and energy infrastructure can also have dramatic impacts on landscape structure and habitat quality. These examples illustrate how human policies and the behaviour they regulate can both directly and indirectly impact wolves and the ecological communities they inhabit. But what social and ecological mechanisms determine wolf management policy?

18.3 PEOPLE AND WOLVES: THE ROLE OF INDIVIDUAL BEHAVIOUR

Policies directed at carnivores are, in part, a function of broad socio-cultural forces (e.g. incentives, sanctions, shared knowledge/beliefs) that impact individuals' attitudes and behaviour – that is, their *tolerance* for wolves (Bruskotter & Fulton 2012; Treves 2012). Though studies examining individual-level correlates of behaviour directed at wolves are few, research generally suggests perceptions of risks (or costs) and benefits associated with wolves and one's affect (or emotional) reaction towards wolves explain the majority of variance across a suite of politically relevant behaviour (Slagle et al. 2012).

Research in the fields of environmental and conservation psychology leads to the general expectation that the physical environment itself also shapes how people think and behave (Clayton & Myers 2009), which in turn, may affect public policy and political clashes over how wildlife populations are managed. We currently know very little about how wolf population dynamics and anthropocentric ecosystem changes affect human attitudes, behaviours and support for management policies. However, we expect that these conditions will impact how individuals perceive wolves, and we anticipate that how they respond behaviourally will be based, at least in part, on those perceptions (Slagle et al. 2012; Bruskotter & Wilson 2014). For example, decreases in wild ungulates may create conditions (e.g. reduced opportunity for harvesting ungulates) that impact hunters' attitudes and behaviour (Ericsson & Heberlein 2003); drought and other environmental disturbances may increase the perceived risks associated with wolves (if such disturbances increase stress on livestock or hunted ungulates, for example); spatial distribution and density of wolves relative to human communities may impact how wolves are collectively perceived (i.e. as novelty or nuisance, Ericsson & Heberlein 2003), all of which potentially heighten individuals' perceptions of risks associated with wolves, and lead people to advocate for (or take) actions intended to reduce wolf populations (e.g. sign petitions, attend legislative hearing or poach) (Slagle et al. 2012). In such cases, the management of wolves is likely to focus on reducing negative interactions rather than on increasing positive interactions, which limits opportunities for achieving conservation-related goals (Frank 2016).

Though wolf–human interactions are often categorized as negative (i.e. as conflicts, risks or impacts), this characterization discounts the fact that positive interactions also arise from the presence of wolves, such as increased viewing opportunities, tourism, spiritual happiness or satisfaction, and other psychological benefits. Additional positive outcomes include limiting the abundance of ungulates with populations so large they cause significant damage to human property, as well as reducing disease transmission, and providing a number of other ecological services (Ripple et al. 2014; Gilbert et al. 2016). The perceived potential beneficial impacts of wolves on human communities (e.g. their symbolic value and the value of their ecological services) were used to justify wolf restoration, and continue to be used as justification for wolf preservation (Duffield et al. 2008; Mech 2012).

The picture that emerges from this brief summary is one of a reciprocal relationship, whereby carnivore populations under various environmental conditions impact humans and their perceptions, which, in turn, affect human actions towards the species – including the policies we put in place to manage carnivores. Reciprocal relationships are central features of social-ecological systems (Figure 18.1) as well as a fundamental concept in ecology: density regulation within and between species which also includes trophic cascade effects such as mesopredator release. Importantly, the relationships between ecological conditions, human perceptions and public policy are likely to be moderated by a variety of factors. For example, perceptions may be influenced by the social or interest groups one belongs to (Lute et al. 2014); likewise, the relationship between perception and public policy is likely to be impacted by the level of influence that individual actors and special interest groups have with policy-makers. For example, in the western USA, hunting and agricultural interest groups have long held disproportionate influence with state wildlife management agencies (Nie 2004). Such influence mediates the influence of individual perceptions on public policy.

18.4 ECOLOGICAL IMPACTS OF WOLVES AND PEOPLE: WHO IS THE TOP CARNIVORE?

Wolf reintroduction to Yellowstone and central Idaho created an opportunity to study the direct and indirect effects of predation in an ecological system where human presence is minimal. One question of some importance is: how do wolves impact populations of their prey? Temporal (but not spatial) correlational data show a strong negative association between wolf abundance and elk abundance in the Northern Range of Yellowstone National Park (Peterson et al. 2014). In addition, when wolves were absent from Yellowstone, growing elk populations degraded plant communities and required additional human intervention (in the form of mass culls and increased human hunting). From this one might infer causation; however, such inference is too simplistic. Indeed, evidence suggests that hunter harvest of elk and natural fluctuations in climate provide a better explanation of changes in elk density across the Northern Range (Vucetich et al. 2005). In other words, changes in the densities of wolf prey species can be influenced by a variety of factors – predator abundance, abiotic factors (such as climate), human-caused mortality, disease and potentially competitors like bison (*Bison bison*).

The extent to which wolves and other top carnivores indirectly impact ecological communities is the subject of intense debate among ecologists (Peterson et al. 2014). There appears to be general consensus that top carnivores precipitate a variety of effects on ecological communities (e.g. mesopredator release or vegetative response to reduced herbivore densities; Estes et al. 2011). However, such effects are likely to be conditional – that is, to depend on other factors (Peterson et al. 2014), including human-caused mortality. Indeed, where wolf populations are held artificially low by high rates of human hunting (legal and illegal), we should not expect wolves to have strong direct effects on ungulate populations nor strong indirect ecological effects (Mech 2012). Consequently, we would expect the indirect effects of wolves on lower trophic levels (i.e. trophic cascades), such as vegetation, to be geographically limited to those regions where humans allow wolves to attain ecologically functional densities (Mech 2012), such as protected areas. However, whether wolves are exerting a strong indirect effect on lower trophic levels is very difficult to detect. In part this is because their effect on lower trophic levels may shift over time in a single region, due to dynamic changes in both biotic and abiotic factors, making the attribution of causation extremely difficult. However, the fact that indirect effects of wolves (and other carnivores) might be negligible or difficult to detect under some circumstances does not significantly influence most arguments in favour of wolf presence in the landscape.

Our collective experience with wolves in the western USA reveals that even in some of the wildest parts of North America – places where human population densities are extremely low and human impact on ecological systems appears minimal – human communities and the policies we enact have both direct and indirect impacts on carnivores and the ecological communities they occupy. Viewed from this perspective, the ecological effects proximately associated with large carnivores are *ultimately* determined by human communities, our collective tolerance for these species and the policies we put in place to manage and conserve them. Because these ecological effects are closely tied to human communities, in the following section we describe some broad-scale social and economic forces that appear to be fundamentally changing how human communities interact with and perceive large carnivores, like wolves, and consequently how we manage them. The effects of these forces are felt across the American West and also other large regions occupied by both people and wolves, such as in Europe (Chapron et al. 2014). Exploring the effects of these social and economic forces enables us to anticipate likely future scenarios of human–wolf interactions.

18.5 SOCIAL AND ECONOMIC FORCES SHAPING HUMAN–WOLF INTERACTIONS

Previously, we have discussed interactions between humans and wolves at relatively fine degrees of resolution on both spatial and temporal scales. However, it is important to note that the socio-economic and biophysical conditions underlying those interactions have gradually changed over the last few decades, with implications for understanding future patterns of human–wolf interactions. One such major change involves economic modernization, which is fundamentally altering how human societies perceive and behave towards wildlife generally, and carnivores, specifically (Bruskotter et al. 2017; Chapter 1). Research indicates that economic modernization is driving a shift in societal values from survival-based, *materialist* values towards those that emphasize self-expression or *post-materialism* (Inglehart & Welzel 2005). Accordingly, one way modernization may affect perceptions of wolves is by promoting more egalitarian or *mutualist* value orientations in which nature and wildlife are viewed as an important part of individuals' moral communities rather than *domination* value orientations, which are expressed in human actions that attempt to subdue nature and put natural resources to instrumental human uses (Figure 18.3; Manfredo et al. 2009, 2016; Teel & Manfredo 2010; Chapter 2). Manfredo et al.

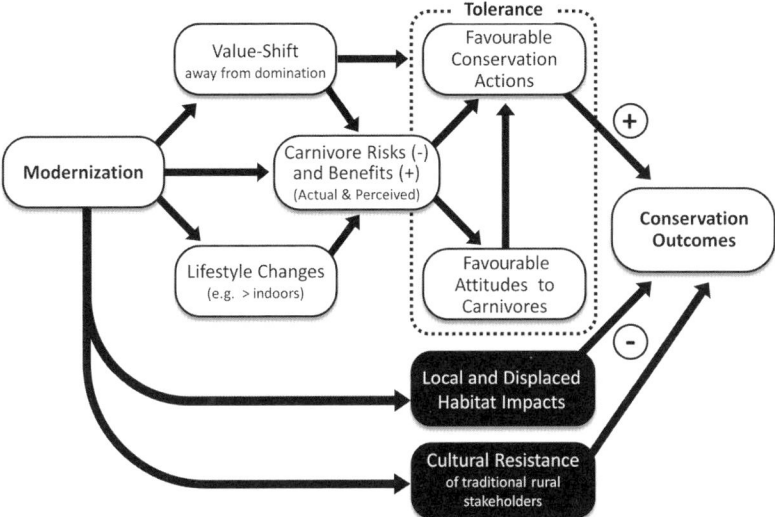

Figure 18.3 Conceptual model outlining how modernization can impact the conservation of large carnivores. Arrows represent proposed causal influences. Figure adapted from Bruskotter et al. (2017).

(2009) found that modernization forces (i.e. urbanization, increasing affluence, education) explained much of the geographic variation in wildlife value orientations across states in the western USA. For example, in the northern Rocky Mountain States of Idaho, Montana and Wyoming, the percentage of respondents with a domination value orientation ranged from 43.8 per cent to 48.6 per cent compared to the 17.9–18.9 per cent of respondents who expressed a mutualist value orientation. In contrast, the percentages of respondents in nearby Washington and Oregon with domination or mutualist value orientations were approximately equal (~33–37 per cent). Value orientations, in turn, were strongly correlated with people's judgements regarding the trade-offs between wildlife protection and human interests (Dietsch et al. 2016; Manfredo et al. 2016), including judgements about carnivores (Dietsch et al. 2016). For example, respondents in Idaho classified as having a domination value orientation were more likely to accept a *reduction in the number of wolves to produce more elk and deer for hunting* than those respondents with a mutualist value orientation (Teel & Manfredo 2010). The high human population growth rate in the Inter-mountain West region, with many of the in-migrants from other regions throughout the USA (Hansen et al. 2002), is likely facilitating changes in the wildlife value orientations of people in the region.

In addition to having an impact on value orientations, modernization might act to physically separate human populations from wolves, ultimately reducing risks associated with these animals (Figure 18.3; Bruskotter et al. 2017). For example, by moving jobs and people out of rural areas into cities, agricultural mechanization reduces the risks associated with wolves at a societal level, although livestock losses to wolves can still be quite significant at local levels or on specific ranches (Muhly & Musiani 2009). Furthermore, modernization is associated with technological innovation and the subsequent proliferation of modern technological conveniences (e.g. air-conditioning, passenger cars, electricity, internet) that serve to promote indoor professions and lifestyles, which can further reduce the risks associated with wolves. Recent work has found that variation among nations in large carnivore conservation outcomes was related to modernization forces believed to reduce the risks associated with large carnivores (Bruskotter et al. 2017). However, some aspects of modernization (e.g. increased affluence, habitat modification) have long been viewed as important drivers of biodiversity loss and species endangerment (Czech et al. 2000). Certainly growth of urban areas, the expansion of subdivisions and the proliferation of transport, energy and recreational infrastructure contributes to habitat loss and fragmentation (Woodroffe 2000). Furthermore, the relationships between modernization and value orientations or behaviours towards carnivores are likely heterogeneous over broad geographic areas. For example, in contrast to our example in the American West, some cultures in India and Mongolia apparently have traditional rural attitudes that are more tolerant of carnivores than the emerging modern attitudes (Athreya et al. 2013).

Societal values do not change uniformly across societies. While the majority of the urban populations in the USA appears to increasingly embrace mutualistic and post-materialistic values, more traditional materialist values are retained among some – especially those in rural communities. These communities are associated with the activities (agriculture, ranching, hunting, rural lives) that potentially face the greatest costs of wolf conservation. All of these changes might drive cultural resistance to outside influences, such as nature-protective policies, and lead to intergroup conflict in human communities by associated social and economic changes (e.g. increasing unemployment in certain areas and sectors) that impact livelihoods, and ultimately, human well-being (Nie 2001; Skogen et al. 2017).

Wolves can be an important symbol of outside influences, which can motivate interested groups to respond in dramatically different ways (Bruskotter et al. 2009; Lute et al. 2014). According to social identity theory, intergroup conflict is driven, in large part, when out-group members take actions that threaten the identity of in-group members (Tajfel & Turner 1979; Tajfel 1982). Broad social changes due to economic modernization might aggravate intergroup conflict, with group members adopting prototypical attitudes and beliefs reinforced through interactions with in- and out-group members (Schneider 2004). Those who strongly identify with groups and regularly interact with in-group members are likely to express different views about the wolf-related risks and benefits to those who identify with *outside* groups. Indeed, tensions over impositions from external groups have exacerbated conflicts over wolves in the United States, as farmers and ranchers feel that the animals are imposed upon them by remote, urban governments and elites unconcerned with the costs incurred by farming communities. In Yellowstone National Park, wolf reintroduction was seen by some rural communities as the controlling, domineering, intrusive federal government overriding the freedom and self-determination of local people (Scarce 1998), and exerting external control over people's private property (Wilson 1997). Wolf-related impacts and human–human conflicts over wolf management can lead to illegal killing of wolves, lawsuits, ballot initiatives and political action to weaken conservation goals in general (Nie 2003; Treves et al. 2016; Carter et al. 2017). By accounting for the various effects of broad social and economic forces on human–wolf interactions as well as the dynamic feedbacks between humans and wolves (Figure 18.4), managers can tailor actions that best accommodate the priorities and goals of a diverse constituency including in those areas where wolves are expected to expand in the future.

18.6 NAVIGATING THE CHALLENGES OF WOLF MANAGEMENT

The political landscape for wolves is in a near-constant flux and defies simple solutions and sustainable outcomes. Although many factors affect wolf policy development (Clark et al. 1996; Primm & Clark 1996), a social-ecological systems framework can be a useful heuristic tool for identifying key levers at the individual, group and societal levels (Figure 18.4) that influence, and are influenced by, decision-making

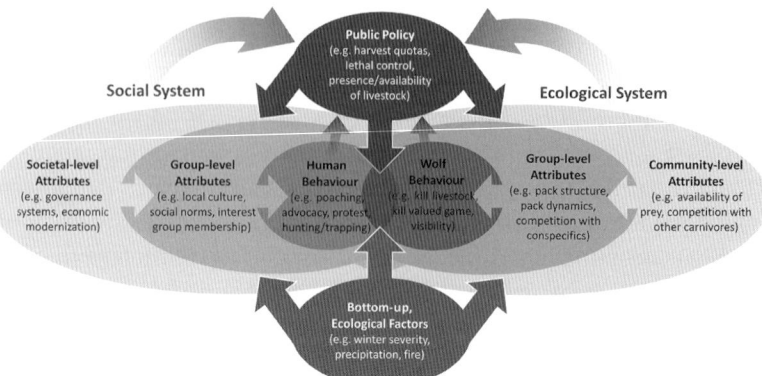

Figure 18.4 A depiction of the social-ecological system of human–wolf interactions in the Northern Rockies, USA. Although not shown here, it is also important to note that behaviours of individual humans and wolves are related to individual-level attributes. For humans, these can include attitudes, emotions, perceived risk, social identity and social trust. For wolves, these can include size, age, health and sex. For the sake of simplicity, other important model actors (i.e. livestock, wild ungulates) and interactions (e.g. human modification of ecological factors) are not depicted in this model.

about wildlife, such as wolves. In the following, we provide a brief and non-exhaustive treatment of some of these levers.

One lever that strongly affects how carnivore conservation policies and practices are created and implemented is political power within governance systems (Clark & Rutherford 2014; Lute & Gore 2014). Configurations of power, wealth and culture in the organizations charged with carnivore management can lead to considerable partiality for a relatively narrow set of special interests (Mattson 2014). To counterbalance the disproportionate effects of a narrow set of special interests, some authors suggest that different degrees of power over various species or resources should lie at different hierarchical levels of government depending on their ecological, social and economic characteristics (Linnell 2015). For example, overall conservation goals could be defined at the national level, while state and local actors are empowered to develop specific, locally adapted policies and practices that are constrained by those broader-level goals and limitations (Redpath et al. 2017). This model is not unlike the model of the ESA, in which federal scientists determine the recovery goals and work with state or local interests to achieve those goals. However, setting overall goals for wolf conservation at regional or national scales, especially in situations where costs are disproportionately borne at the local scale, will often be controversial (Skogen et al. 2017). Indeed, as noted earlier, the extent to which voters in urban areas are seen as dictating wolf policy in rural communities has been a consistent source of tension in the greater Yellowstone region (Wilson 1997; Nie 2001).

On the other hand, in the USA, state wildlife management agencies receive substantial funding from fees and excise taxes paid by consumptive users (i.e. hunters, trappers, gun owners; Williams 2010) and therefore may be more likely to implement wolf policies that maximize benefits (i.e. hunting opportunities) for those users rather than for those of non-consumptive users (i.e. wildlife tourists). For that reason, some authors suggest creating opportunities for both consumptive and non-consumptive users of wolves to meaningfully participate in, and fund, wildlife management programmes, as doing so may lead to greater compromise at local and regional levels (Olson et al. 2015). For example, the state of Montana considered issuing a wolf stamp that would have been available for purchase to anyone, and used to fund non-lethal management of wolves; this would have empowered non-consumptive users to help fund wolf conservation efforts aimed at reducing conflicts. However, successful examples of these types of initiatives appear to be rare as illustrated by *Teaming with Wildlife*, a failed initiative started in the 1990s that proposed taxing outdoor recreational equipment to support non-game species conservation (Spidalieri 2012).

Balancing these disparate priorities among stakeholder groups is a major challenge for management agencies, particularly as stakeholder values change and diversify (Smith et al. 2016). The field of conflict resolution offers useful insights on how to reconcile differences among people (Maser & Pollio 2011). Practitioners advocate the use of deliberative, participatory processes (e.g. collaborative learning, structured decision-making), which assist stakeholders in separating empirical premises (and supporting factual information) from values (desired outcomes). These processes also facilitate joint exploration of consequences of different actions (Maxwell et al. 2015). Participatory processes, characterized by bottom-up representation and legitimization, have also proven successful at negotiating outcomes that are viewed as acceptable, although they work best at small local levels (Daniels & Walker 2001; Young et al. 2013), and there is still a long way to go to fully understand when, and how, they bring benefits (López-Bao et al. 2017; Sterling et al. 2017). Ensuring that participatory processes improve decision-making necessitates that information flows across management levels (e.g. local to national and vice versa), between sectors and that some form of upward and downward accountability exists (Linnell 2015).

Facilitating information exchange, however, does not resolve the issue that occurs when wildlife management priorities differ across jurisdictions. This challenge becomes especially relevant when wildlife, especially wolves, cross jurisdictional boundaries between agencies or states (Smith et al. 2016). For example, wolves are subjected to high levels of hunting immediately upon leaving Yellowstone National Park, where they are protected, to follow prey species that migrate outside the park in the autumn. This high mortality pressure affects pack structure and wolf populations inside the park and conflicts with the mission of the National Park Service of maintaining *naturalness* and providing non-consumptive benefits to park visitors. In such cases, recognizing and negotiating the different missions and management approaches of state and federal agencies, while considering the ecological scales within which wolves exist, are important steps in developing coordinated management across jurisdictions that mitigates public opposition. One suggested coordinated approach involves delineating a transition zone directly adjacent to Yellowstone National Park, to enable wolves from the park to temporarily migrate outside the park without being exposed to liberal harvests near the park boundary (Smith et al. 2016).

Even where mechanisms exist to coordinate management approaches across agencies, rapid social or environmental changes (e.g. disturbances such as fire), political volatility and uncertainties regarding anticipated social and ecological outcomes are likely to challenge agencies' capacity to adapt policies and practices. The adaptive capacity of institutions has been the focus of much research (Brown 2003). A full treatment of institutional fit is beyond the scope of this chapter, but we highlight one approach that is especially relevant to human–wolf coexistence: adaptive management. It focuses on understanding (through experimentation of different management actions) and responding (through iterative modification of strategies) to rapid and unpredictable changes in management contexts when systematic monitoring data are available (Keith et al. 2011). Adaptive management programmes can enhance institutional capacity to learn what drives human–carnivore impacts and human–human conflicts and adjust those policies and practices that may be ineffective. For example, adaptive management can also be applied to funding mechanisms whereby the cost-benefit ratio of wolf presence can be adjusted. This can enable revenue from non-consumptive users, such as wolf-watching tourists, to supplement funding for non-lethal methods for deterring wolves from depredating livestock (Olson et al. 2015).

So far we have focused on those landscapes where humans and wolves are most likely to interact and implicitly assumed that the geographic boundaries of protected areas remain unchanged in the future. However, looking ahead to what habitat wolves need and prioritize protecting those areas would allow wolves to occupy the ecological niches they once filled, thereby providing various ecosystem services and benefits on a larger landscape. By giving wolves the space to roam, it would maintain the *wildness* of the western USA amid growing human pressures and simultaneously reduce the likelihood of negative interactions between humans and wolves in the future. Achieving this longer term, strategic vision would likely require innovative public-private institutional arrangements (see Chapter 8 and Chapter 14). For example, a form of adaptive co-management might be needed that provides incentives and enhanced shared learning across different levels of organization (e.g. private landowners, local organizations and federal agencies) and across geographic space (Armitage et al. 2009). In this case, the boundaries of the SES should be carefully re-delineated (e.g. to encompass a landscape extending beyond the Northern Rockies, USA) to sufficiently accommodate the different sets of human and non-human actors that might accrue benefits or costs from newly established wolf populations (Martín-López et al. 2017). This example highlights the importance of changing how we study human–wolf interactions to address the complex and dynamic challenges of wolf management across a gradient of human-influenced landscapes.

18.7 CHANGING HOW WE STUDY HUMAN–WOLF INTERACTIONS

Much can be gained by integrating the disparate fields of research related to wolves. For example, the inclusion of social science can help reveal how actor groups, social networks, governance structures, power relations or ethics limit or enhance coexistence between human societies and wolf populations. Likewise, social scientists can learn from the ecological sciences, for instance, by examining how ecological disturbances impact human attitudes and values. Viewing wolves and humans as fellow actors within an SES (Figure 18.4) dramatically alters how one thinks about the ecological effects that have become so controversial. For example, viewing human populations as just another member of the ecological community changes the foci of the debate surrounding trophic cascades. Rather than attributing changes in the ecological community to wolves and other top carnivores, we are forced to step backwards along the causal chain and ask – *what factors affect wolf populations, and thereby determine if, when and where trophic cascades will take place?* The answer largely centres on human-induced wolf mortality – a mechanism that is itself highly regulated. Similar questions should be asked for the ways in which human activities directed towards habitat and prey will also modulate wolf-centric trophic cascades. This type of recognition does not challenge the traditional thinking of ecologists in the least; rather, it seeks to better frame humanity's role within the larger realm of ecological processes. Without such recognition, our fundamental understanding of such processes, and especially, our ability to predict when, where and under what conditions they will occur, will suffer.

Of course, analytically disentangling these interactions and feedbacks is a major challenge. For example, it may take several decades (or more) for ecological systems to reorganize after human intervention (Gunderson & Holling 2002); yet our political systems move much faster (e.g. every 2–4 years), challenging our ability to predict ecological dynamics, evaluate management actions and adapt our interventions to achieve desired results. Fortunately, tools that allow scientists to systematically link social and ecological systems are emerging. One example is quantitative models of SESs that incorporate both social and ecological mechanisms of change. Mechanistic models contrast with statistical models that optimize model parameters through the use of correlative data and dependencies between multiple factors which are usually not appropriate to extrapolate the results to unprecedented conditions (Stillman et al. 2015). Rather, mechanistic modelling uses first principles (e.g. animals seek to maximize individual fitness or humans seek to maximize individual utility or well-being) to construct equations and algorithms representing behaviours and interactions that provide the basis for understanding and predicting patterns of interest (Stillman et al. 2015). Mechanistic models have been used separately in the social and ecological sciences to help understand how and why systems change (Drechsler et al. 2007). For example, individual-based ecological models have incorporated adaptive animal-movement ecology in changing landscapes, and investigated wildlife population persistence as a bottom-up process emerging from individual variations and events (Grimm &

Railsback 2005). The models have thus been used to quantitatively examine critical habitats from the individual- to population-level (McLane et al. 2011). Furthermore, by representing different modes of human decision-making, agent-based models (similar in concept to individual-based ecological models) have become powerful tools in ecological economics, land-use science, political science and natural resource management (Filatova et al. 2013). However, despite their potential utility, mechanistic SES models for wildlife conservation are much less common. In part, this is because they require a great deal of social and ecological data that are compatible with each other (that is, collected at a comparable level across different scales) in order to represent real-world systems. For example, where wildlife research is generally spatially explicit, social science research is usually spatially implicit. Even those social science studies that do spatially represent their data (Teel & Manfredo 2010; Dietsch et al. 2016) do so at spatial scales (e.g. state or county boundaries) that do not match well with ecological data, such as individual animal movements that are not bounded by geopolitical borders, although some recent research is beginning to tackle these scale mismatches (Behr et al. 2017). Being one of the most-studied organisms in the world, the grey wolf is therefore an excellent candidate for which to construct an SES mechanistic model. Still, more information on the social system is urgently needed to fully parameterize such models.

Another challenge to understanding human–wolf interactions is the uncertainty associated with our knowledge of social and ecological processes and how policies might affect those processes. When unaccounted or inadequately communicated, uncertainties can lead to inappropriate expectations in the public about the benefits and costs of wildlife, detrimentally affect wildlife populations and in general diminish the efficacy of governance structures. For example, recent studies are calling attention to the uncertainties surrounding legal and illegal hunting rates of wolves, and demonstrate the significant implications these unaccounted uncertainties have on wolf populations and management goals (Creel et al. 2015; Treves et al. 2017). A number of approaches (e.g. integrated assessment models, optimization algorithms and multicriteria decision analysis tools) now exist to systematically account for uncertainties in environmental decision-making (Ascough et al. 2008). One such approach is management strategy evaluation (MSE), a relatively recent method for systematically assessing multiple outcomes of different strategies. MSE uses simulation models to test the

future effects of alternative management procedures on species population dynamics (Milner-Gulland et al. 2010; Bunnefeld et al. 2011).
Unlike other modelling approaches, MSE incorporates various forms
of uncertainty, including process, measurement and structural uncertainty (e.g. resource user compliance with regulations). By engaging
different stakeholder groups, MSE can explicitly include a range of
realistic human behaviours in shared landscapes and facilitate knowledge exchange. Due to the transparency of the MSE approach, for
example, previous studies have shown that it can significantly reduce
the time and effort various stakeholder groups need to reach agreement
on management decisions (Bunnefeld et al. 2011). For those reasons,
MSE offers an exciting way to link research (e.g. from mechanistic SES
models) and policy for wolf conservation in the dynamic landscapes of
the American West.

18.8 RECOMMENDATIONS AND FUTURE DIRECTIONS

We list below some take-home messages that emerge when using an
SES framework to understand and manage human–carnivore interactions, like those between humans and wolves in the Northern Rockies
of the United States.

- Viewing carnivores and humans as interdependent fellow actors
  within an SES is useful for understanding the causes and consequences of ecosystem change, and identifying sources of human–
  wildlife conflicts. A good starting point for doing so is to evaluate how
  gradients of anthropogenic landscapes and activities (e.g. livestock
  grazing, hunting and policies and practices governing these activities)
  affect the populations and functional roles of large carnivores, like
  wolves, at various scales.
- Whether coexistence can be achieved (or what coexistence might look
  like), however, rests not only on the biophysical capacity of a landscape to be shared by humans and wildlife, but also on the capacity
  for human societies to adjust to and accept some level of conflict with
  carnivores (Carter & Linnell 2016; Frank 2016). Human tolerance
  (and intolerance) for carnivores is a function not only of human
  perception, but also social norms and structures, all of which are
  undergoing changes due to broad social and economic forces, such as
  modernization. Increasing our knowledge of how this broad suite of

human factors interacts to regulate human–carnivore coexistence remains an important challenge to sustainably sharing landscapes.

- Because both social and ecological factors shape policy towards carnivores, an SES perspective can highlight ways to navigate the polarizing and challenging issues surrounding carnivore management. In particular, an SES approach could assist in identifying the underlying mechanisms exacerbating human–human conflict over carnivore management, and ways to ameliorate these conflicts, such as through participatory processes or redistribution of costs and benefits between stakeholder groups.
- A large carnivore–human SES is data-rich and could form the basis for successful conservation strategies and outcomes using adaptive management procedures focused on coexistence.
- Through continuing integration of social and ecological sciences and use of ever-advancing computational tools, more and more insights will emerge that help stakeholders and decision-makers maximize positive interactions between humans and carnivores, while minimizing the negative impacts. Maximizing positive interactions will foster a shift towards the neutral to positive side of the conflict-to-coexistence continuum.

The increasing human footprint on planet Earth necessitates a more holistic view of ecological systems – one that explicitly incorporates, rather than ignores, human actors. An important step in learning to live with carnivores is the recognition that even in places where the human footprint is as light as in Yellowstone National Park, carnivores are fundamentally and importantly impacted by people. In a geological epoch dominated by human impacts (the Anthropocene), carnivore persistence is likely to depend upon our ability to coexist with these animals in landscapes altered, and managed, by people. It is time for ecology to recognize and embrace humans, our social systems and institutions as key actors and attributes of ecological systems. Doing so will help reframe human–carnivore interactions from conflict to coexistence.

18.9 References

Armitage, D. R., Plummer, R., Berkes, F., Arthur, R. I., Charles, A. T., Davidson-Hunt, I. J., Diduck, A. P., Doubleday, N. C., Johnson, D. S., Marschke, M., McConney, P., Pinkerton, E. W. & Wollenberg, E. K. (2009). Adaptive comanagement for social-ecological complexity. *Frontiers in Ecology & the Environment*, 7, 95–102.

Ascough, J. C., Maier, H. R., Ravalico, J. K. & Strudley, M. W. (2008). Future research challenges for incorporation of uncertainty in environmental and ecological decision-making. *Ecological Modelling*, 219, 383–99.

Athreya, V., Odden, M., Linnell, J. D. C., Krishnaswamy, J. & Karanth, U. (2013). Big cats in our backyards: Persistence of large carnivores in a human dominated landscape in India. *PLoS ONE*, 8(3), e57872.

Bangs, E. & Shivik, J. (2001). Managing wolf conflict with livestock in the northwestern United States. *Carnivore Damage Prevention News*, 3, 2–5.

Barnosky, A. D., Matzke, N., Tomiya, S., Wogan, G. O. U., Swartz, B., Quental, T. B., Marshall, C., McGuire, J. L., Lindsey, E. L., Maguire, K. C., Mersey, B. & Ferrer, E. A. (2011). Has the Earth's sixth mass extinction already arrived? *Nature*, 471, 51–7.

Behr, D. M., Ozgul, A. & Cozzi, G. (2017). Combining human acceptance and habitat suitability in a unified socio-ecological suitability model: A case study of the wolf in Switzerland. *Journal of Applied Ecology*, 54, 1919–29.

Bergstrom, B. J., Arias, L. C., Davidson, A. D., Ferguson, A. W., Randa, L. A. & Sheffield, S. R. (2014). License to kill: Reforming federal wildlife control to restore biodiversity and ecosystem function. *Conservation Letters*, 7, 131–42.

Borg, B. & Brainerd, S. (2015). Impacts of breeder loss on social structure, reproduction and population growth in a social canid. *Journal of Animal Ecology*, 84, 177–87.

Bradley, E. H., Robinson, H. S., Bangs, E. E., Kunkel, K., Jimenez, M. D., Gude, J. A. & Grimm, T. (2015). Effects of wolf removal on livestock depredation recurrence and wolf recovery in Montana, Idaho, and Wyoming. *Journal of Wildlife Management*, 79(8), 1337–46.

Brainerd, S. & Andrén, H. (2008). The effects of breeder loss on wolves. *The Journal of Wildlife Management*, 72, 89–98.

Brown, K. (2003). Integrating conservation and development: A case of institutional misfit. *Frontiers in Ecology & the Environment*, 1(9), 479–87.

Bruskotter, J. T. & Fulton, D. C. (2012). Will hunters steward wolves? A comment on Treves and Martin. *Society & Natural Resources*, 25(1), 97–102.

Bruskotter, J. T., Toman, E., Enzler, S. A. & Schmidt, R. H. (2010). Are gray wolves endangered in the Northern Rocky Mountains? A role for social science in listing determinations. *BioScience*, 60(11), 941–8.

Bruskotter, J. T., Vaske, J. J. & Schmidt, R. H. (2009). Social and cognitive correlates of Utah residents' acceptance of the lethal control of wolves. *Human Dimensions of Wildlife*, 14(2), 119–32.

Bruskotter, J. T., Vucetich, J. A., Enzler, S., Treves, A. & Nelson, M. P. (2014). Removing protections for wolves and the future of the U.S. Endangered Species Act (1973). *Conservation Letters*, 7(4), 401–7.

Bruskotter, J. T., Vucetich, J. A., Karns, G., Manfredo, M. J., Wolf, C., Ard, K., Carter, N. H., Lopez-Bao, J., Gehrt, S. & Ripple, W. J. (2017). Modernization, risk and conservation of the world's largest carnivores. *BioScience*, 67(1), 646–55.

Bruskotter, J. T. & Wilson, R. S. (2014). Determining where the wild things will be: Using psychological theory to find tolerance for large carnivores. *Conservation Letters*, 7, 158–65.

Bunnefeld, N., Hoshino, E. & Milner-Gulland, E. J. (2011). Management strategy evaluation: A powerful tool for conservation? *Trends in Ecology & Evolution*, 26(9), 441–7.

Cardillo, M., Purvis, A., Sechrest, W., Gittleman, J. L., Bielby, J. & Mace, G. M. (2004). Human population density and extinction risk in the world's carnivores. *PLoS Biology*, 2(7), e197.

Carter, N. H. & Linnell, J. D. C. (2016). Co-adaptation is key to coexisting with large carnivores. *Trends in Ecology & Evolution*, 31(8), 575–8.

Carter, N. H., López-Bao, J., Bruskotter, J. T., Gore, M. L., Chapron, G., Johnson, A., Epstein, Y., Shrestha, M., Frank, J., Ohrens, O. & Treves, A. (2017). A conceptual framework for understanding illegal killing of large carnivores. *Ambio*, 46(3), 251–64.

Carter, N. H., Riley, S. J. & Liu, J. (2012). Utility of a psychological framework for carnivore conservation. *Oryx*, 46(4), 525–35.

Carter, N. H., Viña, A., Hull, V., McConnell, W. J., Axinn, W., Ghimire, D. & Liu, J. (2014). Coupled human and natural systems approach to wildlife research and conservation. *Ecology & Society*, 19(3), art. 43.

Chapron, G., Kaczensky, P., Linnell, J. D. C., von Arx, M., Huber, D., Andrén, H., López-Bao, J. V., Adamec, M., Álvares, F., Anders, O., Balčiauskas, L., Balys, V., Bedő, P., Bego, F., Blanco, J. C., Breitenmoser, U., Brøseth, H., Bufka, L., Bunikyte, R., Ciucci, P., Dutsov, A., Engleder, T., Fuxjäger, C., Groff, C., Holmala, K., Hoxha, B., Iliopoulos, Y., Ionescu, O., Jeremić, J., Jerina, K., Kluth, G., Knauer, F., Kojola, I., Kos, I., Krofel, M., Kubala, J., Kunovac, S., Kusak, J., Kutal, M., Liberg, O., Majić, A., Männil, P., Manz, R., Marboutin, E., Marucco, F., Melovski, D., Mersini, K., Mertzanis, Y., Mysłajek, R. W., Nowak, S., Odden, J., Ozolins, J., Palomero, G., Paunović, M., Persson, J., Potočnik, H., Quenette, P.-Y., Rauer, G., Reinhardt, I., Rigg, R., Ryser, A., Salvatori, V., Skrbinšek, T., Stojanov, A., Swenson, J. E., Szemethy, L., Trajçe, A., Tsingarska-Sedefcheva, E., Váňa, M., Veeroja, R., Wabakken, P., Wölfl, M., Wölfl, S., Zimmermann, F., Zlatanova, D. & Boitani, L. (2014). Recovery of large carnivores in Europe's modern human-dominated landscapes. *Science*, 346, 1517–19.

Chapron, G. & Treves, A. (2016). Blood does not buy goodwill: Allowing culling increases poaching of a large carnivore. *Proceedings of the Royal Society B*, 283 (1830), 20152939.

Clark, S. & Rutherford, M. (2014). *Large Carnivore Conservation: Integrating Science and Policy in the North American West*. Chicago: University of Chicago Press.

Clark, T. W., Curlee, A. P. & Reading, R. P. (1996). Crafting effective solutions to the large carnivore conservation problem. *Conservation Biology*, 10(4), 940–8.

Clayton, S. & Myers, O. (2009). *Conservation Psychology: Understanding and Promoting Human Care of Nature*. Cambridge: Cambridge University Press.

Creel, S., Becker, M., Christianson, D., Dröge, E., Hammerschlag, N., Hayward, M. W., Karanth, U., Loveridge, A., Macdonald, D. W., Matandiko, W.,

M'Soka, J., Murray, D. L., Rosenblatt, E. & Schuette, P. (2015). Questionable policy for large carnivore hunting. *Science*, 350, 1473–5.

Creel, S. & Rotella, J. (2010). Meta-analysis of relationships between human offtake, total mortality and population dynamics of gray wolves (*Canis lupus*). *PLoS ONE*, 5(9), e12918.

Crooks, K. & Soulé, M. (1999). Mesopredator release and avifaunal extinctions in a fragmented system. *Nature*, 400, 563–6.

Czech, B., Krausman, P. R. & Devers, P. K. (2000). Economic associations among causes of species endangerment in the United States. *BioScience*, 50(7), 593–601.

Daniels, S. & Walker, G. (2001). *Working through Environmental Conflict: The Collaborative Learning Approach*. Westport, CT: Praeger.

De Groot, R. S., Wilson, M. A. & Boumans, R. M. (2002). A typology for the classification, description and valuation of ecosystem functions, goods and services. *Ecological Economics*, 41(3), 393–408.

Dietsch, A. M., Teel, T. L. & Manfredo, M. J. (2016). Social values and bio-diversity conservation in a dynamic world. *Conservation Biology*, 30(6), 1212–21.

Drechsler, M., Grimm, V., Mysiak, J. & Wätzold, F. (2007). Differences and similarities between ecological and economic models for biodiversity conser-vation. *Ecological Economics*, 62(2), 232–41.

Duffield, J., Neher, C. J. & Patterson, D. A. (2006). *Wolves and People in Yellowstone: Impacts on the Regional Economy*. University of Montana, Final Report for Yellowstone Park Foundation.

Duffield, J. W., Neher, C. J. & Patterson, D. A. (2008). Wolf recovery in Yellowstone Park: Visitor attitudes, expenditures, and economic impacts. *Yellowstone Science*, 16(1), 20–5.

Eberhardt, L. L. & Breiwick, J. M. (2010). Trend of the Yellowstone grizzly bear population. *International Journal of Ecology*, art. ID 924197.

Ericsson, G. & Heberlein, T. A. (2003). Attitudes of hunters, locals, and the general public in Sweden now that the wolves are back. *Biological Conser-vation*, 111(2), 149–59.

Estes, J. A, Terborgh, J., Brashares, J. S., Power, M. E., Berger, J., Bond, W. J., Carpenter, S. R., Essington, T. E., Holt, R. D., Jackson, J. B. C., Marquis, R. J., Oksanen, L., Oksanen, T., Paine, R. T., Pikitch, E. K., Ripple, W. J., Sandin, S. A., Scheffer, M., Schoener, T. W., Shurin, J. B., Sinclair, A. R. E., Soulé, M. E., Virtanen, R. & Wardle, D. A. (2011). Trophic downgrading of planet Earth. *Science*, 333(6040), 301–6.

Filatova, T., Verburg, P. H., Parker, D. C. & Stannard, C. A. (2013). Spatial agent-based models for socio-ecological systems: Challenges and prospects. *Environmental Modelling & Software*, 45, 1–7.

Folke, C., Carpenter, S. R., Walker, B., Scheffer, M., Chapin, T. & Rockström, J. (2010). Resilience thinking: Integrating resilience, adaptability and trans-formability. *Ecology & Society*, 15(4), art. 20.

Frank, B. (2016). Human–wildlife conflicts and the need to include tolerance and coexistence: An introductory comment. *Society & Natural Resources*, 29 (6), 738–43.

Gehrt, S. D., Riley, S. P. D. & Cypher, B. L. (2010). *Urban Carnivores: Ecology, Conflict, and Conservation*. Baltimore, MD: Johns Hopkins University Press.

Gilbert, S. L., Sivy, K. J., Pozzanghera, C. B., Dubour, A., Overduijn, K., Smith, M. M., Zhou, J., Little, J. M. & Prugh, L. R. (2016). Socioeconomic benefits of large carnivore recolonization through reduced wildlife–vehicle collisions. *Conservation Letters*, 10(4), 431–9.

Grimm, V. & Railsback, S. (2005). *Individual-Based Modeling and Ecology*. Princeton, NJ: Princeton University Press.

Gunderson, L. & Holling, C. S. (2002). *Panarchy: Understanding Transformations in Human and Natural Systems*. Washington, DC: Island Press.

Hansen, A., Rasker, R., Maxwell, B., Rotella, J., Johnson, J., Parmenter, A., Langner, U., Cohen, W., Lawrence, R. & Kraska, M. (2002). Ecological causes and consequences of demographic change in the new west. *BioScience*, 52, 151–62.

Inglehart, R. & Welzel, C. (2005). *Modernization, Cultural Change, and Democracy: The Human Development Sequence*. New York: Cambridge University Press.

Jimenez, M. D. & Becker, S. A. (2015). *Northern Rocky Mountain Wolf Recovery Program 2014 Interagency Annual Report*. US Fish and Wildlife Service, Idaho Department of Fish and Game, Montana Fish, Wildlife & Parks, Wyoming Game and Fish Department, Nez Perce Tribe, National Park Service, Blackfeet Nation, Confederated Salish and Kootenai Tribes, Wind River Tribes, Confederated Colville Tribes, Spokane Tribe of Indians, Washington Department of Fish and Wildlife, Oregon Department of Fish and Wildlife, Utah Department of Natural Resources and USDA Wildlife Services. USFWS, Ecological Services, 585 Shepard Way, Helena, Montana, 59601, pp. 1–12.

Keith, D. A., Martin, T. G., McDonald-Madden, E. & Walters, C. (2011). Uncertainty and adaptive management for biodiversity conservation. *Biological Conservation*, 144(4), 1175–8.

Levin, S., Xepapadeas, T., Crépin, A.-S., Norberg, J., de Zeeuw, A., Folke, C., Hughes, T., Arrow, K., Barrett, S., Daily, G., Ehrlich, P., Kautsky, N., Mäler, K.-G., Polasky, S., Troell, M., Vincent, J. R. & Walker, B. (2012). Social-ecological systems as complex adaptive systems: Modeling and policy implications. *Environment & Development Economics*, 18(2), 111–32.

Liberg, O., Chapron, G., Wabakken, P., Pedersen, H. C., Hobbs, N. T. & Sand, H. (2011). Shoot, shovel and shut up: Cryptic poaching slows restoration of a large carnivore in Europe. *Proceedings of the Royal Society of London Series B*, 270, 91–8.

Linnell, J. D. C. (2015). Defining scales for managing biodiversity and natural resources in the face of conflicts. In S. M. Redpath, R. J. Guitiérrez, K. A. Wood & J. C. Young, eds., *Conflicts in Conservation: Navigating towards Solutions*. Cambridge: Cambridge University Press, pp. 208–18.

Linnell, J. D. C., Swenson, J. E. & Anderson, R. (2001). Predators and people: Conservation of large carnivores is possible at high human densities if management policy is favourable. *Animal Conservation*, 4, 345–9.

Liu, J., Dietz, T., Carpenter, S. R., Alberti, M., Folke, C., Moran, E., Pell, A. N., Deadman, P., Kratz, T., Lubchenco, J., Ostrom, E., Ouyang, Z., Provencher,

W., Redman, C. L., Schneider, S. H. & Taylor, W. W. (2007). Complexity of coupled human and natural systems. *Science*, 317(5844), 1513–16.

López-Bao, J. V., Chapron, G. & Treves, A. (2017). The Achilles heel of participatory conservation. *Biological Conservation*, 212, 139–43.

Lute, M. L., Bump, A. & Gore, M. L. (2014). Identity-driven differences in stakeholder concerns about hunting wolves. *PLoS ONE*, 9(12), e114460.

Lute, M. L. & Gore, M. L. (2014). Knowledge and power in wildlife management. *The Journal of Wildlife Management*, 78(6), 1060–8.

Manfredo, M. J., Teel, T. L. & Dietsch, A. M. (2016). Implications of human value shift and persistence for biodiversity conservation. *Conservation Biology*, 30(2), 287–96.

Manfredo, M. J., Teel, T. L. & Henryk. L. (2009). Linking society and environment: A multilevel model of shifting wildlife value orientations in the western United States. *Social Science Quarterly*, 90(2), 407–27.

Martín-López, B., Palomo, I. & García-Llorente, M. (2017). Delineating boundaries of social-ecological systems for landscape planning: A comprehensive spatial approach. *Land Use Policy*, 66, 90–104.

Maser, C. & Pollio, C. A. (2011). *Resolving Environmental Conflicts*. Boca Raton, FL: CRC Press.

Mattson, D. J. (2014). State-level management of a common charismatic predator: Mountain lions in the west. In S. Clark & M. Rutherford, eds., *Large Carnivore Conservation: Integrating Science and Policy in the North American West*. Chicago and London: University of Chicago Press, pp. 29–64.

Maxwell, S. L., Milner-Gulland, E. J., Jones, J. P. G., Knight, A. T., Bunnefeld, N., Nuno, A., Bal, P., Earle, S., Watson, J. E. M., Rhodes, J. R., Tittensor, D. P., Perrings, C., Stafford-Smith, M., Skjaerseth, J. B., Bull, J., Redpath, S. M., Smead, R., Bunnefeld, N., Fulton, E. A., Daniels, S. E., Walker, G. B., Pouwels, R., Dubash, N. K. & Campbell, L. M. (2015). Environmental science: Being smart about SMART environmental targets. *Science*, 347(6226), 1075–6.

McLane, A. J., Semeniuk, C., McDermid, G. J. & Marceau, D. J. (2011). The role of agent-based models in wildlife ecology and management. *Ecological Modelling*, 222(8), 1544–56.

Mech, D. (2012). Is science in danger of sanctifying the wolf? *Biological Conservation*, 150(1), 143–9.

Mech, L. D. (1995). The challenge and opportunity of recovering wolf populations. *Conservation Biology*, 9, 270–8.

Milner-Gulland, E. J., Arroyo, B., Bellard, C., Blanchard, J., Bunnefeld, N., Delibes-Mateos, M., Edwards, C., Nuno, A., Palazy, L., Reljic, S., Riera, P. & Skrbinsek, T. (2010). New directions in management strategy evaluation through cross-fertilization between fisheries science and terrestrial conservation. *Biology Letters*, 6(6), 719–22.

Muhly, T. B., Hebblewhite, M., Paton, D., Pitt, J. A., Boyce, M. S. & Musiani, M. (2013). Humans strengthen bottom-up effects and weaken trophic cascades in a terrestrial food web. *PLoS ONE*, 8(5), e64311.

Muhly, T. B. & Musiani, M. (2009). Livestock depredation by wolves and the ranching economy in the Northwestern U.S. *Ecological Economics*, 68(8–9), 2439–50.

Nie, M. (2001). The sociopolitical dimensions of wolf management and restoration in the United States. *Human Ecology*, 8, 1–12.

Nie, M. (2004). State wildlife policy and management: The scope and bias of political conflict. *Public Administration Review*, 64, 221–33.

Nie, M. A. (2003). *Beyond Wolves: The Politics of Wolf Recovery and Management*. Minneapolis: University of Minnesota Press.

Obst, C., Hein, L. & Edens, B. (2016). National accounting and the valuation of ecosystem assets and their services. *Environmental & Resource Economics*, 64 (1), 1–23.

Olson, E. R., Stenglein, J. L., Shelley, V., Rissman, A. R., Browne-Nuñez, C., Voyles, Z., Wydeven, A. P. & Van Deelen, T. (2015). Pendulum swings in wolf management led to conflict, illegal kills, and a legislated wolf hunt. *Conservation Letters*, 8(5), 351–60.

Packer, C., Kosmala, M., Cooley, H. S., Brink, H., Pintea, L., Garshelis, D., Purchase, G., Strauss, M., Swanson, A., Balme, G., Hunter, L. & Nowell, K. (2009). Sport hunting, predator control and conservation of large carnivores. *PLoS ONE*, 4(6), e5941.

Peebles, K. A., Wielgus, R. B., Maletzke, B. T. & Swanson, M. E. (2013). Effects of remedial sport hunting on cougar complaints and livestock depredations. *PLoS ONE*, 8(11), e79713.

Peterson, R. O., Vucetich, J. A., Bump, J. M. & Smith, D. W. (2014). Trophic cascades in a multicausal world: Isle Royale and Yellowstone. *Annual Review of Ecology, Evolution, & Systematics*, 45, 325–45.

Primm, S. A. & Clark, T. W. (1996). Making sense of the policy process for carnivore conservation. *Conservation Biology*, 10(4), 1036–45.

Prugh, L. R., Stoner, C. J., Epps, C. W., Bean, W. T., Ripple, W. J., Laliberte, A. S. & Brashares, J. S. (2009). The rise of the mesopredator. *BioScience*, 59(9), 779–91.

Redpath, S. M., Linnell, J. D. C., Festa-Bianchet, M., Boitani, L., Bunnefeld, N., Dickman, A., Gutiérrez, R. J., Irvine, R. J., Johansson, M., Majić, A., McMahon, B. J., Pooley, S., Sandström, C., Sjölander-Lindqvist, A., Skogen, K., Swenson, J. E., Trouwborst, A., Young, J. & Milner-Gulland, E. J. (2017). Don't forget to look down: Collaborative approaches to predator conservation. *Biological Reviews*, 92, 2157–63.

Redpath, S. M., Young, J., Evely, A., Adams, W. M., Sutherland, W. J., Whitehouse, A., Amar, A., Lambert, R. A., Linnell, J. D. C., Watt, A. & Gutiérrez, R. J. (2013). Understanding and managing conservation conflicts. *Trends in Ecology & Evolution*, 28(2), 100–9.

Riley, S. J., Nesslage, G. M. & Maurer, B. A. (2004). Dynamics of early wolf and cougar eradication efforts in Montana: Implications for conservation. *Biological Conservation*, 119(4), 575–9.

Ripple, W. J., Estes, J. A., Beschta, R. L., Wilmers, C. C., Ritchie, E. G., Hebblewhite, M., Berger, J., Elmhagen, B., Letnic, M., Nelson, M. P., Schmitz, O. J., Smith, D. W., Wallach, A. D. & Wirsing, A. J. (2014). Status and ecological effects of the world's largest carnivores. *Science*, 343(6167), 1241484.

Scarce, R. (1998). What do wolves mean? Conflicting social constructions of *Canis lupus* in Bordertown. *Human Dimensions of Wildlife*, 3(3), 26–45.

Schneider, D. (2004). *The Psychology of Stereotyping: Distinguished Contributions in Psychology*. New York: Guilford Publications.

Skogen, K., Krange, O. & Figari, H. (2017). *Wolf Conflicts: A Sociological Study*. New York: Berghahn Books.

Slagle, K., Bruskotter, J. & Wilson, R. (2012). The role of affect in public support and opposition to wolf management. *Human Dimensions of Wildlife*, 17, 44–57.

Smith, D. & Bangs, E. (2009). Reintroduction of wolves to Yellowstone National Park: History, values, and ecosystem restoration. In M. W. Hayward & M. J. Somers, eds., *Reintroduction of Top-Order Predators*. Oxford: Wiley-Blackwell, pp. 92–125.

Smith, D. W., Bangs, E. E., Oakleaf, J. K., Mack, C., Fontaine, J., Boyd, D., Jimenez, M., Pletscher, D. H., Niemeyer, C. C., Meier, T. J., Stahler, D. R., Holyan, J., Asher, V. J. & Murray, D. L. (2010). Survival of colonizing wolves in the Northern Rocky Mountains of the United States, 1982–2004. *Journal of Wildlife Management*, 74(4), 620–34.

Smith, D. W., White, P., Stahler, D., Wydeven, A. P. & Hallac, D. (2016). Managing wolves in the Yellowstone area: Balancing goals across jurisdictional boundaries. *Wildlife Society Bulletin*, 40, 436–45.

Spidalieri, K. (2012). Looking beyond the bang for more bucks: A legislative gift to fund wildlife conservation on its 75th anniversary. *Cleveland State Law Review*, 60(3), 769–98.

Sterling, E. J., Betley, E., Sigouin, A., Gomez, A., Toomey, A., Cullman, G., Malone, C., Pekor, A., Arengo, F., Blair, M., Filardi, C., Landrigan, K. & Porzecanski, A. L. (2017). Assessing the evidence for stakeholder engagement in biodiversity conservation. *Biological Conservation*, 209, 159–71.

Stillman, R. A., Railsback, S. F., Giske, J., Berger, U. & Grimm, V. (2015). Making predictions in a changing world: The benefits of individual-based ecology, *Bioscience*, 65(2), 140–50.

Tajfel, H. (1982). Social psychology of intergroup relations. *Annual Review of Psychology*, 33, 1–39.

Tajfel, H. & Turner, J. (1979). An integrative theory of intergroup conflict. *The Social Psychology of Intergroup Relations*, 33, 33–47.

Teel, T. L. & Manfredo, M. J. (2010). Understanding the diversity of public interests in wildlife conservation. *Conservation Biology*, 24(1), 128–39.

Thrower, J. (2009). Ranching with wolves: Reducing conflicts between livestock and wolves through integrated grazing and wolf management plans. *Journal Land Resources & Environmental Law*, 319, 327–8.

Treves, A. (2012). Tolerant attitudes reflect an intent to steward: A reply to Bruskotter and Fulton. *Society & Natural Resources*, 25, 103–4.

Treves, A., Artelle, K., Darimont, C. T. & Parsons, D. R. (2017). Mismeasured mortality: Correcting estimates of wolf poaching in the United States. *Journal of Mammalogy*, 98(5), 1256–64.

Treves, A., Krofel, M. & McManus, J. (2016). Predator control should not be a shot in the dark. *Frontiers in Ecology and the Environment*, 14(7), 380–8.

Vucetich, J. A., Smith, D. W. & Stahler, D. R. (2005). Influence of harvest, climate and wolf predation on Yellowstone elk, 1961–2004. *Oikos*, 111(2), 259–70.

Walker, B. & Salt, D. (2006). *Resilience Thinking: Sustaining Ecosystems and People in a Changing World*. Washington, DC: Island Press.

Wielgus, R. B. & Peebles, K. A. (2014). Effects of wolf mortality on livestock depredations. *PLoS ONE*, 9(12), e113505.

Williams, S. (2010). Wellspring of wildlife funding: How hunter and angler dollars fuel wildlife conservation. *The Wildlife Professional*, Fall, 35–8.

Wilson, M. A. (1997). The wolf in Yellowstone: Science, symbol, or politics? Deconstructing the conflict between environmentalism and wise use. *Society & Natural Resources*, 10(5), 453–68.

Woodroffe, R. (2000). Predators and people: Using human densities to interpret declines of large carnivores. *Animal Conservation*, 3(2), 165–73.

Young, J. C., Jordan, A., Searle, K. R., Butler, A., Simmons, P. & Watt, A. D. (2013). Framing scale in participatory biodiversity management may contribute to more sustainable solutions. *Conservation Letters*, 6(5), 333–40.

19

Planning for Coexistence in a Complex Human-Dominated World

SILVIO MARCHINI, KATIA M. P. M. B. FERRAZ,
ALEXANDRA ZIMMERMANN, THAÍS GUIMARÃES-
LUIZ, RONALDO MORATO, PEDRO L. P. CORREA AND
DAVID W. MACDONALD

The interdisciplinary field of research focusing on human–wildlife conflict (HWC) and ways to turn conflict into coexistence, although relatively new, is developing fast. In the last 20 years, the number of scientific publications addressing HWC has increased almost exponentially (Figure 19.1). Initially focused on the ecological and economic aspects of wildlife damage (Woodroffe et al. 2005), the emphasis of this literature has been gradually shifting to the human dimensions of HWC. Examples of issues that have been addressed recently include the roles of cognition and feelings such as attitudes (Kansky & Knight 2014), values (Manfredo et al. 2016), risk perception (Decker et al. 2010) and emotions (Jacobs 2012) about single species or groups of related species (e.g. large carnivores), models to predict tolerance (Bruskotter & Wilson 2014; Kansky et al. 2016) and behaviours (Marchini & Macdonald 2012) towards the species involved, and approaches to understanding conflicts between groups of people over wildlife management (Madden & McQuinn 2014). There has been, indeed, considerable progress in the understanding of the ecology and economics of wildlife damage and of the drivers of tolerance and hostility towards wildlife at small scales (i.e. individual to household to community level). Nonetheless, this understanding has not translated significantly into management and policy at larger scales. In the meantime, HWC is escalating in the world in general, and in developing countries in particular (Hoare 2015; Marchini & Crawshaw Jr 2015; Manral et al. 2016).

Two important factors behind this *research-implementation gap* are complexity and limited resources. Whereas researchers have understandably focused mostly on local, single-species issues and individual-

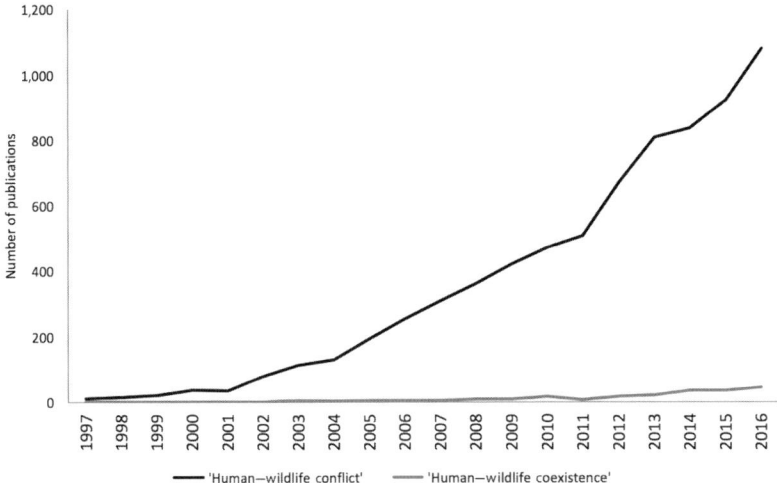

Figure 19.1 Google Scholar search for *human–wildlife conflict* and *human–wildlife coexistence*. Custom range search was used to generate the number of results for each year, from 1997 to 2016 (on 15 December 2017).

level responses to them, managers and policy-makers working at larger scales (e.g. in local to national government agencies), and in an increasingly globalized world, have to deal with problems that typically involve (1) multiple species, including endangered and abundant, native and exotic, wild and domestic, (2) a growing set of stakeholders holding ever stronger interests in wildlife governance and different views on the costs and benefits of different situations, (3) patterns and processes that take place at broader spatial and temporal scales and at higher hierarchical levels of decision (e.g. institutional rather than individual) than what has been usually examined, (4) alternative management actions with varying urgency, coverage and cost-effectiveness and (5) diverse forms of financial and human resources constraints.

We argue that the bridge between HWC research and the implementation of large-scale human–wildlife coexistence is good planning. In this chapter, we address the potential application of strategic planning, combined with the growing fields of scientific modelling and data science, to inform decisions regarding the conflict-to-coexistence continuum (Frank 2016) and propose a framework for integrating data and stakeholders – planners, researchers, modellers, policy-makers, managers and citizens – in the process of planning for coexistence.

19.1 WHAT IS PLANNING AND WHY WE NEED IT

Planning is the process of identifying a course of action in a systematic manner to achieve objectives by utilizing the available resources competently in a cost-effective way (Mintzberg & Quinn 1996). Planning is often depicted as an iterative, cyclic process comprised of the following fundamental steps (Figure 19.2):

- Situation assessment, where a description and understanding of the current situation is developed by addressing the questions *where are we* and *why are we here?*
- Decision-making, which involves establishing what the project will aim to achieve, defining the agreed vision and short-term activities that must be completed to ensure that longer-term goals are met, and determining what needs to be done to achieve the desired results, including how results will be monitored and evaluated. The guiding questions are *where do we want to get* and *how do we get there?*
- Implementation, which involves putting into practice the previous planning work through the development and implementation of specific work plans while ensuring sufficient resources, capacity and partners.
- Monitoring and evaluation, where data collected during and after implementation are analysed in order to measure success, usually at different levels (e.g. output, outcome and impact). The guiding questions are *are we getting there?* and *have we got there?* At this step, revisiting the decision-making and implementation steps should be considered, thus closing the cycle.

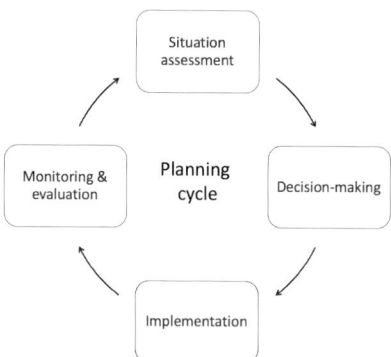

Figure 19.2 Planning cycle and its fundamental steps.

Covering the entire planning cycle, *learning and communication* are powerful means for helping those responsible for the planning and implementation if others can learn from their experiences and can provide feedback (IUCN Species Survival Commission 2017). Learning and communication are relevant for project and organization staff, and also externally for other planners, funders and donors, the conservation community, governments and civil society more broadly.

The steps of the planning cycle do not fit a standard format. Their definition and labels vary between sources and users, and different planning approaches emphasize different steps in the cycle. Nonetheless, the underlying principles remain constant and in all cases the planning cycle provides a logical framework to enable managers to tackle the issues in a systematic way, and so ensure that no major issues are left unaddressed.

Systematic efforts to plan for the conservation of biodiversity were initiated in the mid-1970s in response to the limited funding and ad hoc way in which protected areas had generally been established (Groves 2003; Knight et al. 2013). At approximately the same time but independently, The Nature Conservancy (TNC) in the United States and the United Nations Educational, Scientific and Cultural Organization (UNESCO) started to develop their methods to identify high-priority sites for conservation purposes (Groves 2003). Concurrently with these efforts, Australian scientists began to work on establishing a biologically representative network of nature reserves throughout the country (and established themselves as the leading thinkers in systematic planning for the conservation of biological diversity; Groves 2003).

Ever since, there has been an increasing application of techniques using computational tools for informing decisions about conservation in the face of limited financial resources (Knight et al. 2013; Text Box 19.1). These approaches contribute to the development of plans for the implementation and continued application of conservation actions with the aim of reducing biodiversity declines in a transparent and socially responsible manner (Margules & Pressey 2000; Ball et al. 2009). The term *conservation planning* has been used in such studies, as the ultimate goal of planning is to conserve endangered species and/or their habitat. Conservation planning has indeed evolved in two different and mostly independent fronts: species-focused and ecosystem- or area-based planning.

Species conservation planning is intended to generate a blueprint for saving a species or group of species, across all or part of the species'

Text Box 19.1 Software Used in Conservation Planning

The types of software available to assist conservation planners' decision-making have evolved and diversified. Below is a brief description of the leading software used in spatial planning and species-conservation planning.

- **Marxan** (marxan.net) provides decision support to a range of conservation planning problems, including the design of new reserve systems, reporting on the performance of existing reserve systems and developing multiple-use zoning plans for natural resource management.
- **Zonation** (helsinki.fi/en/researchgroups/metapopulation-research-centre/software) identifies areas important for retaining habitat quality and connectivity for multiple species.
- **C-Plan** (marxan.net/cplan) is a conservation decision support software that links with Geographic Information Systems to map options for achieving explicit conservation targets. It acts as a graphical user interface for Marxan and can generate Marxan data sets from C-Plan data sets.
- **Vortex** (vortex10.org), unlike the above, is not primarily a decision support software, but an individual-based population viability analysis (PVA) model (Lacy 2000). PVA is used to estimate the future likelihood of a population's extinction and indicate the urgency of alternative management efforts, and to identify key life stages or processes that should be the focus of those efforts.
- **Ramas** (ramas.com) uses species-specific information such as the current population size, survival rate and fecundity, maps of the species' habitat and population structure to predict the future size, structure and spatial distribution of the species. A sensitivity analysis tool allows identifying the most important data requirements.
- **Maxent** (biodiversityinformatics.amnh.org/open_source/maxent/) for modelling species niches and distributions by applying a machine-learning technique called maximum entropy modelling. From a list of species presence locations and a set of environmental predictors, the model expresses a probability distribution where each grid cell has a predicted suitability of conditions for the species (Phillips & Dudík 2008).
- **Miradi** (miradi.org) is a user-friendly program that allows nature conservation practitioners to design, manage, monitor and learn from their projects to more effectively meet their conservation goals – both species-focused and area-based planning – following a process laid out in the Open Standards for the Practice of Conservation (OS).

range. An example of this approach is the Species Conservation Strategy of the International Union for Conservation of Nature (IUCN) Species Survival Commission (SSC), whose species planning cycle comprises two phases (IUCN 2017): (1) plan (preparing, collecting information, building a vision and setting goals, analysing threats and setting objectives, planning actions), and (2) implement/learn/adapt. The SSC

Conservation Planning Specialist Group (CPSG) conceived a broader approach to species conservation planning, the One Plan Approach, that emphasizes the importance of incorporating *ex situ* collection plans and species management programmes to produce a single, comprehensive conservation plan for a species (Byers et al. 2013). The IUCN Species Conservation Strategy process has been adopted by several countries members of the Convention for Biodiversity Conservation (CBD). Brazil, for example, has established species conservation action plans for over 50 per cent of the species listed as endangered on its National Red List of Endangered Species.

Area-based conservation planning approaches, mostly referred to as *systematic conservation planning* (note: other approaches are systematic too), focus on the strategic spatially explicit allocation of conservation resources so as to ensure that all natural features are represented in formally protected areas whilst maximising the cost-effectiveness of conservation activities (Pressey & Nicholls 1989). Prioritization is based on the principle of complementarity, i.e. the idea that individual areas within a protected area network should complement, rather than duplicate, one another in terms of the natural features they protect (Justus & Sarkar 2002). The modelling methods have evolved rapidly since the development of the first simple algorithms used to identify complementary candidates of protected areas for planning a region (e.g. Pressey & Nicholls 1989). Socio-economic considerations such as implementation costs (Naidoo et al. 2006), willingness and capacity to effectively collaborate and implement conservation action (Knight et al. 2010) and temporal analyses that account for changes in land-use over time (e.g. Wilson et al. 2007; Cushman et al. 2017), have also been included to improve cost-effectiveness.

In both species-focused and area-based conservation planning, modelling has been a pivotal tool. Scientific modelling refers to the generation of a conceptual or mathematical representation – a model – of a real system that is too complex for us to comprehend in its entirety. A model is a simplification of reality as it includes those elements that have broad effect and omits those minor elements that are not relevant to the given level of abstraction. Where sufficient and empirically verified knowledge of a system is available, modelling can provide a means for *sensitivity analysis*, i.e. testing how the system may respond to changes in different factors, and assess the likely responses of the system to alternative possible future management (Bunnefeld et al. 2011). Qualitative predictions of expected trends and system dynamics can be generated even when the knowledge about the system is too limited for modelling to provide robust quantitative predictions (Heinonen & Travis 2015). There is, indeed, a trade-off between precision and generality: while tactical models aim to make predictions of a specific system, but demand data of high quantity and quality, strategic models are less demanding, but generate generic insights that raise concerns about model validation and the usefulness of model outputs (Heinonen & Travis 2015). Sophisticated software has been developed to assist planners to map and model increasingly complex and realistic conservation scenarios (Text Box 19.1).

Despite progress, conservation planning is still an imperfect science that has emphasized the analytical tools for informing decisions, but has been limited in taking stakeholder collaboration into account (Knight et al. 2013). Conservation decisions affect stakeholders, necessitating their involvement in the decision-making processes (Knight et al. 2006). Furthermore, conservation planning has placed more importance on ecological considerations than on social ones (Knight et al. 2013). HWC, however, is essentially a social issue. Even though the conceptual and technical advances in conservation planning are potentially useful for those working on human–wildlife conflict and coexistence issues, the planning to turn conflict into coexistence at large scales will require a broader approach that places more emphasis on the human dimension and whose goals go beyond saving endangered species and ecosystems.

19.2 PLANNING FOR COEXISTENCE

The concept of *coexistence* is relatively new to wildlife science (Madden 2004; Woodroffe et al. 2005; Dickman et al. 2011). Compared to the fast-growing importance given to *human–wildlife conflict*, the expression that refers to the opposite end of the conflict-to-coexistence continuum (Frank 2016), *human–wildlife coexistence*, has had to date only a very modest presence in the scientific literature (Figure 19.1). Its meaning has been taken for granted by authors until even more recently, when confusion between coexistence and similar concepts such as co-occurrence, tolerance, acceptability and stewardship prompted a debate about terminology (Bruskotter et al. 2015, Carter & Linnell 2016; Frank 2016). Regardless of the various definitions of *coexistence*, we posit that the term conveys two fundamental and complementary meanings in the context of planning. One of them is literal: to coexist, to exist together or to live together. Coexistence implies coinciding in space and time and, furthermore, interacting. Coexistence has, therefore, an explicit spatial-temporal ecological dimension. The second meaning, perhaps less explicit and more subjective, concerns how people live together with wildlife. As opposed to conflict, coexistence suggests some level of sustainable harmony and conformity (Dickman et al. 2011), that results in a relatively non-hostile reaction from the human side towards the wildlife and the humans involved. Coexistence has, therefore, also an attitudinal-behavioural – or psychological – dimension. In sum, coexistence has two fundamental elements: *inter-action* (ecological process over a defined space and time) and *re-action* (human attitudinal and behavioural response to that interaction). As a note, in the conservation policy and planning literature the term coexistence has been used to refer to relationships between groups of people (e.g. sport, commercial, and personal-use fishing groups in Alaska, in Loring 2016) and institutions (e.g. marine renewable energy projects and marine protected areas in Europe, in Kyriazi et al. 2016).

Likewise, HWC has ecological and human dimensions. Human–wildlife conflict has been defined as 'when the behaviour of wild animal species poses a direct and recurring threat to the livelihood or safety of a community and, in response, persecution of the species ensues' (Zimmermann et al. 2010, p. 129), but not every species that has a negative impact on, or is negatively impacted by, humans is treated under the label of HWC. While non-charismatic nuisance animals continue to be managed through the traditional animal damage control approach (e.g. lethal control), HWC has been increasingly used to frame issues involving high-profile species (Macdonald et al. 2013; Redpath et al. 2013). The difference between the small rodents and snakes, frogs or invertebrates that are simply controlled as pests versus the carnivores, primates, mega-herbivores and birds addressed in the growing HWC literature is not necessarily the magnitude of the damage they cause or their conservation status, but rather the fact that the animals in the latter group can divide opinions and feelings among broad sectors of society (Macdonald et al. 2013; Marchini 2014). To some people, in some contexts, usually local people in rural contexts, these animals can be hated and feared as much as any pest. But to a growing number of other people, in other contexts, usually urban and distant from the problem, they are highly regarded for their commercial, recreational, ecological, cultural, scientific, spiritual, aesthetic or simply existence value. Although wildlife damage and preventive or retaliatory killing of wildlife – both ecological interactions – are important elements of HWC, what ultimately defines HWC are the clashes between groups of people who hold differing interests towards these animals and their management (Herda-Rapp & Goedeke 2005; Peterson et al. 2010; Marchini 2014; Redpath et al. 2015).

Since HWC is fundamentally about disagreements and resentment between groups of people (over wildlife), and because the field of HWC is increasingly dealing with species that do not entail a significant conservation focus, such as abundant and exotic wildlife in highly human-modified environments (e.g. urban coyotes, capybaras in residential condominiums and wild boars in agricultural fields), and even domestic animals (e.g. free-ranging dogs and cats in protected areas), we argue that *planning for coexistence*, rather than *conservation planning*, may be a phraseology that more helpfully directs thinking to the approaches most likely to be fruitful. Instead of emphasizing the maximization of wildlife populations of endangered species or the cost-effectiveness of protected areas, planning for coexistence would concentrate attention on improving human–wildlife interactions, while attending also to the well-being of the people affected and ensuring just treatment to the wildlife involved (Vucetich et al. 2018).

An additional reason to favour the term coexistence over conservation in this context is that it is arguably more effective in communicating management and policy objectives, and consequently engaging the public. The framing of messages is an important component of effective behaviour change (Tversky & Kahneman 1981; see Chapter 16). There has been increasing support among conservationists for the idea that the path towards effective environmental communication is not via negativity, but through hope and optimism (Bull et al. 2017). The Conservation Optimism (conservationoptimism.com), Earth Optimism (earthoptimism.si.edu) and Ocean Optimism (www.oceanoptimism.org) movements are evidence of this trend. The term *coexistence* refers equally to the two sides of the interaction, and does not suggest any negative reaction from the human side. Accordingly, the public may understand planning for coexistence as having a more balanced and impartial focus on the interaction itself – between humans and wildlife, and also between people or institutions with different interests towards wildlife – than explicitly favouring in a top-down fashion one side of the interaction, the wildlife side, as is perhaps the perception of conservation.

Planning for coexistence, like any other strategic planning, will follow the adaptive management components of situation assessment,

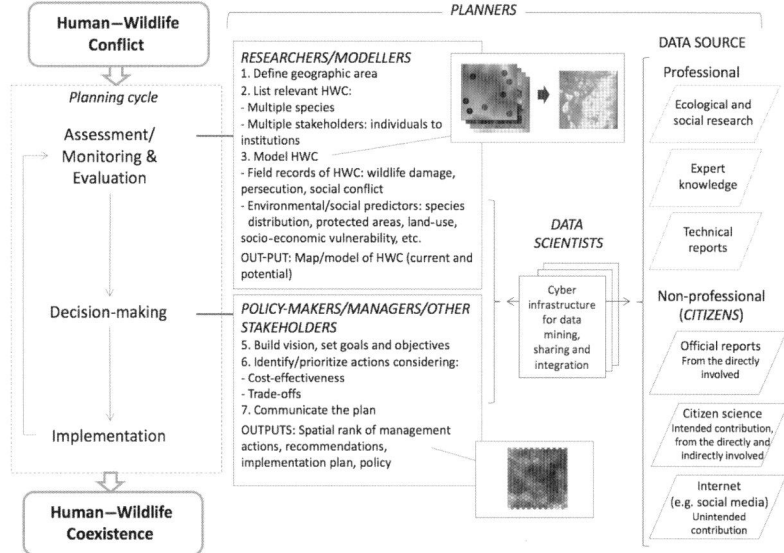

Figure 19.3 Planning for coexistence: framework for integrating stakeholders (in capital italic letters) and data.

decision-making, implementation and evaluation (Figure 19.3). What distinguishes it from conservation planning and from how HWC has been traditionally addressed is rather the breadth of its scope. Planning for coexistence proposes a broader approach to HWC at the following levels:

- **Goal**. Whereas the ultimate goal of conservation planning is to save species and/or ecosystems, in planning for coexistence the emphasis is on achieving that outcome by improving the relationships between people and wildlife, and among stakeholders, including conservationists and animal welfare groups. This implies safeguarding both the viability of wildlife populations and the well-being of individual animals. Planning for coexistence, as we use the term, therefore, encompasses conservation planning, and goes beyond it. As a note, while conservation problems can be, and have been, resolved (e.g. species are saved from extinction and natural areas are converted into effectively protected areas), planning for coexistence should consider that HWC is often difficult or impossible to resolve. In fact, Redpath et al. (2013) stress that no conflict in conservation has ever been completely resolved in the sense that (social) conflict is eliminated, although there have been varying degrees of success at managing them. Nonetheless, planning will be vital in adapting to the changes that are already under way, and preparing for future HWC, which can help mitigate the risks and hassle societies will face from sharing the space with wildlife.

- **Target**. In addition to species of conservation concern, planning for coexistence involves abundant native and exotic wildlife, and even domestic species. In fact, coexistence may require, in some cases, management strategies aiming at reducing wildlife population size.
- **Level of decision**. While some research has emphasized, often for entirely pragmatic and understandable reasons, individual-level determinants of HWC, human–wildlife coexistence requires attention to multiple scales of decision-making, from intra-personal determinants of behaviour (e.g. attitudes) to farm-level management to region-level policies.
- **Spatial scale**. The majority of studies and experiments in HWC management are rather localized. To achieve large-scale coexistence, planning needs to up-scale these analyses and interventions to cover the entire areas of regions, countries or continents where HWC occurs.
- **Methodology**. Assessing HWC at large scales, making decisions that account for the interests of a wide range of stakeholders, from local individuals to global institutions, and implementing strategies that require continued and lasting collaboration between these stakeholders, will require novel combinations of methods to communicate to and engage with stakeholders. Furthermore, planning for coexistence is likely to be, in general, a data-intensive science and as such is likely to rely heavily on modelling and require techniques of data management (acquisition, mining, storage, sharing, integration and visualization) that go beyond hitherto conventional methodologies.

19.3 ASSESSING THE SITUATION: UP-SCALING WITH AN EMPHASIS ON THE HUMAN SIDE

Assessing HWC at large scales involves the challenge of knowing where a particular event of wildlife damage has occurred, and also where else it is likely to occur. By taking known mapped locations of the event and correlating that with a wide set of existing environmental data (e.g. vegetation, land-use, density of roads, proximity to protected areas), predictive distribution models can inform the variation in the likelihood of event occurrence. For example, species distribution models (SDM), also called environmental niche models (ENM), allow mapping the extent and suitability of habitat for a species (Guisan & Tuiller 2005; Elith & Leathwick 2009). There is a variety of methods for distribution modelling; one of the most commonly used is implemented in the software MaxEnt (Phillips & Dudik 2008, Text Box 19.1). These modelling approaches have been used for predicting fire risk (Fonseca et al. 2016), roadkills (Santos et al. 2015) and deforestation (Souza & De Marco 2014), and can be used to predict the risk of wildlife damage such as crop-raiding, livestock predation and the spread of zoonotic disease.

Sitati et al. (2003), for example, demonstrated that crop-raiding and human deaths and injuries caused by elephants in Trans-Mara District, Africa, were correlated with a variety of underlying spatial variables. Likewise, Wilson et al. (2013) found that in Assam, India, over 90 per cent of all elephant crop-raiding occurred within 700 m of forest fragments and tea plantations, which encompasses the areas that tend to be inhabited by the poorest and most disadvantaged communities. As for conflicts involving carnivores, a review of the literature revealed the utility of spatial predation risk modelling as a technique for identifying high-risk priority areas (Miller 2015). The landscape features associated with carnivore attacks related to the species (carnivore and prey), environment, human infrastructure and management interventions. Risk of livestock predation by jaguars along the Trans-Amazon Highway in Brazil, for example, is related with distance to deforestation and cattle density (Carvalho et al. 2015). Regarding the spread of diseases, Pigott et al. (2014) used SDM to predict the transmission of Ebola in twenty-two countries across Central and West Africa based on vegetation, elevation, temperature, evapotranspiration and suspected reservoir bat distributions.

The output from risk models can successfully be applied to managing wildlife damage. The usefulness for conservation, however, will depend upon factors beyond the ecology and economics of damage. The ultimate motivator of retaliatory persecution may not be the actual impact of a species on human safety or property, but rather the cultural and social perceptions of the potential threat that wildlife pose when attacking people, raiding crops and killing livestock (Dickman 2010; Marchini & Macdonald 2012; Dickman et al. 2013). In conflicts between people and carnivores, for example, the perceived impacts often exceed the actual evidence (Zimmermann et al. 2005; Marchini & Macdonald 2012). In addition, factors not directly related to the impacts that wildlife have on human livelihoods (e.g. perceived social status of hunters in the community and thrill of the chase) may also be involved in the persecution of large carnivores. Although actual ecological interactions may, in many cases, play a central role in HWC, the vital element to be assessed if we are to plan for coexistence is, therefore, people's thoughts and actions.

Attitudes, behaviours and the relationships between humans as individuals or groups, however, are more difficult to map and model than wildlife damage. Indeed, few HWC studies have looked at the integration of spatial analyses and human dimensions (Hemson et al. 2009; Carter et al. 2014; Behr et al. 2017). Nonetheless, relevant human behaviours in HWC contexts, such as retaliatory killing, could conceivably be linked to spatially explicit biophysical, social and economic factors such as the distribution and density of multiple wildlife species, proximity to protected areas, density of roads, ease of access, institution presence (or vacuum), predominant land-uses and the policies that govern them. Besides, a person's thoughts and actions towards wildlife and wildlife management can be determined by processes occurring far away and at higher hierarchical levels (Adams 2015). Indeed, human–wildlife conflicts are the result of decisions of a wide range of interacting stakeholders, from households to global corporations, reflecting the working of markets, mediated by the intervention and regulation of the state, and the lobbying of non-governmental actors (Adams 2015). This complexity of the human side of HWC raises the issue of data acquisition in planning for coexistence and in how to inform decisions regarding the conflict-to-coexistence continuum.

19.3.1 Getting the Necessary Data

Assessing the situation in order to plan for coexistence involves the use of ecological and human dimensions data (Figure 19.3). However, while remote sensing, radio-tracking, camera trapping, drones and other technologies have immensely increased capacity to collect ecological data, traditional social research approaches – in-person interviews, self-completed questionnaires or media content analysis – often can make only a relatively modest contribution to the overall data set. Alternative approaches to data collection and sharing may represent promising cost-effective ways to overcome this discrepancy and help fill the gaps in social data. Below we present examples of data sources that can support coexistence planning.

19.3.2 Expert Opinion

Whenever action is required before uncertainties can be resolved, and empirical data are scarce or unavailable and difficult or expensive to collect, expert knowledge can be the best or only source of information (Kuhnert et al. 2010). Commonly used methods for expert elicitation are the Delphi process and its variants, including the Nominal Group Technique (MacMillan & Marshall 2006). These are group elicitation processes in which the experts are first asked for independent input on some parameter of interest. These estimates are collated, and then shared with the whole group, along with the rationale and evidence that each expert used. The experts are then allowed to revise their estimates, if they so desire, to reflect the insights that arose from the group. This process can be repeated a number of times, until the experts are comfortable with their responses. Amit and Jacobson (2017), for example, applied this technique to identify and rank benefits, barriers and motivations of ranchers in Costa Rica to adopt preventive husbandry practices that limit livestock depredation by large carnivores, such as jaguars (*Panthera onca*) and pumas (*Puma concolor*).

Although the use of expert knowledge in applied ecology and conservation is not new (Burgman 2005), the focus on the quantification of uncertainties using probability is relatively new. Structured elicitation techniques aim at extracting and quantifying individual judgement about uncertainty (McBride & Burgman 2012). These techniques increase the reliability of expert opinions and help a planner decide whether expert opinion is useful, how an elicitation should be conducted, and what the most relevant method for elicitation is, facilitating informed decision-making in a cost-effective manner (Sutherland & Burgman 2015).

19.3.3 Technical Reports and Official Records

Ranchers, farmers and other people affected by HWC can be important sources of data on HWC when they report their problems to governmental agencies and non-governmental organizations. These institutions may compile records in the form of claims for relief and compensation and the corresponding response given to each case. Governmental agencies may compile also reports from technical studies that evaluate the impact of infrastructure projects, and from wildlife research and management projects under their authorization. Governmental agencies may hold, indeed, a wealth of information on both the ecological and human dimensions of HWC. However, lack of standard-ization, data gaps, limited use of computational tools for data sharing (especially in developing countries) and poor communication between sectors inside and outside the government, such as academia and non-profit, pose challenges to the use of this resource.

19.3.4 Citizen Science

Citizen science is a method that relies on volunteers to collect data for research investigations (Couvet et al. 2008; Silvertown et al. 2013; Buesching et al. 2015). Mobile applications for smartphones, tablets and other gadgets have dramatically increased the opportunities and capacity in citizen science. Smartphones can automate data collection and incorporate many important data-gathering functions, such as cap-turing images, audio and text, into a single tool that can inform the date, time and geographic coordinates associated with an observation. Citizen science has traditionally been used in ornithological studies (e.g. wikiaves.com.br), but it has also started to gain traction in human–wildlife interaction issues as well. Examples include a project that maps hotspots of conflicts with coyotes in suburban New York by asking citizens to report their experiences and encounters with wildlife (Weckel et al. 2010) and mobile applications to monitor and report poaching and other illegal activities (smartconservationtools.org), roadkill (-sistemaurubu.com.br/us) and zoonotic diseases (biodiversidade.ciss.fio-cruz.br). These data are used to generate maps and feed predictive models. Similar applications could conceivably be used to record the behavioural and social components of HWC as well: preventive and retaliatory persecution, and conflicts of interest between stakeholders.

19.3.5 Internet Data

User-generated content produced via various social media platforms and search engines contains spatially explicit information on human–wildlife interactions, and on people´s thoughts and actions (Text Box 19.2). Unlike in citizen science, this information is not intended to be used by scientists or planners, but it can. Every minute, users all over the world *like* over 4 million posts on Facebook and almost 2 million photos on Instagram, send about 350,000 tweets on Twitter and upload 300 hours of new video on YouTube (Infographic 2015). Part of this content (relatively small, of course, but still substantial in absolute

Text Box 19.2 Social Big Data Analytics Tools

Big data is an evolving term that describes any voluminous amount of structured, semi-structured and unstructured data that has the potential to be mined for information (Mattmann 2013). These data come from sensors used to gather climate, traffic and flight information, transaction records, mobile phone GPS signals, posts to social media sites, digital pictures and videos, to name a few sources. Social big data comes from joining the efforts of two domains: big data and social media. The following are some tools used in social big data analytics (Batrinca & Treleaven 2015):

- **Scientific programming tools**. R, MATLAB and Mathematica for statistical, numeric and symbolic programming respectively; Python for natural language detection, title and content extraction, query matching and sentiment analysis; Apache UIMA analyses large volumes of unstructured information in order to discover knowledge that is relevant to an end-user.
- **Business toolkits**. SAS Sentiment Analysis for scraping content sources, including websites, social media and internal organizational text sources; RapidMiner for data mining and machine learning procedures including data loading and transformation, data pre-processing and visualization, modelling, evaluation and deployment.
- **Social media monitoring tools.** Social Mention provides social media alerts similarly to Google Alerts; Trackur are online reputation monitoring tools that track what is being said on the Internet; Google Trends shows how often a particular search-term input compares to the total search volume.
- **Text analysis tools.** Many options including Python Natural Language Toolkit, GATE, NVivo, IBM LanguageWare, IBM SPSS Text Analytics for Surveys, STATISTICA Text Miner and WordStat.
- **Data visualization tools.** SAS Visual Analytics, Tableau Desktop and Microsoft Power BI.

terms) is about wildlife related-issues and can be relevant for conservation and management purposes.

Social media data have been extensively used for marketing purposes and in numerous fields of science as well, from identifying crime hotspots (Malleson & Andresen 2015) to measuring foreign policy dynamics (Zeitzoff et al. 2015) to monitoring public health (Khoury & Ioannidis 2014). Internet data can be particularly relevant when the research question involves large spatial scales, access to informants is difficult and expensive (e.g. they are scattered over large and/or remote areas) and reliability is an issue (e.g. informants are unlikely to tell the truth in a face-to-face interview). Tenkanen et al. (2017), for example, assessed the usability of Instagram, Flickr and Twitter for visitor monitoring in protected areas, while El Bizri et al. (2015) used YouTube to detect the occurrence of sport hunting in Brazil and evaluate the opinions of hunters and internet users on sport hunting. Search engines can also be the source of useful information. Google Trends, for example, has emerged as a source of information to investigate how social trends evolve over space and time (see Figure 19.1). Knowing how the level of interest in conservation topics varies over space and time, using Google search volume as a proxy, can help support targeted conservation science communication (Nghiem et al. 2016).

Particularly promising new tools in big data analytics are opinion mining and sentiment analysis, which deal with the computational treatment of opinion, sentiment and subjectivity in text (Text Box 19.2). Sentiment analysis involves algorithms based on natural-language-processing machine learning that can analyse the text of social media such as reviews, forum discussions, blogs, micro-blogs, Twitter and social networks to understand the meaning behind a person's statement (Akcora 2010). These algorithms can create a quantified score of the public's feelings and allow the monitoring of opinions. Sentiment analysis could be used to map and monitor stakeholders' feelings (i.e. strongly negative to strongly positive) and opinions about both species and management actions, contributing to the assessment of the cost-effectiveness of the actions (e.g. the stronger the negative feelings, the higher social cost of implementation).

19.4 INTEGRATING STAKEHOLDERS AND DATA

Planning for large-scale coexistence and to inform the conflict-to-coexistence continuum involves integrating large and diverse data sets, engaging various stakeholders and making decisions about problems that are dynamic and may be rapidly changing (Yurco et al. 2017). Recent advances in data technologies can contribute to such integration and velocity. A cyber-infrastructure (Campbell 2003), for example, is a research environment that supports advanced data processing (acquisition, storage, management, integration, mining and visualization) and other computing and information processing services distributed over the Internet beyond the scale of a single institution. Examples of these infrastructures are the Global Biodiversity Information Facility (www .gbif.org) and the Data Observation Network for Earth (www.dataone .org). For planning purposes, cyber-infrastructure could be a technological and sociological solution to the problem of efficiently connecting data and stakeholders (individuals and institutions).

All scientific evidence, together with gaps and uncertainty, is then taken into account in the structured decision-making around the trade-offs involved in finite resource allocation. Accounting for cost-effectiveness is key to planning for large-scale human–wildlife coexistence. Planning for coexistence must incorporate cost-benefit analyses that are based not only on the objective and tangible terms of ecology and economy (McManus et al. 2015), but that take also into account the subjectivity of intangible affective, social and cultural costs and benefits as perceived by each stakeholder.

Turning conflict into large-scale coexistence will require communication, engagement and long-term commitment among the stakeholders. The process can be facilitated by the establishment of dialogues in which planning products are shared or co-created with relevant stakeholders, the creation of *fora* by means of which knowledge of what works and does not, and why, can be exchanged across regions, and by the mobilization of public and private sectors to implement effective actions. Gaining acceptance in the political arena may be particularly important. Converting planning products into instruments of public policy such as national action plans, combined with lists of endangered species and other forms of regulation, may be necessary to ensure stakeholder commitment to the plan. At the international level, the recognition of the importance of upscaling the analysis and management of HWC so as to engage and integrate various stakeholders into a data-intensive planning for coexistence is new, but it is growing. The establishment of the Task Force on Human–Wildlife Conflict (www.hwctf.org) is an evidence of this trend. The task force was established in 2016 by the IUCN SSC and its aims include to support their network – the IUCN SSC Conservation Planning Specialist Group and national governments, for example – in addressing HWC by providing interdisciplinary guidance and expert support, and a platform for the exchange of best practice.

When grounded in the available science and widespread stakeholder participation, the planning framework described above appears promising in providing legitimate, effective and fair mechanisms for local society to coexist with wildlife in an ever more complex and human-dominated world.

19.5 RECOMMENDATIONS AND FUTURE DIRECTIONS

The key to turning HWC into large-scale coexistence is good planning. Given the complexity of HWC and the typical scarcity of financial and human resources, it is imperative to (1) assess the multiple dimensions of HWC, with the proper emphasis on the human reactions and their determinants at multiple levels, from both top-down and bottom-up angles, (2) make decisions that account for the interests of the stakeholders, from local individuals to global institutions, (3) implement strategies that guarantee continued and lasting collaboration between these stakeholders and (4) increase professional capacity in planning. This is arguably more important in HWC than in other conservation issues, as poorly handled HWC mitigation makes the situation worse, creating even deeper divides.

- Because HWC increasingly involves species that do not entail a significant conservation focus, such as abundant and exotic wildlife in highly human-modified environments, and because HWC is fundamentally about disagreements and resentment between people (over wildlife), *planning for coexistence* as we define it, rather than conservation planning, is a more inclusive, productive approach. Planning for coexistence has an emphasis on improving the well-being of the people affected while ensuring just treatment to the wildlife involved. As a result, it would be more effective in communicating and engaging with society.
- Social data are typically scarce relative to ecological and environmental data. Nonetheless, planning for coexistence should become a data-intensive science, and novel mechanisms for data sharing and sources of data such as citizen science and big data (e.g. social media) need to be explored.
- Modelling is a valuable tool for planning for large-scale coexistence. It reduces complexity to help us understand where and why conflicts and coexistence occur and how the systems are likely to respond to alternative management actions. A challenge ahead is to incorporate social and behavioural data in spatial modelling.
- Planning for coexistence is a truly interdisciplinary field, and interdisciplinary training is still limited among wildlife professionals. Capacity building, including the development of technical and framework guidance materials, resources, tools and training as needed by those working on human–wildlife conflict and coexistence issues, should be a priority.

19.6 References

Adams, W. M. (2015). The political ecology of conservation conflicts. In S. M. Redpath, R. J. Gutiérrez, K. A. Wood & J. C. Young, eds., *Conflicts in*

Conservation: Navigating towards Solutions. Cambridge: Cambridge University Press, pp. 64–75.

Akcora, C. G. (2010). *Using Microblogs for Crowdsourcing and Public Opinion Mining*. Buffalo, NY: State University of New York.

Amit, R. & Jacobson, S. K. (2017). Stakeholder barriers and benefits associated with improving livestock husbandry to prevent jaguar and puma depredation. *Human Dimensions of Wildlife*, 22(3), 246–66.

Ball, I. R., Possingham, H. P. & Watts, M. (2009). Marxan and relatives: Software for spatial conservation prioritisation. In A. Moilanen, K. A. Wilson & H. P. Possingham, eds., *Spatial Conservation Prioritisation: Quantitative Methods and Computational Tools*. Oxford: Oxford University Press, pp. 185–95.

Batrinca, B. & Treleaven, P. C. (2015). Social media analytics: A survey of techniques, tools and platforms. *AI & Society*, 30(1), 89–116.

Behr, D. M., Ozgul, A. & Cozzi, G. (2017). Combining human acceptance and habitat suitability in a unified socio-ecological suitability model: A case study of the wolf in Switzerland. *Journal of Applied Ecology*, 54, 1919–29.

Bruskotter, J. T., Singh, A., Fulton, D. C. & Slagle, K. (2015). Assessing tolerance for wildlife: Clarifying relations between concepts and measures. *Human Dimensions of Wildlife*, 20(3), 255–70.

Bruskotter, J. T. & Wilson, R. S. (2014). Determining where the wild things will be: Using psychological theory to find tolerance for large carnivores. *Conservation Letters*, 7(3), 158–65.

Buesching, C. D., Slade, E., Newman, C., Riutta, T., Riordan, P. & Macdonald, D. W. (2015). Many hands make light work – but do they? A critical evaluation of citizen science. In D. W. Macdonald & R. E. Feber, eds., *Wildlife Conservation on Farmlands*. 2 vols. *Conflict in the Countryside*. Oxford: Oxford University Press.

Bull, J. W., Verissimo, D. & Milner-Gulland, E. J. (2017). When a ripple becomes a flood – why we didn't sign Ripple et al.'s 'World Scientists' Warning to Humanity: A Second Notice'. *Conservation Optimism*. Available from https://conservationoptimism.com/blog (accessed December 2017).

Bunnefeld, N., Hoshino, E. & Milner-Gulland, E. J. (2011). Management strategy evaluation: A powerful tool for conservation?. *Trends in Ecology & Evolution*, 26(9), 441–7.

Burgman, M. (2005). *Risks and Decisions for Conservation and Environmental Management*. Cambridge: Cambridge University Press.

Byers, O., Lees, C., Wilcken, J. & Schwitzer, C. (2013). The 'One Plan Approach': The philosophy and implementation of CBSG's approach to integrated species conservation planning. *WAZA Magazine*, 14, 2–5.

Campbell, J. (2003). NBII Enterprise Architecture – Section 2 – Business Architecture. NBII Report Programme.

Carter, N. H. & Linnell, J. D. (2016). Co-adaptation is key to coexisting with large carnivores. *Trends in Ecology & Evolution*, 31(8), 575–8.

Carter, N. H., Riley, S. J., Shortridge, A., Shrestha, B. K. & Liu, J. (2014). Spatial assessment of attitudes toward tigers in Nepal. *Ambio*, 43(2), 125–37.

Carvalho, E. A., Zarco-González, M. M., Monroy-Vilchis, O. & Morato, R. G. (2015). Modeling the risk of livestock depredation by jaguar along the Trans-Amazon highway, Brazil. *Basic & Applied Ecology*, 16(5), 413–19.

Couvet, D., Jiguet, F., Julliard, R., Levrel, H. & Teyssedre, A. (2008). Enhancing citizen contributions to biodiversity science and public policy. *Interdisciplinary Science Reviews*, 33(1), 95–103.

Cushman, S. A., Macdonald, E. A., Landguth, E. L., Malhi, Y. & Macdonald, D. W. (2017). Multiple-scale prediction of forest loss risk across Borneo. *Landscape Ecology*, 32, 1581–98.

Decker, D. J., Evensen, D. T., Siemer, W. F., Leong, K. M., Riley, S. J., Wild, M. A., Castle, K. T. & Higgins, C. L. (2010). Understanding risk perceptions to enhance communication about human–wildlife interactions and the impacts of zoonotic disease. *ILAR Journal*, 51(3), 255–61.

Dickman, A. J. (2010). Complexities of conflict: The importance of considering social factors for effectively resolving human–wildlife conflict. *Animal Conservation*, 13(5), 458–66.

Dickman, A. J., Macdonald, E. A. & Macdonald, D. W. (2011). A review of financial instruments to pay for predator conservation and encourage human–carnivore coexistence. *Proceedings of the National Academy of Sciences*, 108(34), 13937–44.

Dickman, A. J., Marchini, S. & Manfredo, M. (2013). The human dimension in addressing conflict with large carnivores. *Key Topics in Conservation Biology*, 2, 110–26.

El Bizri, H., Morcatty, T., Lima, J. & Valsecchi, J. (2015). The thrill of the chase: Uncovering illegal sport hunting in Brazil through YouTube™ posts. *Ecology & Society*, 20(3), art. 30.

Elith, J. & Leathwick, J. R. (2009). Species distribution models: Ecological explanation and prediction across space and time. *Annual Review of Ecology, Evolution, & Systematics*, 40, 677–97.

Fonseca, M. G., Aragão, L. E. O., Lima, A., Shimabukuro, Y. E., Arai, E. & Anderson, L. O. (2016). Modelling fire probability in the Brazilian Amazon using the maximum entropy method. *International Journal of Wildland Fire*, 25(9), 955–69.

Frank, B. (2016). Human–wildlife conflicts and the need to include tolerance and coexistence: An introductory comment. *Society & Natural Resources*, 29 (6), 738–43.

Groves, C. R. (2003). *Drafting a Conservation Blueprint: A Practitioner's Guide to Planning for Biodiversity*. Washington, DC: Island Press.

Guisan, A. & Thuiller, W. (2005). Predicting species distribution: Offering more than simple habitat models. *Ecology Letters*, 8(9), 993–1009.

Heinonen, J. P. M. & Travis, J. M. J. (2015). Modelling conservation conflicts. In S. M. Redpath, R. J. Gutiérrez, K. A. Wood & J. C. Young, eds., *Conflicts in Conservation: Navigating towards Solutions*. Cambridge: Cambridge University Press, pp. 195–211.

Hemson, G., Maclennan, S., Mills, G., Johnson, P. & Macdonald, D. (2009). Community, lions, livestock and money: A spatial and social analysis of attitudes to wildlife and the conservation value of tourism in a human–carnivore conflict in Botswana. *Biological Conservation*, 142(11), 2718–25.

Herda-Rapp, A. & Goedeke, T. L. (2005). *Mad about Wildlife: Looking at Social Conflict over Wildlife*. Leiden: Brill Academic.

Hoare, R. (2015). Lessons from 20 years of human–elephant conflict mitigation in Africa. *Human Dimensions of Wildlife*, 20(4), 289–95.

Infographic. (2015).The Data Explosion in 2014 Minute by Minute. Available from http://aci.info/2014/07/12/the-data-explosion-in-2014-minute-by-minute-info graphic (accessed November 2018).

IUCN Species Survival Commission. (2017). Guidelines for Species Conservation Planning. Version 1.0. Gland, Switzerland.

Jacobs, M. H. (2012). Human emotions toward wildlife. *Human Dimensions of Wildlife*, 17(1), 1–3.

Justus, J. & Sarkar, S. (2002). The principle of complementarity in the design of reserve networks to conserve biodiversity: A preliminary history. *Journal of Biosciences*, 27(4), 421–35.

Kansky, R., Kidd, M. & Knight, A. T. (2016). A wildlife tolerance model and case study for understanding human wildlife conflicts. *Biological Conservation*, 201, 137–45.

Kansky, R. & Knight, A. T. (2014). Key factors driving attitudes towards large mammals in conflict with humans. *Biological Conservation*, 179, 93–105.

Khoury, M. J. & Ioannidis, J. P. (2014). Big data meets public health. *Science*, 346 (6213), 1054–55.

Knight, A. T., Cowling, R. M. & Campbell, B. M. (2006). An operational model for implementing conservation action. *Conservation Biology*, 20, 408–19.

Knight, A. T., Cowling, R. M., Difford, M. & Campbell, B. M. (2010). Mapping human and social dimensions of conservation opportunity for the scheduling of conservation action on private land. *Conservation Biology*, 24(5), 1348–58.

Knight, A. T., Rodrigues, A. S., Strange, N., Tew, T. & Wilson, K. A. (2013). Designing effective solutions to conservation planning problems. *Key Topics in Conservation Biology*, 2, 362–83.

Kuhnert, P. M., Martin, T. G. & Griffiths, S. P. (2010). A guide to eliciting and using expert knowledge in Bayesian ecological models. *Ecology Letters*, 13(7), 900–14.

Kyriazi, Z., Maes, F. & Degraer, S. (2016). Coexistence dilemmas in European marine spatial planning practices: The case of marine renewables and marine protected areas. *Energy Policy*, 97, 391–9.

Lacy, R. C. (2000). Structure of the VORTEX simulation model for population viability analysis. *Ecological Bulletins*, 48, 191–203.

Loring, P. A. (2016). Toward a theory of coexistence in shared social-ecological systems: The case of Cook inlet salmon fisheries. *Human Ecology*, 44(2), 153–65.

Macdonald, D. W., Boitani, L., Dinerstein, E., Fritz, H. & Wrangham, R. (2013). Conserving large mammals. *Key Topics in Conservation Biology*, 2, 277–312.

MacMillan, D. C. & Marshall, K. (2006). The Delphi process: An expert-based approach to ecological modelling in data-poor environments. *Animal Conservation*, 9(1), 11–19.

Madden, F. (2004). Creating coexistence between humans and wildlife: Global perspectives on local efforts to address human–wildlife conflict. *Human Dimensions of Wildlife*, 9(4), 247–57.

Madden, F. & McQuinn, B. (2014). Conservation's blind spot: The case for conflict transformation in wildlife conservation. *Biological Conservation*, 178, 97–106.

Malleson, N. & Andresen, M. A. (2015). The impact of using social media data in crime rate calculations: Shifting hot spots and changing spatial patterns. *Cartography and Geographic Information Science*, 42(2), 112–21.

Manfredo, M. J., Teel, T. L. & Dietsch, A. M. (2016). Implications of human value shift and persistence for biodiversity conservation. *Conservation Biology*, 30(2), 287–96.

Manral, U., Sengupta, S., Hussain, S. A., Rana, S. & Badola, R. (2016). Human wildlife conflict in India: A review of economic implication of loss and preventive measures. *Indian Forester*, 142(10), 928–40.

Marchini, S. (2014). Who's in conflict with whom? Human dimensions of the conflicts involving wildlife. In L. M. Verdade, M. C. Lyra-Jorge & C. I. Piña, eds., *Applied Ecology and Human Dimensions in Biological Conservation*. New York: Springer Press, pp. 189–209.

Marchini, S. & Crawshaw Jr, P. G. (2015). Human–wildlife conflicts in Brazil: A fast-growing issue. *Human Dimensions of Wildlife*, 20(4), 323–8.

Marchini, S. & Macdonald, D. W. (2012). Predicting ranchers' intention to kill jaguars: Case studies in Amazonia and Pantanal. *Biological Conservation*, 147 (1), 213–21.

Margules, C. R. & Pressey, R. L. (2000). Systematic conservation planning. *Nature*, 405(6783), 243–53.

Mattmann, C. A. (2013). Computing: A vision for data science. *Nature*, 493 (7433), 473–5.

McBride, M. F. & Burgman, M. A. (2012). What is expert knowledge, how is such knowledge gathered, and how do we use it to address questions in landscape ecology? In A. H. Perera, C. A. Drew & C. J. Johnson, eds., *Expert Knowledge and its Application in Landscape Ecology*. New York: Springer, pp. 11–38.

McManus, J. S., Dickman, A. J., Gaynor, D., Smuts, B. H. & Macdonald, D. W. (2015). Dead or alive? Comparing costs and benefits of lethal and non-lethal human–wildlife conflict mitigation on livestock farms. *Oryx*, 49(4), 687–95.

Miller, J. R. (2015). Mapping attack hotspots to mitigate human–carnivore conflict: Approaches and applications of spatial predation risk modeling. *Biodiversity & Conservation*, 24(12), 2887–911.

Mintzberg, H. & Quinn, J. B. (1996). *The Strategy Process: Concepts, Contexts, Cases*. London: Prentice Hall.

Naidoo, R., Balmford, A., Ferraro, P. J., Polasky, S., Ricketts, T. H. & Rouget, M. (2006). Integrating economic costs into conservation planning. *Trends in Ecology & Evolution*, 21(12), 681–7.

Nghiem, L. T., Papworth, S. K., Lim, F. K. & Carrasco, L. R. (2016). Analysis of the capacity of Google Trends to measure interest in conservation topics and the role of online news. *PLoS ONE*, 11(3), e0152802.

Peterson, M. N., Birckhead, J. L., Leong, K., Peterson, M. J. & Peterson, T. R. (2010). Rearticulating the myth of human–wildlife conflict. *Conservation Letters*, 3(2), 74–82.

Phillips, S. J. & Dudík, M. (2008). Modeling of species distributions with Maxent: New extensions and a comprehensive evaluation. *Ecography*, 31(2), 161–75.

Pigott, D. M., Golding, N., Mylne, A., Huang, Z., Henry, A. J., Weiss, D. J., Brady, O. J., Kraemer, M. U., Smith, D. L., Moyes, C. L. & Bhatt, S. (2014). Mapping the zoonotic niche of Ebola virus disease in Africa. *Elife*, 3, e04395.

Pressey, R. L. & Nicholls, A. O. (1989). Efficiency in conservation evaluation: Scoring versus iterative approaches. *Biological Conservation*, 50(1–4), 199–218.

Redpath, S. M., Bhatia, S. & Young, J. (2015). Tilting at wildlife: Reconsidering human–wildlife conflict. *Oryx*, 49(2), 222–5.

Redpath, S. M., Young, J., Evely, A., Adams, W. M., Sutherland, W. J., White-house, A., Amar, A., Lambert, R. A., Linnell, J. D., Watt, A. & Gutierrez, R. J. (2013). Understanding and managing conservation conflicts. *Trends in Ecology & Evolution*, 28(2), 100–9.

Santos, S. M., Marques, J. T., Lourenço, A., Medinas, D., Barbosa, A. M., Beja, P. & Mira, A. (2015). Sampling effects on the identification of roadkill hotspots: Implications for survey design. *Journal of Environmental Management*, 162, 87–95.

Silvertown, J., Buesching, C. D., Jacobson, S. K. & Rebelo, T. (2013). Citizen science and nature conservation. *Key Topics in Conservation Biology*, 2, 127–42.

Sitati, N. W., Walpole, M. J., Smith, R. J. & Leader-Williams, N. (2003). Predicting spatial aspects of human–elephant conflict. *Journal of Applied Ecology*, 40 (4), 667–77.

Souza, R. A. & De Marco, P. (2014). The use of species distribution models to predict the spatial distribution of deforestation in the western Brazilian Amazon. *Ecological Modelling*, 291, 250–9.

Sutherland, W. J. & Burgman, M. A. (2015). Use experts wisely. *Nature*, 526 (7573), 317–18.

Tenkanen, H., Di Minin, E., Heikinheimo, V., Hausmann, A., Herbst, M., Kajala, L. & Toivonen, T. (2017). Instagram, Flickr, or Twitter: Assessing the usability of social media data for visitor monitoring in protected areas. *Scientific Reports*, 7(1), 17615.

Tversky, A. & Kahneman, D. (1981). The framing of decisions and the psychology of choice. *Science*, 211 (4481), 453–8.

Vucetich, J. A., Burnham, D., Macdonald, E. A., Bruskotter, J. T., Marchini, S., Zimmermann, A. & Macdonald, D. W. (2018). Just conservation: What is it and should we pursue it? *Biological Conservation*, 221, 23–33.

Weckel, M. E., Mack, D., Nagy, C., Christie, R. & Wincorn, A. (2010). Using citizen science to map human–coyote interaction in suburban New York, USA. *Journal of Wildlife Management*, 74(5), 1163–71.

Wilson, K. A., Underwood, E. C., Morrison, S. A., Klausmeyer, K. R., Murdoch, W. W., Reyers, B., Wardell-Johnson, G., Marquet, P. A., Rundel, P. W., McBride, M. F. & Pressey, R. L. (2007). Conserving biodiversity efficiently: What to do, where, and when. *PLoS Biology*, 5(9), 223.

Wilson, S., Davies, T. E., Hazarika, N. & Zimmermann, A. (2013). Understanding patterns of human–elephant conflict in Assam: An analysis to inform mitigation strategies. *Oryx*, 49(01), 140–9.

Woodroffe, R., Thirgood, S. & Rabinowitz, A. (2005). *People and Wildlife, Conflict or Coexistence?* Cambridge: Cambridge University Press.

Yurco, K., King, B., Young, K. R. & Crews, K. A. (2017). Human–wildlife interactions and environmental dynamics in the Okavango Delta, Botswana. *Society & Natural Resources*, 30(9), 1112–26.

Zeitzoff, T., Kelly, J. & Lotan, G. (2015). Using social media to measure foreign policy dynamics: An empirical analysis of the Iranian–Israeli confrontation (2012–13). *Journal of Peace Research*, 52(3), 368–83.

Zimmermann, A., Baker, N., Inskip, C., Linnell, J. D. C., Marchini, S., Odden, J., Rasmussen, G. & Treves, A. (2010). Contemporary views of human-carnivore conflicts on wild rangelands. In J. T. du Toit, R. Kock & J. C. Deutsch, eds., *Wild Rangelands: Conserving Wildlife While Maintaining Livestock in Semi-Arid Ecosystems*. Chichester: John Wiley & Sons, Ltd, pp. 129–51.

Zimmermann, A., Walpole, M. J. & Leader-Williams, N. (2005). Cattle ranchers' attitudes to conflicts with jaguar *Panthera onca* in the Pantanal of Brazil. *Oryx*, 39(4), 406–12.

20

Human–Wildlife Interactions

Multifaceted Approaches for Turning Conflict into Coexistence

JENNY A. GLIKMAN, BEATRICE FRANK AND
SILVIO MARCHINI

In their concluding chapter, Woodroffe et al. (2005) asked the question *which management strategies and policies can encourage human–wildlife coexistence?* Their recommendations included technical solutions, economic incentives, legal protection and community involvement. In practice, lethal and non-lethal methods are used to dissuade wildlife from consuming crops and livestock, while economic and legal forces are directed at people to keep them from persecuting wildlife. The authors acknowledged that engaging with local people is a key component of any strategy to resolve human–wildlife conflict. Their book – *People and Wildlife, Conflict or Coexistence?* – focused on wildlife damage and retributive killing, providing recommendations to mitigate the mutual impacts of such negative interaction. The book set the tone for what has become increasingly evident: human–wildlife conflict (HWC) requires multifaceted approaches to be understood and managed. From its initial focus on the ecology and economics of wildlife damage, the study and mitigation of HWC has gradually expanded its scope to incorporate also the human dimensions of the whole spectrum of human–wildlife relationships, including positive interactions. During the preparation of this volume, several books and peer-reviewed articles were published on subjects like acceptance and tolerance towards wildlife and human–wildlife coexistence (e.g. Broekhuis et al. 2017; Hill et al. 2017; Pooley et al. 2017). When considering solutions to HWC, it has been increasingly recognized that much HWC are in reality human–human conflict over wildlife and its management (Madden 2004; Fraser-Celin et al. 2018). Even the use of the term *conflict* itself has been challenged among academics and practitioners (e.g. Hill 2017), the question being: should we just keep on framing human–wildlife interactions as conflicts or should we move to a more inclusive and broad definition that better encompasses the diverse relationship humans have with and over wildlife? We propose that human–wildlife interactions, rather than HWC, may be a terminology that better communicates the entire spectrum of relevant relationships between people and wildlife, and among humans over wildlife, from conflict to coexistence. Indeed, it is needed to first understand what type of interaction is taking place, and then determine whether it is a conflict, for whom and over what.

We embarked on this book with the intention of catalyzing a paradigm shift from discourses on HWC to dialogues promoting human–wildlife interactions and coexistence. Yet a change in name from conflict to coexistence will not be enough. Beyond writing and theorizing about human–wildlife interactions what is required is acting and fostering actual change, one that favours a shift from conflict to coexistence. Findings within this book indicate that in some cases this shift is already in its course. Successful projects focused on positive interactions between wildlife and humans are showcased throughout this book, from beehive fences that protect crops from elephants (Chapter 11) to 'Lion Guardians' monitoring large carnivores in order to protect local communities in Southern Kenya (Chapter 17). Many of these successful projects and positive interactions are now shared and implemented in other geographical areas. For example, traditional warriors are engaged in conservation programmes and in protecting and monitoring lions in Samburu, Kenya, with the Warrior Watch (i.e. Ewaso Lions, ewasolions.org/conservation/warrior-watch/) and in Tanzania with the Lion Defenders (i.e. Ruaha Carnivore Project, www.ruahacarnivoreproject.com/lion-defenders-6/). These organizations have come together and founded The Pride Alliance to conserve lions across Africa (pridelionalliance.org/our-strategies). These initiatives represent promising changes in the way of addressing human–wildlife conflicts and seeking coexistence.

As stressed in the conflict-to-coexistence continuum, there are many factors influencing human willingness to coexist with wildlife that include: 'the species involved, the location in which the species is encountered, and the person's culture ... the sociocultural background, the types of conservation law enforcement, economic benefits, and other aspects of societies living with wildlife' (Frank 2016, pp. 740–1). What we learnt throughout the advancement of this book is that while every single chapter provides unique insight towards building coexistence in specific settings, each of them offers lessons that can be shared across socio-political settings, scales and geographical areas.

Figure 20.1 Word cloud revealing the major themes of the book (generated by worditout.com). The more common a word, the larger and darker it is in the picture.

Besides the greater attention to social rather than ecological issues, and to coexistence rather than conflict (Figure 20.1), three salient trends emerge across this book: (1) the emphasis on the psychological and emotional dimensions, (2) the importance of scaling up analysis and action in terms of target (from single to multiple species), space (from local to global) and level of decision (from individual to institutions) and (3) the urgent need to incorporate and encourage other disciplines to embrace and further push the paradigm shift towards coexistence.

20.1 DO NOT FEAR THE FEELINGS: FOCUSING ON THE ACTUAL DRIVERS OF HUMAN BEHAVIOUR

Traditionally, the inclusion of humans and the so-called human dimensions of wildlife conservation/management has focused on the rational – or cognitive component – of human mental dispositions towards wildlife (Fulton et al. 1996; Hrubes et al. 2001). Following the scientific revolution, everything needed to be understood through an objective lens and studied with scientifically rigorous approaches, including humans (Chapter 1). Yet humans' reactions to and interactions with wildlife – from fear of snakes to appreciation and wonder about charismatic species such as whales, bears and many others – are often not explained by rational decision-making processes (Chapter 4). Over time and through extended research, the knowledge on cognitive systems has been integrated and expanded to include the non-rational side – the affective component – of attitudes in predicting human behaviours towards wildlife. To further these streams of research, recent studies have focused on emotions and their role in determining humans' feeling for and response to wildlife. Several chapters in our book stress that the affective component of attitudes and emotions plays a determining role in human–wildlife interactions. Specifically, positive emotions towards a species (Chapters 1, 4, 5, 9) and the perception that a species poses low risk or that such risk can be controlled favours coexistence between humans and wildlife (Chapters 3, 5).

Additional factors have been proven to enable coexistence and foster positive affect and emotions towards wildlife, such as trust and respect (Chapters 2, 17) and incurring in benefits from coexisting with wildlife (Chapters 5, 9). It is also important to recognize and understand that the concerns related to vulnerable others (Chapter 3) and the moral bounds of the encounter (Chapter 15) shape ethical judgements towards a species by a person or identity group, and influence perceptions about the benefits a person or group may gain by coexisting with wildlife. Better understanding these underlying drivers and the possible hidden impacts of non-rational mental dispositions towards wildlife is key to establish the reasons behind people's willingness to tolerate wildlife and their ability to perform certain specific behaviours.

Only the tip of the iceberg has been explored when it comes down to feelings and emotions towards wildlife, especially in relation to coexistence. However, we are now aware that affect and emotions need to be included when engaging the public, stakeholders and other key players in discussing wildlife decision-making. While examples of this inclusion exist for some species (e.g. carnivores), it is currently non-existent for others (e.g. pest species, invasive species) due to limited studies focusing on affect and emotions, and lack of longitudinal research on coexistence. Future work is required to address this gap by including emotions and positive mental disposition as core research components – from questions on willingness to coexist with species, to inquiries about feelings towards specific species, to workshops tailored at fostering tolerance for wildlife and trust between groups having opposing views about wildlife conservation and management. General appraisal criteria developed in other disciplines about affect and emotions, such as novelty, valence, goals, agency and norms can be used to further prompt research in this domain (Chapter 4). Such research would allow pushing the definition of coexistence further and helping to determine why, when and how people are willing to share the landscape with wildlife.

20.2 SCALE UP: ADDING SPACE TO THE EQUATION

In addition to the recognition of the importance of the psychological realm discussed above, the last decade has also seen a growing understanding that, what ultimately determines the conflict or coexistence between people and wildlife, is beyond the local, single-species issues and individual-level responses to them. Theory and practice have indeed moved 'down' from the individual-level models to explore intra-personal factors such as emotions and values. Nonetheless, there is also a need to scale 'up' to consider the full range of spatial scales, from local to global, and to incorporate the corresponding hierarchy of decision-making levels, from individual to societal.

Human–wildlife interactions have a spatial dimension, where the territories of species might overlap. The establishment of exclusive places for humans and for wildlife results in defining human–wildlife interactions as a transgression, i.e. people's understanding of when a species has crossed a person's idea of the right 'place' for it to be (Chapters 1, 15). It is therefore fundamental to acknowledge that socially constructed landscapes – what is referred to as *taskscapes*, for example – can influence people's mental disposition towards conservation and wildlife issues (Chapter 7). Indeed, human behaviours such as retaliatory killing could conceivably be linked to spatially explicit biophysical, social and economic factors. Furthermore, a person's thoughts and actions towards wildlife and wildlife management can be determined by processes occurring far away and at higher hierarchical levels (Chapters 14, 18, 19). The socio-ecological system approach proposes that the interactions between wildlife and humans are the result of nested, multiscalar and multilevel hierarchical processes. These processes, in part, will determine if, where and when interactions take place, and how the interactions are interpreted by the human and non-human actors (Chapter 14). Human–wildlife conflicts or coexistence are arguably the result of decisions of a wide range of interacting stakeholders, from households to global corporations, reflecting the working of markets, mediated by the intervention and regulation of the state, and the lobbying of non-governmental actors (Chapter 19). The key to turning HWC into large-scale coexistence is, therefore, the assessment of the multiple dimensions of human–wildlife interaction, at the right spatial scales and with the proper emphasis on the human reactions and their determinants at multiple levels, from both top-down and bottom-up angles (Chapters 18, 19). This up-scaling in analysis and action will benefit from novel mechanisms for data sharing and sources of data such as citizen science and big data analytics (Chapter 19). Furthermore, if coexistence is to be achieved at large scales, conservationists will need to overcome interdisciplinary boundaries and expand the information basis upon which we make decisions, going beyond the usual realms of ecology and economics to incorporate the perspectives from multiple disciplines, the social sciences in particular.

20.3 THINK OUTSIDE THE BOX: OVERCOMING DISCIPLINARY SILOS

Traditional wildlife management and conservation consists of experts leading research projects employing biological or classic social (e.g. sociology, economics and anthropology) perspectives[1] and, in a minority of cases, the integration of these two sciences. More recently, conservationists have started exploring a broader array of social phenomena (e.g. markets, policy) and incorporating applied social science disciplines (e.g. education, law and philosophies) to embrace and further push the paradigm shift of conservation towards coexistence (Bennett et al. 2017). Innovative approaches have been borrowed from other social science disciplines (e.g. communication and marketing, and development), as showcased through the conservation marketing approach for the conservation of the cockatoo in the Philippines (Chapter 16) and the business approach used in the Lion Guardians project in Kenya to foster long-term coexistence between local communities and these carnivores (Chapter 17). Applying socio-ecological and socio-environment system approaches to urban wildlife (Chapter 6) and wolves' conservation (Chapter 18), as well as planning for coexistence (Chapter 19), demonstrate the necessity of incorporating new interdisciplinary[2] and

---

[1] Defined as those that have emerged from specific disciplines; see Bennet et al. (2017).
[2] Integrating the approaches from two or more disciplines for answering one research question (each discipline can affect the research output of the other.)

transdisciplinary[3] perspectives for predicting interactions between humans and wildlife (e.g. Behr et al. 2017). Conservation criminology (Gore 2017), for example, is an interdisciplinary field of research focusing on the environmental exploitation and risks at the intersection of human and natural systems. Within this, the Illegal Wildlife Trade programme (www.illegalwildlifetrade.net/) uses an array of disciplines (e.g. public health, computer science), across species (e.g. saiga, sharks) and countries (e.g. Singapore, Indonesia), to address consumers' demand towards illegal and unsustainable wildlife trade. Regarding the transdisciplinary fields, an example is the ethnoprimatology, which uses an anthropological holistic approach to understand primates' interfaces, including the human (Dore et al. 2017; Fuentes et al. 2017). Specifically, the integration of qualitative social data (e.g. ethnography, participatory observation) and quantitative biological data (e.g. market survey, ecological transect) allows a more nuanced understanding of conservation issues (Setchell et al. 2017). Many other fields of research are working at the interface between multiple and diverse disciplines, offering to human–wildlife interaction studies lessons learned and pathways on how to move closer to coexistence endeavours. Indeed, to advance the shift in focus, and work on increasing coexistence rather than mitigating conflicts, conservationists need to overcome disciplinary silos studies by better integrating interdisciplinary and transdisciplinary perspectives. To foster more interdisciplinary and transdisciplinary approaches, conservationists need to include new frameworks and adopt a holistic lens for better understanding and including interactions from both a human and a wildlife perspective (Parathian et al. 2018). From the human perspective, the focus needs to be on the social drivers and hidden costs of conservation (e.g. Bond & Mkutu 2018), with an emphasis on how insights from environmental justice and ethics are used to foster long-lasting coexistence between people and wildlife (Vucetich et al. 2018). Future research should focus on discourses around the intrinsic values and the deep-rooted meaning of nature and wildlife that will help the integration of humans in nature. From a wildlife perspective, increasing studies on the behavioural responses of wildlife to different scenarios can help mitigate some

---

[3] Integration of knowledge that crosses many disciplinary boundaries to create a holistic approach; output is created as a result of disciplines integrating to become something completely new.

conflicts (e.g. Blackwell et al. 2017; Gaynor et al. 2018). Indeed, humans are modifying the natural landscape in a fast pace, with wildlife failing to adapt its behaviours and needs accordingly. As such, it is necessary to understand the type and speed of the reaction that individual animals have to specific stimuli to design appropriate mitigation strategies (Greggor et al. 2016). In addition, understanding the cognition of wildlife, the cues used for navigating the territory and intra- and interspecies social learning can assist planning for coexistence (Barrett et al 2018). As stressed in the previous section of this chapter, expanding the focus of studies to a wider spectrum of species other than the charismatic, mega fauna animals, might help in identifying new factors influencing human–wildlife interactions, creating new opportunities and crafting novel solutions in reducing barriers for an improved coexistence.

Another fundamental silos challenge to overcome is how conservationists as well as media present and frame conservation issues, and how such information sharing impacts the general public (Dayer et al. 2017). Indeed, the media influences what the public gets to know by deciding which stories to cover and how to portray those stories. The wording of the narrative can indeed deeply influence people's perceptions and mental dispositions towards a species. When certain carnivores (e.g. wolves, leopards, coyotes) are involved, most often, the representation of the animals and the focus of the story is negative (Alexander & Quinn 2012; Crown & Doubleday 2017). Outcomes from social sciences can support the delivery of messages that best resonate with the target audiences, as demonstrated in studies on the design of experimental risk communication looking at concerns related to vulnerable others (Chapter 3), and in the process of designing an intervention based on the specific needs of the selected target audience (Chapter 16). It is key to address the emotional component of perceived wildlife risk by focusing on what people are really afraid of, such as losing a child to a predator. Addressing such a worry can result in a greater effect towards coexistence than just focusing on the factual aspect that the risk to be attacked might be low. In addition, when promoting the establishment of a specific behaviour (e.g. vegetable plots that did not require forest destruction), it is fundamental to pay attention to the needs of the selected target audience (e.g. farming), while identifying benefits (e.g. economic revenues) and reducing barriers (e.g. acquiring new skills) of the behaviour that is being encouraged, which will increase effectiveness of the behaviour's acquisition (Chapter 16).

20.4 A UNIFIED FRAMEWORK: THE CONFLICT-TO-COEXISTENCE CONTINUUM

In this book, we use the conflict-to-coexistence continuum as a leading theme to help reasoning around key terms such as coexistence and tolerance, and to better understand context-specific, multidimensional and complex human–wildlife interactions. The continuum illustrates what human–wildlife interactions look like when we use a more balanced and impartial focus on the interaction itself, one that includes both negative to positive sides and considers equally wildlife and humans. This approach is expected to facilitate a more cohesive understanding of what coexistence means. Based on the key insights provided by the team of experts in this book, we have identified three main points of improvement in the domain of human–wildlife coexistence discourses needed for the conflict-to-coexistence continuum framework to move forward: (1) define and operationalize the term coexistence, (2) consider the multidimensional domains of coexistence and (3) embrace the dynamic nature of coexistence. Below we briefly discuss these three main points.

20.4.1 Coexistence Is More Than a Positive Mindset

Coexistence is complex as it depends upon the interplay of intrinsic deep-rooted and often *irrational* psychological human traits (Chapters 2–5; Chapter 15), as well as from extrinsic factors – which span from biophysical settings to socio-political processes (Chapters 8, 12, 14). This is further complicated by the fact that coexistence does not necessarily imply a positive interaction with wildlife. This term places an emphasis on some 'level of harmony or relatively low hostile reactions between human and wildlife' (Chapter 19, p. 421). Thus coexistence represents a delicate interplay between positive and negative aspects of human–wildlife and human–human interactions. The turning point from conflict to coexistence is, therefore, more blurred than initially defined in the continuum and may shift based on how people define human–wildlife interactions, as a glass half full (coexistence) or half empty (conflict). Defining coexistence from the perspective of the individuals and groups with whom we are working on wildlife-related discourses is instrumental to really be able to evaluate if coexistence has been reached and to what degree. Such an approach will also help in furthering the understanding of what coexistence means in different settings, with different species, across culture and socio-political contexts.

20.4.2 Coexistence Is Multidimensional

The conflict-to-coexistence continuum has mostly focused on expanding the approach to human–wildlife interactions so as to address their whole complexity from conflict to coexistence. It has mostly helped exploring individuals' dispositions towards wildlife. Yet altogether the contributed chapters in this book reveal the need to expand the analysis and management of human–wildlife interactions also across hierarchical scales of decision-making, both below and above the individual level. By moving towards a system approach that considers the interconnectedness of a broader whole – from intra-personal factors such as values and emotions to individual behaviour to interpersonal and intergroup relationships to community and institutional forces, from both bottom-up and top-down perspectives – it will be possible to understand the processes that are essential to effectively engage all parties. Such an understanding will help communities move towards a more desirable future, one that better fulfils human and wildlife needs. We are currently engaging different audiences (e.g. academics from different disciplines, practitioners across continents) to further conceptualize and define the terminology around coexistence and to visualize it beyond the linear continuum dimension. While some have been comfortable in keeping the conflict-to-coexistence continuum as a linear framework projection, others have imagined this approach as a series of nesting circles that represent the different level of scales from individual, to community to larger landscape (e.g. similar to the *panarchy* cycles; see Holling 2001). Future research in this domain will allow us to further question the conflict-to-coexistence visual projection, thus continue to explore how, when and for whom human–wildlife interactions are defined as mainly coexistence experiences.

20.4.3 Coexistence Is Not an End Point

We fully acknowledge that coexistence is not just a fixed point at the end of the conflict-to-coexistence continuum, but a process that encompasses two fundamental elements: (1) the well-being of the people interacting with wildlife and (2) the ethical handling of the wildlife involved in the interaction (Chapter 19). Being a process, coexistence is not an unalterable, predetermined or locked state. It is regenerative, as every time human–wildlife and human–human interactions destabilize or disrupt the status quo, a new level of organization and expression of coexistence emerges – a new continuum is forged. As a matter of fact, coexistence is a dynamic condition, just like the environment and the socio-political processes by which it is influenced.

20.5 LOOKING FORWARD TOWARDS HUMAN–WILDLIFE COEXISTENCE

We believe that focusing on coexistence rather than conflict will allow researchers and practitioners to pursue a desired, more optimistic future, one that better fits human and wildlife long-term needs. The shift from conflict to coexistence is a key step in achieving conservation goals. The next necessary endeavour is to propel human well-being discourses in planning for coexistence by including themes such as equity and pluralism (Loring 2016). While equity is about the fair treatment of individuals and groups and the impartial sharing of cost and benefits of wildlife conservation (Law et al. 2018), pluralism fosters acceptance between competing groups by recognizing that different stakeholders hold different values (e.g. cultural diversity) and acknowledging others' right to access wildlife (Loring 2016). Mechanisms for equity and pluralism (e.g. policies, harvest quotas, multi-sector engagement) can be used to better include marginal views and redistributing the costs of living with wildlife to a broader constituency – hence better consider wildlife and human well-being in current and future contexts and across society. These two themes can help develop processes and outcomes based on more ethical decision-making mechanisms. As perfectly summarized by Green et al. (2015, p. 386) 'It follows that positive change (i.e. conservation success) is defined by individuals and groups from diverse belief systems, social structures and backgrounds who hold equally diverse values for biodiversity.' Not including these two themes in coexistence discourses can hinder the full potential of planning for coexistence (Law et al. 2018). A future step for the conflict-to-coexistence continuum will be to explore the role played by equity and pluralism in fostering coexistence. We are also aware that other important factors of inclusive conservation might be currently missing in the conflict-to-coexistence continuum framework – a gap we hope to address with other experts in future research and projects focused on coexistence.

We envision the conflict-to-coexistence continuum as an adaptive and dynamic process with feedback loops that will allow to explore human–wildlife interactions in a system that contains multilevels of spatial scales, time and social organizations. There is similarity with the *panarchy* framework for social-ecological systems of resilience (Holling 2001), where the adaptive dynamic system cycles change, stabilize and grow and regenerate with continuous feedback among cycle phases. We hope that this book allows *the system* to move into the new phase of human–wildlife interactions, one that has as its ultimate goal the coexistence between humans and wildlife.

20.6 References

Alexander, S. M. & Quinn, M. S. (2012). Portrayal of interactions between humans and coyotes (*Canis latrans*): Content analysis of Canadian print media (1998–2010). *Cities and the Environment (CATE)*, 4(1), art. 9.

Barrett, L. P., Stanton, L. & Benson-Amram, S. (2018). The cognition of 'nuisance' species. *Animal Behaviour*, DOI: 10.1016/j.anbehav.2018.05.005.

Behr, D. M., Ozgul, A. & Cozzi, G. (2017). Combining human acceptance and habitat suitability in a unified socio-ecological suitability model: A case study of the wolf in Switzerland. *Journal of Applied Ecology*, 54, 1919–29.

Bennett, N. J., Roth, R., Klain, S. C., Chan, K., Christie, P., Clark, D. A., Cullman, G., Curran, D., Durbin, T. J. & Epstein, G. (2017). Conservation social science: Understanding and integrating human dimensions to improve conservation. *Biological Conservation*, 205, 93–108.

Blackwell, B. F., DeVault, T. L., Fernandez-Juricic, E., Gese, E. M., Gilbert-Norton, L. & Breck, S. W. (2017). No single solution: Application of behavioural principles in mitigating human–wildlife conflict. *Animal Behaviour*, 120, 245–54.

Bond, J. & Mkutu, K. (2018). Exploring the hidden costs of human–wildlife conflict in Northern Kenya. *African Studies Review*, 61(1), 33–54.

Broekhuis, F., Cushman, S. A. & Elliot, N. B. (2017). Identification of human–carnivore conflict hotspots to prioritize mitigation efforts. *Ecology & Evolution*, 7(24), 10630–9.

Crown, C. A. & Doubleday, K. F. (2017). Man-eaters in the media: Representation of human–leopard interactions in India across local, national, and international media. *Conservation & Society*, 15(3), 304–12.

Dayer, A. A., Williams, A., Cosbar, E. & Racey, M. (2017). Blaming threatened species: Media portrayal of human–wildlife conflict. *Oryx*, 1–8.

Dore, K. M., Rile, E. P. & Fuentes, A. (2017). *Ethnoprimatology: A Practical Guide to Research at the Human–Nonhuman Primate Interface*. Cambridge: Cambridge University Press.

Frank, B. (2016). Human–wildlife conflicts and the need to include tolerance and coexistence: An introductory comment. *Society & Natural Resources*, 29 (6), 738–43.

Fraser-Celin, V.-L., Hovorka, A. J. & Silver, J. J. (2018). Human conflict over wildlife: Exploring social constructions of African wild dogs (Lycaon pictus) in Botswana. *Human Dimensions of Wildlife*, DOI: 10.1080/10871209 .2018.1443528.

Fuentes, A., Riley, E. P. & Dore, K. M. (2017). Ethnoprimatology matters: Integration, innovation, and intellectual generosity. In K. M. Dore, E. P. Riley & A. Fuentes, eds., *Ethnoprimatology: A Practical Guide to Research at the Human–Primate Interface*. New York: Cambridge University Press, pp. 297–302.

Fulton, D. C., Manfredo, M. J. & Lipscomb, J. (1996). Wildlife value orientations: A conceptual and measurement approach. *Human Dimensions of Wildlife*, 1 (2), 24–47.

Gaynor, K. M., Hojnowski, C. E., Carter, N. H. & Brashares, J. S. (2018). The influence of human disturbance on wildlife nocturnality. *Science*, 360(6394), 1232–5.

Gore, M. L. (2017). *Conservation Criminology*. Chichester: John Wiley & Sons.

Green, S. J., Armstrong, J., Bogan, M., Darling, E., Kross, S., Rochman, C. M., Smyth, A. & Veríssimo, D. (2015). Conservation needs diverse values, approaches, and practitioners. *Conservation Letters*, 8, 385–7.

Greggor, A. L., Berger-Tal, O., Blumstein, D. T., Angeloni, L., Bessa-Gomes, C., Blackwell, B. F., St Clair, C. C., Crooks, K., de Silva, S., Fernández-Juricic, E., Goldenberg, S. Z., Mesnick, S. L., Owen, M., Price, C. M., Saltz, D., Schell, C. J., Suarez, A. V., Swaisgood, R. R., Winchell, C. S. & Sutherland, W. J. (2016). Research priorities from animal behaviour for maximising conservation progress. *Trends Ecology & Evolution*, 31, 953–64.

Hill, C. (2017). Introduction. Complex problems: Using a biosocial approach to understanding human–wildlife interactions. In C. M. Hill, A. D. Webber & N. E. C. Priston, eds., *Understanding Conflicts about Wildlife: A Biosocial Approach*. Oxford: Berghahn, pp. 1–14.

Hill, C. M., Webber, A. D. & Priston, N. E. C. (2017). *Understanding Conflicts about Wildlife: A Biosocial Approach*. Oxford: Berghahn.

Holling, C. S. (2001). Understanding the complexity of economic, ecological, and social systems. *Ecosystems*, 4, 390–405.

Hrubes, D., Ajzen, I. & Daigle, J. (2001). Predicting hunting intentions and behavior: An application of the theory of planned behavior. *Leisure Sciences*, 23(3), 165–78.

Law, E. A., Bennett, N. J., Ives, C. D., Friedman, R., Davis, K. J., Archibald, C. & Wilson, K. A. (2018). Equity trade-offs in conservation decision making. *Conservation Biology*, 32(2), 294–303.

Loring, P. A. (2016). Toward a theory of coexistence in shared social-ecological systems: The case of Cook inlet salmon fisheries. *Human Ecology*, 44(2), 153–65.

Madden, F. (2004). Can traditions of tolerance help minimize conflict? An exploration of cultural factors supporting human–wildlife coexistence. *Policy Matters*, 13, 234–41.

Parathian, H. E., McLennan, M. R., Hill, C. M., Frazão-Moreira, A. & Hockings, K. J. (2018). Breaking through disciplinary barriers: Human–wildlife interactions and multispecies ethnography. *International Journal of Primatology*, DOI:10.1007/s10764-018-0027-9.

Pooley, S., Barua, M., Beinart, W., Dickman, A., Holmes, G., Lorimer, J., Loveridge, A., Macdonald, D., Marvin, G. & Redpath, S. (2017). An interdisciplinary review of current and future approaches to improving human–predator relations. *Conservation Biology*, 31, 513–23.

Setchell, J. M., Fairet, E., Shutt, K., Waters, S. & Bell, S. (2017). Biosocial conservation: Integrating biological and ethnographic methods to study human–primate interactions. *International Journal of Primatology*, 38(2), 401–26.

Vucetich, J. A., Burnhamb, D., Macdonald, E. A., Bruskotter, J. T., Marchini, S., Zimmermann, A. & Macdonald, D. W. (2018). Just conservation: What is it and should we pursue it? *Biological Conservation*, 221, 23–33.

Woodroffe, R., Thirgood, S. & Rabinowitz, A. (2005). *People and Wildlife, Conflict or Coexistence?* Cambridge: Cambridge University Press.

Index